Modern Applications o[illegible]
Chemistry

Modern Applications of Cycloaddition Chemistry

Edited by

Paolo Quadrelli
Department of Chemistry, University of Pavia, Pavia, Italy

ELSEVIER

Elsevier
Radarweg 29, PO Box 211, 1000 AE Amsterdam, Netherlands
The Boulevard, Langford Lane, Kidlington, Oxford OX5 1GB, United Kingdom
50 Hampshire Street, 5th Floor, Cambridge, MA 02139, United States

British Library Cataloguing-in-Publication Data
A catalogue record for this book is available from the British Library

Library of Congress Cataloging-in-Publication Data
A catalog record for this book is available from the Library of Congress

ISBN: 978-0-12-815273-7

Publisher: Susan Dennis
Acquisition Editor: Emily McCloskey
Editorial Project Manager: Sara Pianavilla
Production Project Manager: Nilesh Kumar Shah
Cover Designer: Christian Bilbow

Cover picture by Maria Kjær Waag

Typeset by MPS Limited, Chennai, India

Contents

List of Contributors

Misal Giuseppe Memeo Department of Chemistry, University of Pavia, Pavia, Italy

Daniele Merli Department of Chemistry, University of Pavia, Pavia, Italy

Chiara Milanese Department of Chemistry, University of Pavia, Pavia, Italy

Mattia Moiola Department of Chemistry, University of Pavia, Pavia, Italy

Paolo Quadrelli Department of Chemistry, University of Pavia, Pavia, Italy

Foreword

The discovery and development of cycloaddition chemistry, namely Diels–Alder cycloaddition and Huisgen 1,3-dipolar cycloaddition, has helped transform the broad field of organic chemistry. These methodologies have enabled synthetic chemists to design strategies to construct complex molecular architecture with unprecedented control over the number of bonds that can be transformed in a single reaction, stereochemistry, and functional group compatibility. The impact and associated evolution of cycloaddition chemistry are particularly noteworthy, with a number of elegant variations emerging since the pioneering reports. Using an easy-to-read platform and carefully selected examples, this book effectively illustrates how cycloaddition chemistry can help solve modern synthetic and mechanistic problems in organic chemistry. Coupled with the recent understanding of the power of cycloaddition chemistry, an important point highlighted in this book is the development of applications outside traditional organic chemistry. The authors elegantly bring to light this underlying shift in strategy that has enabled cycloaddition chemistry to become a valuable tool for a wide spectrum of researchers.

As mentioned by the authors in their preface, cycloaddition reactions were, are, and prominently will remain the most valuable tools for organic synthesis, and their widespread application beyond organic synthesis makes *Modern Applications of Cycloaddition Chemistry* timely as our scientific community becomes more and more interdisciplinary. It is my hope that this book, dedicated to highlighting how cycloaddition chemistry has helped transform fields ranging from chemical biology to energy harvesting, will inspire new developments in an even wider range of fields. I would like to congratulate Professor Quadrelli, Dr. Memeo, Professor Merli, and Professor Milanese on their wonderful role in bringing this exciting topic to a wider audience.

Javier Read de Alaniz
University of California, Santa Barbara, CA, United States
February 15, 2019

Chapter 1

Cycloaddition reactions for antiviral compounds

Paolo Quadrelli and Mattia Moiola
Department of Chemistry, University of Pavia, Pavia, Italy

Cycloaddition: general aspects

Cycloaddition reactions, whose applications are the main topic of the present book and which were classically classified within the group of pericyclic reactions, have been exhaustively studied by many research groups during the last 50 years. These investigations have been pursued both on the basis of experimental research and upon theoretical methodologies based on ab initio calculations.[1] The evolution of theory and the use of increasingly powerful computers and software increased the level of theory applied to more complex molecular systems. The electron density analysis, as proposed by Domingo's Molecular Electron Density theory,[2] promises to go into deep of cycloaddition reaction mechanisms, and the methods seem to offer novel opportunities with respect to the studies based on simple molecular orbital (MO) interactions, as well as on the Frontier MO (FMO) theory, typically employed in the last century. New theoretical investigations, based on accurate experimental studies, built new, solid, reactivity models for these relevant types of organic reactions that allow a profound rationalization; the new findings imply strong and significant repercussions on cycloaddition reactions and applications in modern organic chemistry and synthesis, specifically. The matter is quite complex, and more and more discussions will appear in literature in the next years as the level of complexity of the theoretical investigation increases with the development of new computational clusters.

The complex matter regarding mechanisms and potential synthetic applications of cycloaddition reactions has been and is still constantly discussed in literature as the level of complexity of theoretical investigations and solution demands increases. The synthesis of molecules with high level of complexity and increasing importance, mainly in the medicinal chemistry field, places the topic of cycloaddition in front of the dichotomy: to be useful and

Modern Applications of Cycloaddition Chemistry. DOI: https://doi.org/10.1016/B978-0-12-815273-7.00001-0

participate in the pivotal step of a synthesis or to be relegated to a simple chemical curiosity, not applicable to solve tremendous needs in modern organic chemistry. Several books and reviews[3] have been published from the late 1960s to nowadays dealing with general and more specific aspects of cycloaddition chemistry.

Cycloaddition reactions are a mature topic of organic chemistry but abundantly used in different areas where chemistry represents the pivotal step for reaching the desired target. Books dedicated to cycloaddition reactions deal typically with the synthesis of heterocycles, general organic synthesis applications, specific applications in natural product synthesis, or they describe the use of a class of organic compounds as partners in cycloaddition reactions.[3] In 2016, a book appeared in the Topic in Current Chemistry Collections dealing with the bioorthogonal chemistry edited by Vrabel and Carrell. This represents a first example of a book dedicated to the view of cycloadditions in modern organic chemistry. However, other subjects still demand a general review since in recent years pericyclic reactions were extensively applied to different chemistry areas such as chemical biology, biological processes, catalyzed cycloaddition reactions, delivery of antiviral or anticancer drugs, targeting of biological active molecules, material chemistry with applications in medicine, photovoltaic processes, photochemistry, energy harvesting, and so on. None of these topics, where cycloaddition reactions spread out, has received a unifying overview in a single book that could offer a solid background for future development in these and other subjects. Moreover, the chronological structure of the references regarding a general topic allows the reader to recognize the relevance of specific subjects during the past decades and the evolution of the synthesis with respect to specific active compounds. This book aims to fill the gap, shedding a new light over pericyclic reactions and demonstrating how these valuable tools may elegantly solve synthetic and mechanistic problems in modern organic chemistry.

The Diels–Alder (DA) reactions are one of the general classes of cycloadditions, which include also hetero DA (HDA) reactions, [2 + 2] cycloadditions, and 1,3-dipolar cycloaddition reactions (Fig. 1.1).

Within this frame, both the inter- and intramolecular versions of these reactions are widely used in organic synthesis. In many cases, these reactions occur at room temperature, but harsher reaction conditions are sometimes required to get the final compounds. Moreover, the use of catalysts has been increasingly developed not only in the way to accelerate the reactions but mainly to ensure enantioselective and diasteroselective processes.

Most DA reactions involve a diene bearing electron-donating substituents and an electron-poor dienophile. However, DA exists also in the inverse electron demand version where the substituents located on diene and dienophile display reversed electronic properties with respect to the classical DA

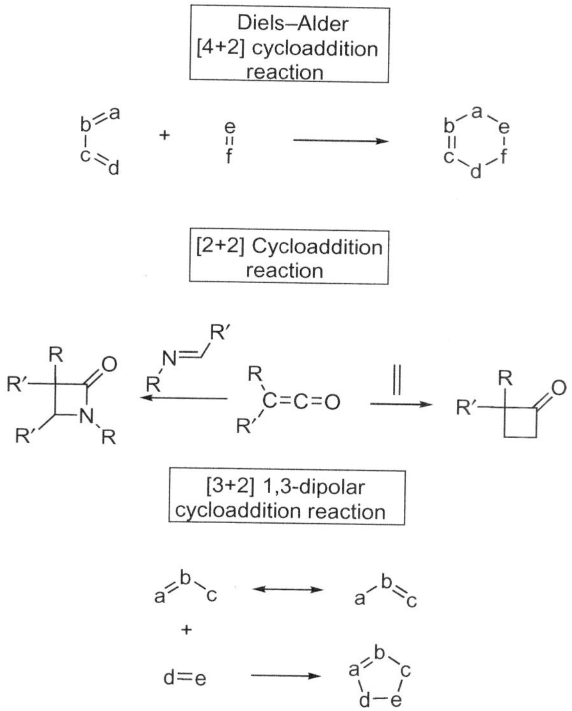

FIGURE 1.1 General schemes for [4 + 2], [2 + 2], and [3 + 2] cycloaddition reactions.

reaction. The two cycloaddends must have complementary electronic character to ensure the usefulness of the DA cycloaddition reactions (Fig. 1.2).

The monumental work of Rolf Huisgen et al. in the 1960s led to the general concept of 1,3-dipolar cycloaddition reactions. These extremely valuable synthetic tools permit the preparation of molecules of higher complexity than the reactants. Few reactions rival these processes in synthetic utility, mainly in the field of heterocyclic chemistry. Five-membered heterocycles are the synthetic target of these reactions and their ability to produce heterocycles extends the importance of the 1,3-dipolar cycloaddition to other areas of organic chemistry, not strictly related to the ring formation processes. Moreover, the introduction of new chiral catalysts allowed for the control of new stereocenters derived from the cycloaddition reactions with the valuable stereochemical outcome borne by the newly synthesized molecules. If these molecules are drugs, this is a key point for determining the success of a synthetic approach.

Mechanisms, reactivities, and selectivities of these reactions have been completely rationalized according to modern theoretical investigations.

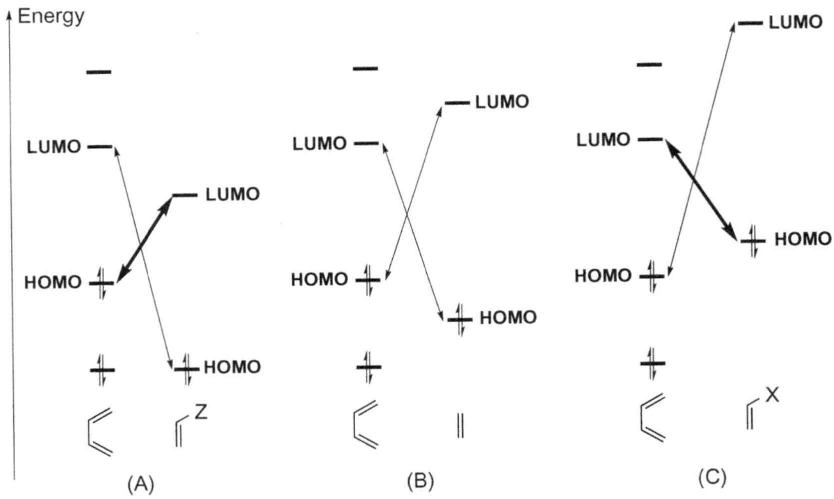

FIGURE 1.2 FMO relative energies. (A) Dienophile with low-energy LUMO; (B) diene + dienophile; (C) dienophile with high-energy HOMO. *FMO*, Frontier molecular orbital.

However, the traditional references to these concepts will still be used to illustrate basic ideas that are in the background of the subjects treated in this book. Figures and schemes will focus reader attention on the pivotal steps of the syntheses highlighting the role of cycloaddition reaction in achieving the best results both in terms of chemical yields and selectivities and in terms of biological activities.

Cycloaddition for antiviral compounds

The virus

The smallest and simplest biological structure, called a virus (from Latin: poison), lacks any type of cellular organization and is forced to live within cells, being intracellular parasites. Viral infections deeply marked human history, as the Spanish conquest of South American territories showed in a paradigmatic picture of death. The development of new and powerful diagnostic methodologies allowed for an increasing capacity to identify new viruses. From the 1960s forward, new frightening hemorrhagic African fevers were discovered and attributed to the filoviruses, while the AIDS appeared in 1981, although HIV was already present in the human population in the 1930s.

Another emergency is the reappearance of viral infections due to virus mutations and resistance to antiviral therapies. This is the case of the influenza viruses; the virus surface proteins have changed and are not recognizable by the human immune system any longer. Sometimes the virus genoma can be exchanged from humans to animals and back to humans, determining new pandemic situations and social alarms. Viral infections are quite detrimental since, contrary to bacteria, viruses cannot be defeated with antibiotics. Until very recently, a small number of antiviral compounds could be used in the pharmacological treatment of these diseases. Luckily, not only are viruses in progress, so are chemical research and synthetic approaches to new and powerful methods for the contrast to viral infections. The demand of antiviral compounds is constantly increasing and chemistry is constantly under examination to furnish novel strategies for a large-scale preparation of active molecules.[4]

Diels–Alder and hetero Diels–Alder reactions and the antivirals

To start discussing the application of DA cycloaddition to the synthesis of value-added molecules, it is worth citing the case of the influenza viruses and specifically the H1N1 virus and how neuraminidase (NA) inhibitors can be prepared with different synthetic approaches, having in common the pivotal step represented by the DA cycloaddition reaction. The research subject offers also a brilliant example of joint efforts from academia and pharmaceutical industries. The influenza infection is known to be a serious health concern, and the cause of morbidity and mortality in many countries. Current options for its treatment and prevention sometimes have severe limitations, underscoring the need for new, effective antiinfluenza agents. Influenza NA is one of the two major surface glycoproteins expressed by influenza viruses. NA promotes the cleavage of the terminal sialic acid residues attached to glycoproteins and glycolipids.[5] Therefore, NA represents a potential target for developing antiinfluenza agents. In fact, in 1999 a merged research group by Gilead Sciences Inc. and Roche Discovery proposed the synthesis of a 1,4,5,6-tetrahydropyridazine derivative and its C-5 epimer, which possess side chains similar to compound **GS4071**. These compounds were synthesized via HDA reaction, and evaluated as influenza NA inhibitors. Both compounds exhibited a μM range of influenza NA inhibitory activity.[6] The strategy followed is based on an HDA cycloaddition reaction between the hydrazone **3** and alkene **4** in the presence of sodium carbonate in acetonitrile to afford a 1:3 mixture of the desired cycloadduct **5** and its regioisomer in 87% yield (Scheme 1.1).

After removal of the Boc group on **5** and saponification, the amino compound **6** could be easily separated from its regioisomer by simple chromatography. The influenza NA inhibitory activity of compound **6** is reported in Scheme 1.1. Compound **6**, having a similar side chain and stereochemistry (*trans*) to **GS4071**, exhibited IC_{50} of 6 μM against influenza A and 62 μM against influenza B. Increased activity was resulted from converting the amino group of **6** into the guanidine functionality. The relatively weak NA inhibitory activity of **6** compared to that of **GS4071** can be explained from

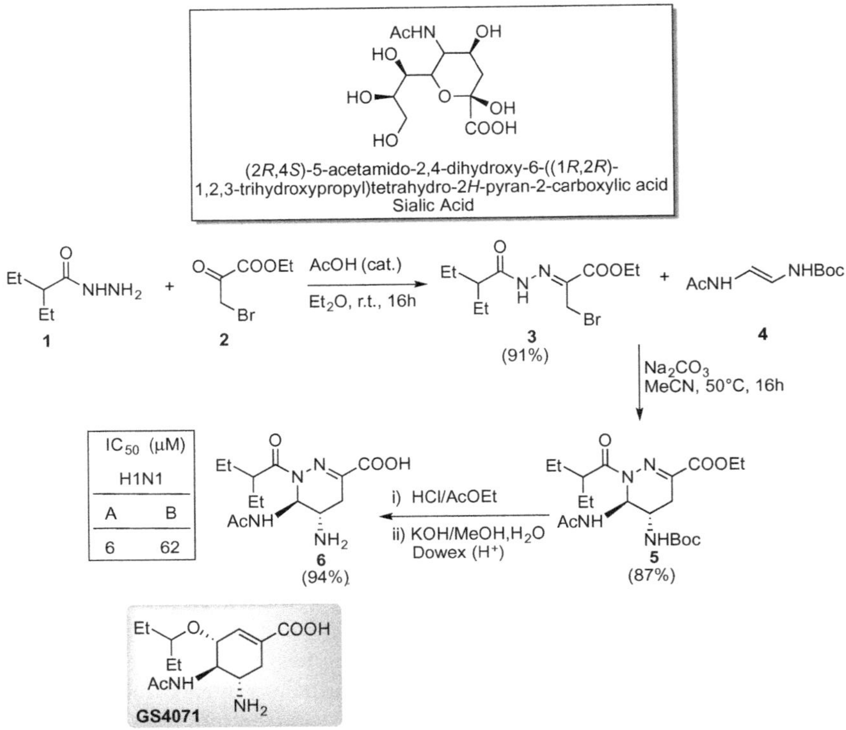

SCHEME 1.1 Synthetic strategy for (5*S*,6*S*)-6-acetamido-5-amino-1-(2-ethylbutanoyl)-1,4,5,6-tetrahydropyridazine-3-carboxylic acid **6**.

conformational and binding studies. These investigations revealed that the NH_2 and NHAc groups of **6** occupy the pseudo-axial positions, rather than the normally preferred pseudo-equatorial positions that are adopted on binding to influenza NA, indicating that an energy penalty is incurred upon the binding. In addition, the X-ray structure of **6** bound to NA reveals a different binding mode for the 3-pentyl side chain as compared with that of **GS4071**. In the case of **GS4071**, the two ethyl groups of the 3-pentyl side chain bind in two different pockets. However, presumably owing to the partly planar nature of the amide bond, the 3-pentyl side chain of **6** poorly fits and exposes the hydrophobic group to water.[6]

A fruitful collaboration between the University of Manchester and a research group of a big chemical company, GlaxoSmithKline, was established in 2001 for the synthesis of new inhibitors of influenza virus sialidases. Structurally related cyclohexene derivatives, Zanamivir **7** and **GS4071** (administered in prodrug form as Oseltamivir), are known to interfere with viral replication by acting as potent inhibitors of virus sialidases. These enzymes, located on the viral surface, cleave the α-glycosidic bond of terminal sialic acid residues. Zanamivir **7** and **GS4071** bind to the active site of a sialidase in an analogous manner. Whereas the trihydroxypropyl side chain binds to a hydrophilic pocket via hydrogen bonding interactions, the pentyloxy function is accommodated in a hydrophobic pocket. The importance of the substituent structures has been demonstrated in several binding studies and structural modification by replacement of selected substituents aimed to tune properly the inhibitory activity. For example, the replacement of the trihydroxypropyl side chain by a dialkylcarbamoyl entity leads to compounds that retain potency against influenza A sialidase but lose effectiveness against influenza B sialidase. Surprisingly, in this series, the guanidino function appears to play no role in binding to influenza A sialidase. Also the location of the C=C double bond seems to be critical and the retrosynthetic analysis has to take into account the generation of regioisomers. The synthesis of two novel compounds, whose inhibition activity was evaluated in comparison with the **GS4071**, is based upon a DA tactic between the nitro acrylamide **8** and the diene **9** (Scheme 1.2).[7]

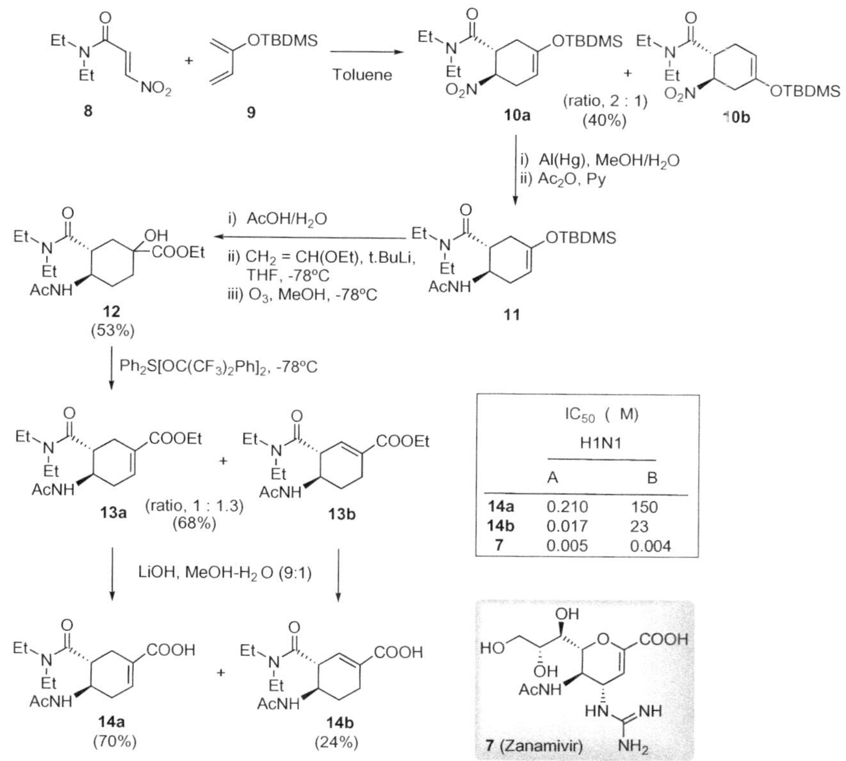

	IC$_{50}$ (M) H1N1 A	B
14a	0.210	150
14b	0.017	23
7	0.005	0.004

SCHEME 1.2 Synthetic strategy for Zanamivir-like compounds **14a,b**.

The reactions afford a separable 2:1 mixture of the regioisomeric cycloadducts **10a,b** in 40% overall yield. The synthesis proceeds by using just the regioisomer **10a** that was reduced by Al(Hg) and acetylated to give **11**, subsequently transformed in three steps into the α-hydroxy ester **12** in 53% yield as a mixture of diastereoisomers (ratio 3:1). Martin's sulfurane dehydration reaction provided a 1:1.3 mixture of compounds **14a** and **14b** (68% yield after chromatography). Finally, hydrolysis of the ester **13a** [LiOH, MeOH–H_2O (9:1), 2 h] was uneventful, leading to the desired acid **14a** in 70% yield. Under similar conditions, the ester **13b** afforded a 4:2:2:1 mixture of products; the most prevalent component [isolated in a near-pure state as a syrup (24% yield) by addition of EtOAc to the mixture and decantation of the liquid] was identified as the required acid **14b**.

The inhibitory activities of the acids **14a,b**, compared with Zanamivir **7**, against influenza A and B sialidases are shown in Scheme 1.2 in the inset. Both compounds are highly selective for influenza A sialidase by a factor of over 700 for **14a** and over 1300 for **14b**. Against influenza A sialidase, the acid **14b** is about 3 times less active than Zanamivir **7** whereas the acid **14a** is about 40 times less active. These findings indicate that the presence of a basic function is not a requirement for high potency. They also reveal that the position of the double bond contributes to, but is not critical for high activity against influenza A sialidase.[7]

The same authors further explored the substituent effects as well as the potentialities of the synthesized compounds in racemic forms by preparing the nitro-cyclohexene **17** by DA cycloaddition of the nitro-olefin **15**a and the diene **16**. The product was obtained in good yields (56%) and submitted to a sequence of well-established chemical conversions to get the final acids **18a,b** in excellent yields (Scheme 1.3).[8]

The inhibitory activities of the acids **18a,b** against influenza A and B sialidases in comparison with **GS4071** are shown in Scheme 1.3. Only compound **18a** showed any appreciable activity. Compared with **GS4071**, it was 70 times less active against influenza A sialidase and 280 times less active

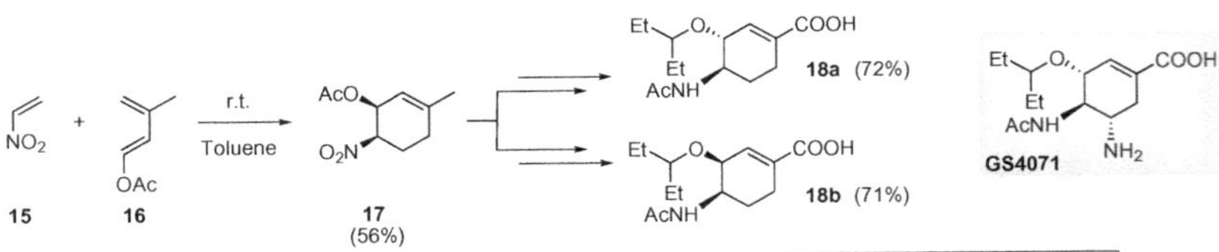

	IC50 (μM) H1N1 A	B
18a	0.071	0.83
18b	172	>700
GS4071	0.001	0.003

SCHEME 1.3 [4 + 2] Cycloaddition reaction strategy for **GS4071**-like compounds.

against influenza B sialidase. These results indicate that the amino group of **GS4071** contributes to but is not obligatory for high potency, although it is important in binding to the sialidase. The poor activity of the acid **18b** demonstrates the inadequacy of the *syn*-arrangement of the pentoxy and acetylamino groups, a relevant structure–activity relationship (SAR) information for future-planned syntheses of potential active analogs. From the synthetic point of view, the excellent *endo*-stereoselectivity realized in the DA reactions is attributable to the presence of the *O*-acetyl substituent in the oxybuta-1,3-diene. In conclusion, the sialidase inhibitory properties of the acids **18a,b** provide new insights into SAR of antiinfluenza agents.

The relevance of the contrast to influenza pandemic is a worldwide concern. As told before, pandemic is attributed to a mutation of influenza virus proteins that allows the virus to evade the human immune system. As result of today's extensive global transport, a local influenza epidemic cannot be restricted to a specific area. Because of the high mutation frequency of influenza virus, an effective antiinfluenza drug should target fundamental molecular processes that are essential and specific for the life cycle of the virus. This principle is based on the hypothesis that structures of fundamental proteins are conserved even in mutant viruses, and NA belongs to these virus proteins. An influenza virion budding from an infected cell binds to a terminal sialic acid residue on the host cell surface glycoprotein with hemagglutinin (HA: Fig. 1.3). NA hydrolytically cleaves the glycosidic bond of sialic acid to release the virus from the host cell surface. This process liberates the budding virion from the infected cell and is essential for spreading the infection. As expected, the active site of NA is highly conserved across the influenza A and B virus strains. Therefore, an NA inhibitor is a prime candidate for broad spectrum antiinfluenza drugs.

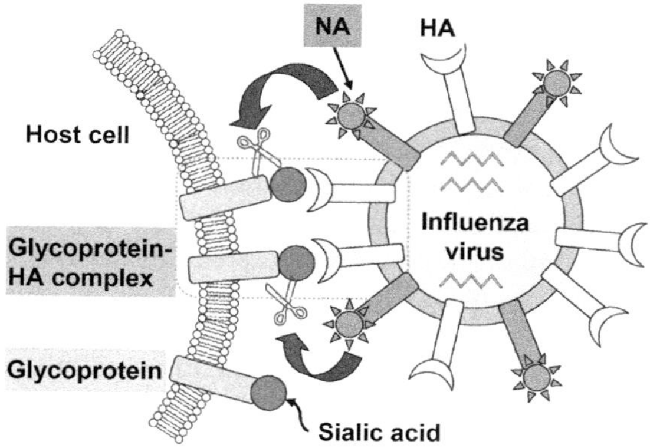

FIGURE 1.3 Schematic representation of an influenza virion budding from a host cell. *From: Shibasaki, M.; Kanai, M. Eur. J. Org. Chem.* ***2008****, 1839–1850.*

Binding studies provided molecular-level structural information for the design of NA inhibitors. Several research groups, both from academia and pharmaceutical industries, have investigated the topic and proposed different synthetic approaches to potential active molecules. Two antiinfluenza drugs, Zanamivir (**7**: GlaxoSmithKlein's Relenza)[9] and Oseltamivir phosphate (**19**: Gilead's Tamiflu, marketed by Roche),[10] inhibit NA by strongly binding its active site. The IC_{50} values of these drugs are in the nanomolar range. Oseltamivir phosphate is an orally active prodrug, active in the corresponding carboxylic acid form, whereas Zanamivir has low bioavailability and is administered by inhalation. These drugs are considered to be effective to treat H5N1 influenza. Although these molecules are relatively small, the development of a practical synthesis that can provide the quantity required worldwide is highly challenging. Some of the literature-proposed synthetic strategies for Oseltamivir phosphate are based on DA strategies.

The Roche purpose of process chemistry is to provide the target compound on a large scale (ton scale) as efficiently as possible without using toxic and hazardous reagents. A successful DA route in this direction is shown in Scheme 1.4. This synthesis started with a racemic DA reaction between furan and ethyl acrylate. The reaction proceeded in the presence of 1 equiv. $ZnCl_2$ affording the thermodynamically more stable *exo*-**20** as major product (ratio: 9:1). Enzymatic resolution of **20** allowed access to the desired (*R*) isomer **21** with 97% ee at 75% conversion (20% yield). Compound **21** was converted into the *endo* aziridine **22** by treatment with diphenylphosphoryl azide through [3 + 2] cycloaddition. After transesterification at the phosphate moiety and aziridine-opening with pentan-3-ol, compound **23** was obtained and submitted to hydrolysis of the phosphoryl amide and hydrochloride formation to give **24**. This latter can be converted into **19** by aziridine formation and introduction of an amino functionality (Scheme 1.4).[11]

In 2006, Corey et al. reported a concise azide-free synthesis of **19** starting from the catalyzed enantioselective DA reaction developed in his group (Scheme 1.5).[12] The DA reaction between butadiene and trifluoroethyl acrylate proceeded at room temperature in a sealed tube without solvent in the presence of 10 mol% triflimide-activated oxazaborolidine catalyst **25**, giving the cyclohexene derivative **26** in 97% yield and with 97% ee. The excellent enantioselectivity was explained by a transition state model, in which butadiene approached the activated dienophile from the *Re*-face opposite the shielding phenyl group of the catalyst. The chiral ligand was recovered efficiently and ammonolysis of **26** followed by conversion into the *N*-Boc-imide **27** (oxalyl chloride and tBuOH) allowed for the successive bromolactamization performed in the presence of LiOtBu and NBS, affording **28**. After elimination of HBr with Diazabicycloundecene (DBU), allylic bromination followed by ethanolysis under basic conditions produced dienyl ester **29**. Regio- and stereoselective bromoamidation of the C3=C4 double bond was achieved by the originally developed method using

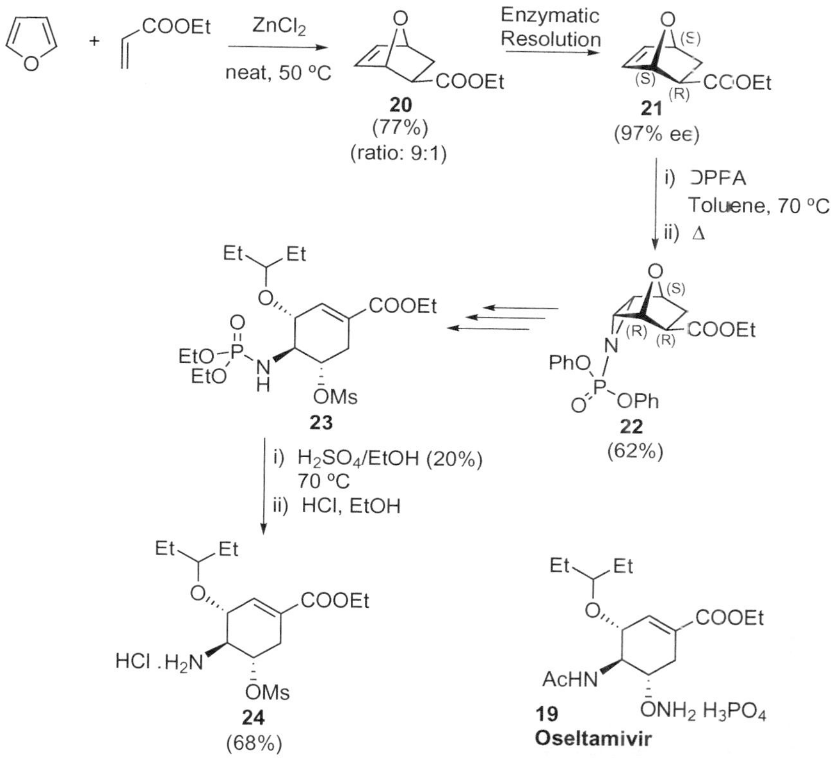

SCHEME 1.4 Synthetic strategy for the ethyl (3*R*,4*S*,5*S*)-4-amino-5-[(methylsulfonyl)oxy]-3-(pentan-3-yloxy)cyclohex-1-ene-1-carboxylate **24**.

N-bromoacetamide in the presence of a catalytic amount of $SnBr_4$ as an activator of the Br^+ donor in acetonitrile as solvent. The reaction proceeded through bromonium ion formation with attractive interactions between the carbonyl oxygen of the Boc group and Br^+, giving **30**.

The conversion of **30** into Oseltamivir phosphate (**19**) was straightforward. Treatment of **30** with tetrabutylammonium hexamethyldisilazide produced an aziridine intermediate; the three-membered ring-opening was accomplished through a Lewis acid-catalyzed (Cu^{2+}) reaction with pentan-3-ol. The cleavage of the Boc group was conducted with H_3PO_4, giving **19**. Besides all the described chemical transformations, it can be clearly seen that the pivotal step of Corey's synthesis is the catalytic enantioselective DA reaction affording the enantiomerically enriched starting compound **26**.

The Shibasaki and Kanai's third-generation synthesis of **19** started from the DA reaction between the (buta-1,3-dien-1-yloxy)trimethylsilane and fumaroyl chloride, run at room temperature over 2 h.[13] After the reaction

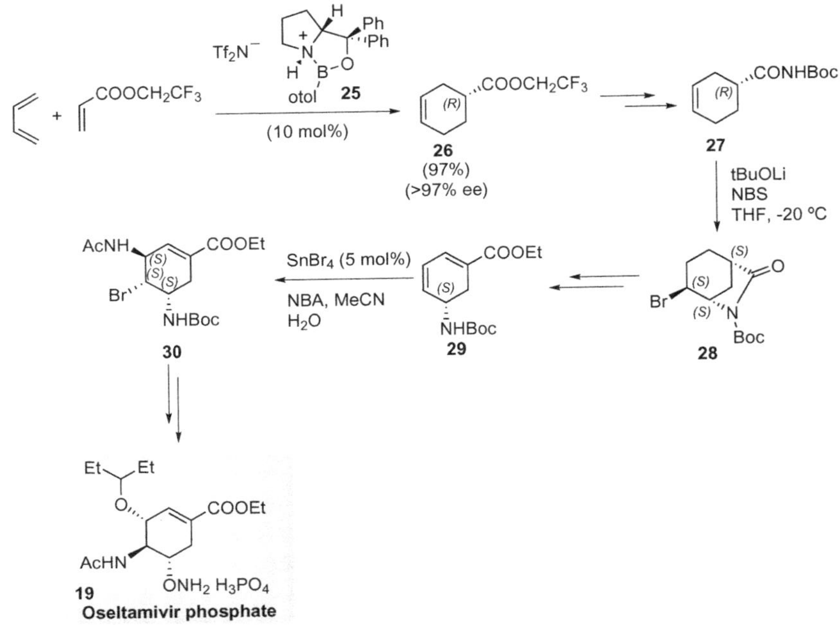

SCHEME 1.5 Synthesis of Oseltamivir phosphate **19**.

was complete, $TMSN_3$ and 4-Dimethylamino pyridine (DMAP) were added to the mixture, and the corresponding acylazide was formed. On quenching of the reaction with HCl, desilylation took place and compound **31** was obtained in 55% yield. Although the DA reaction afforded a 2:1 (*endo/exo*) mixture of diastereomers, the undesired *exo* isomer selectively decomposed during the acidic cleavage of the trimethylsilyl (TMS) ether. A Curtius rearrangement was conducted by heating a tBuOH solution of acylazide **31** at reflux. In this process, two nitrogen atoms at C-4 and C-5 were differentiated, affording **32**. Selective hydrolysis of the cyclic carbonate with LiOH and subsequent *N*-acetylation produced **33**, which was oxidized to enone **34** under modified Moffat conditions with isobutyric anhydride (Scheme 1.6).

The use of the sterically bulky anhydride in this step was essential in order to avoid *O*-acylation. At this stage, the enantiomers were separated by chiral HPLC and the Ni-mediated conjugate addition of TMSCN, followed by α-bromination, elimination of HBr, and stereoselective reduction with LiAl-$(OtBu)_3H$, afforded **35** the same intermediate as in the second-generation synthesis performed by the authors.

Fukuyama's synthesis of **19**, reported in 2007, started with a novel asymmetric DA reaction between the dihydropyridine derivative **36** and acrolein, promoted by MacMillan catalyst (*R*)-5-benzyl-2,2,3-trimethylimidazolidin-4-one (Scheme 1.7), affording the bicyclic aldehyde **37**.[14] Kraus oxidation

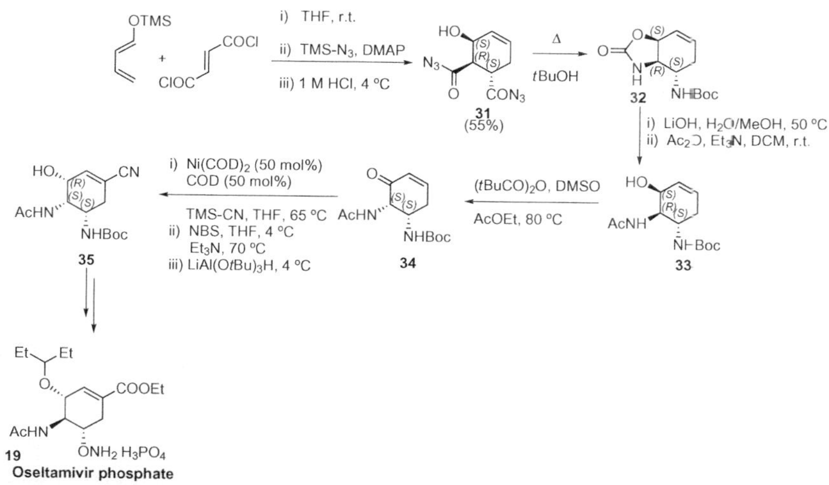

SCHEME 1.6 Synthesis of Oseltamivir phosphate **19** via azido-strategy.

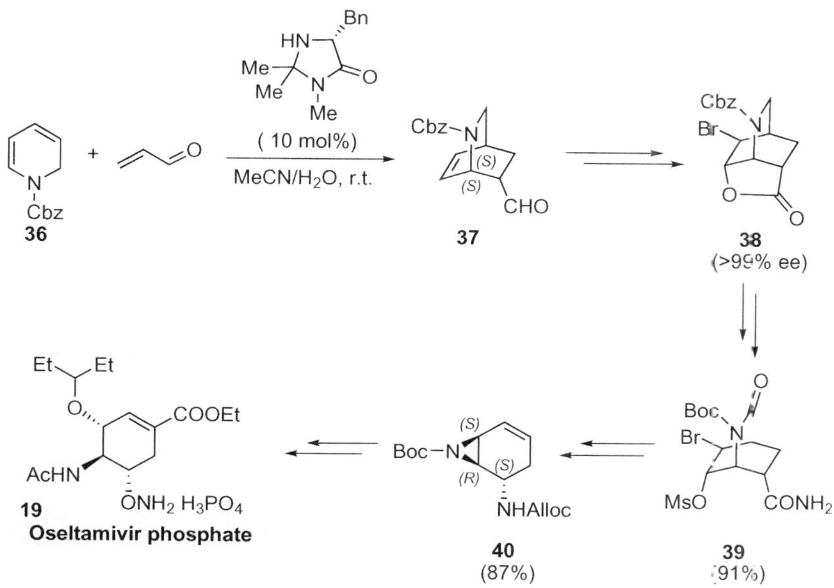

SCHEME 1.7 Synthesis of Oseltamivir phosphate **19**.

followed by bromo-lactonization produced **38** in 26% yield. The moderate yield was attributed to the asymmetric DA reaction, but **38** was easily purified through aqueous acid/base partition and crystallization. Chemically and enantiomerically pure **38** was obtained without column chromatography. The excellent enantioselectivity in the DA reaction can be explained by the model[15] originally proposed by MacMillan: the diene **36** would approach the activated dienophile from the side opposite to the bulky benzyl group located on the catalyst. After the Cbz group of **38** had been exchanged for a Boc group, oxidation with a catalytic amount of $RuO_2 \cdot nH_2O$ and a stoichiometric amount of $NaIO_4$ afforded an imide intermediate. The $RuO_2 \cdot nH_2O$ was recovered for reuse by quenching of the reaction with iPrOH and filtration. Ammonolysis of the lactone followed by *O*-mesylation produced amide **39**, which was subjected to Hofmann rearrangement in the presence of PhI$(OAc)_2$ and allyl alcohol to give an allyl carbamate. Treatment with NaOEt afforded aziridine **40** through ethanolysis of the imide, aziridination, and elimination of HBr in one-pot fashion. An aziridine-opening reaction with pentan-3-ol, cleavage of the Boc group with trifluoroacetic acid (TFA), *N*-acetylation, deprotection of C-5 amine, and phosphate salt formation produced **19**.[16]

Although the yield of the catalytic asymmetric DA reaction requires further improvement, Fukuyama's 14-step synthesis is within the practical range thanks to the following points: (1) the starting material is the inexpensive pyridine, (2) the number of column purifications is minimal, and (3) safe and common reagents are used for synthesis.

Although Oseltamivir (**19**) is a relatively small molecule, developing a practical synthesis that can satisfy worldwide demand in an environmentally friendly and safe way is quite challenging. Roche's current process synthesis relies on naturally occurring shikimic acid as the starting material. There is no doubt, however, that artificial asymmetric catalysts can provide potentially more straightforward starting materials. The synthetic routes developed by three academic groups (the Corey, Shibasaki and Kanai, and Fukuyama groups) support this viewpoint.[17] New synthetic methodologies allow chemists to design conceptually new strategies and, in this sense, improvement in synthetic efficiency is intimately related to progress in synthetic methodologies. Another important research vector is to identify molecules with improved pharmacological effects. Like other drugs, Oseltamivir phosphate is not perfect. The emergence of an Oseltamivir-resistant influenza virus and possible side effects on the human nervous system are two major concerns relating to Oseltamivir. Flexible synthetic routes should allow for the synthesis of analogs in order to address these problems. Hence, organic synthesis can actively contribute to answer biologic questions.

Before a safe and effective vaccine is available to protect the possible pandemic flu, NA inhibitors are the only therapy we have. Because of the oral bioavailability and good toleration, Roche's Tamiflu is recommended as the best choice for therapeutics and prophylaxis of influenza. The spread of the avian virus H5N1 makes the situation of Tamiflu supply and demand increasingly serious. It should be rational to further optimize current manufacturing processes based on increasing the supply of shikimic and quinic acids through improved isolation, fermentation, or synthesis. On the other hand, new synthetic routes that do not involve complex natural products as precursors, but rather cheap and widely available commodity chemicals, should also be developed. Some academic chemists have reported significant results of experimental study. Discovery of new functional molecules is an extremely important task for synthetic chemists.[18]

Regarding the contrast to influenza viruses, an interesting short-step synthesis of a model core structure associated with halenaquinone and related natural compounds[19] is reported for the preparation of the furan-fused tetracyclic compound **43** that has a notable antiviral activity and was concisely synthesized on the basis of *o*-quinodimethane chemistry (Scheme 1.8). Again, the pivotal step is represented by the intramolecular DA reaction of the quinodimethane intermediate **42** with the furan moiety that acts as dienophile.

The Severe Acute Respiratory Syndrome (SARs) of the congeners of **43** were investigated, aiming at the discovery of new candidates for antiviral drugs. As a preliminary study, the antiviral activity was surveyed using the assay method of HA titers. Several derivatives having more potent antiviral activity than the lead compound **43** were obtained and, especially, the Triisopropylsilyl (TIPS) derivative **44** was revealed to have the lowest minimum inhibitory concentration (MIC) value and good therapeutic index.[20] These results prompted the authors to investigate the possibility of revealing dihydrofuran-fused compounds as a new class of antiinfluenza agents

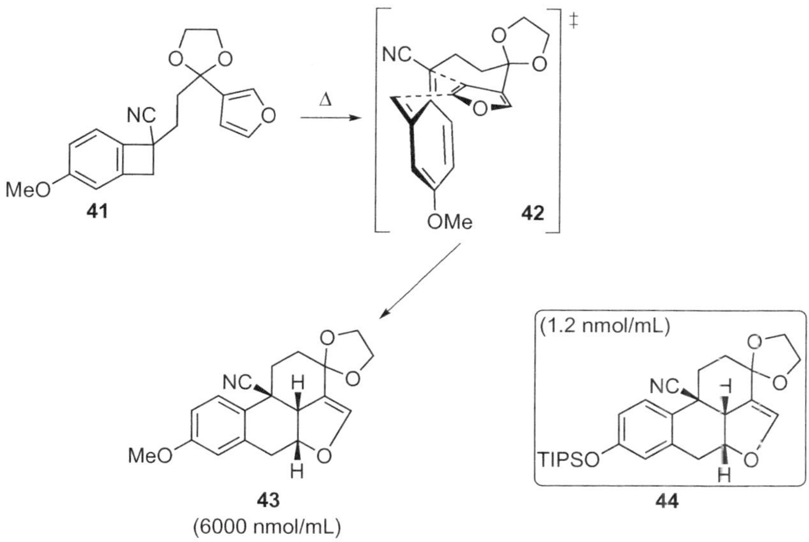

SCHEME 1.8 Intramolecular DA reaction of the quinodimethane intermediate **42**. MIC values for **43** and **44** given in parentheses. *DA*, Diels–Alder.

possessing a novel structural characteristic.[21] The inhibitory activity of the some compounds including **43** and **44** was investigated against viral growth in Madin–Darby canine kidney (MDCK) cells using influenza A/Aichi/2/68 (H3N2 subtype) virus strain at 10 mM drug concentration. The results showed that several compounds inhibited the virus growth and could have potential as new antiinfluenza agents. In particular, some modified structure of **43** exhibited potent activity, suppressing the virus proliferation up to ca. 30% of control.

In summary, influenza being a long-standing health problem, the developments of NA inhibitors promise new hopes for virus treatment. Tamiflu, the phosphate salt of Oseltamivir (**19**), is an orally available NA inhibitor designed to have a cyclohexene scaffold to mimic the structure of the oxonium-like intermediate **47** in the enzymatic cleavage of the terminal sialoside from cell receptors (Scheme 1.9). Many new synthetic procedures of **19** without using shikimic acid have been explored because current industrial manufacturing may have shortages of shikimic acid for global supply during influenza pandemics. In these shikimate-independent syntheses, several different approaches can be utilized to construct the multiple substituted cyclohexene ring: (1) use of existing six-membered rings, (2) construction by DA reactions, (3) construction by intramolecular aldol, Dieckmann and Wittig reactions, (4) construction by tandem Michael–Wittig reactions, (5) construction by olefin metathesis, and (6) construction by Claisen rearrangement. Phosphonate group is generally used as a bioisostere of carboxylate in drug design. The syntheses of Tamiphosphor (**49**) and Zanaphosphor (**50**), which are the phosphonate congeners of **19** and **7**, were accomplished and shown to possess potent inhibitory activities against avian and human

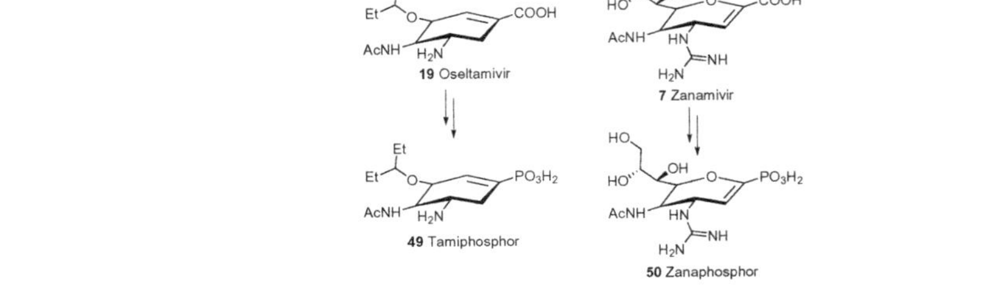

SCHEME 1.9 Enzymatic process toward NA inhibitor syntheses. *NA*, Neuraminidase.

influenza viruses, including the Oseltamivir-resistant strain, according to the NA inhibition, cell-based assay, and mice experiments.

For the first time, phosphonate monoalkyl esters are proved to be real active drugs, but not prodrugs of the parental phosphonic acids. A review covers the design, synthesis, and biological activity of phosphonate compounds as effective antiinfluenza agents.[22]

From the mechanistic point of view, the synthetic approach just reported finds useful applications in other syntheses. In fact, the *o*-quinodimethane intramolecular cycloaddition was applied for the total synthesis of estrane type of steroid.[23] For example, the first asymmetric total synthesis of (+)-chenodeoxycholic acid was achieved via the intramolecular cycloaddition of the *o*-quinodimethane **51** as a key step (Scheme 1.10).[24]

More flexible and efficient routes to both aromatic and nonaromatic steroids were planned on the basis of this type of reaction sequence. In this route, the trans-benzoperhydroindane **53** was set as a key compound for preparing either A-trienic or nonaromatic steroids (**54** or **55**, respectively) by easy manipulation of its benzene ring. In turn, **53** could be accessed by intramolecular cycloaddition of the *o*-quinodimethane **52**, generated in situ by thermolysis of benzocyclobutene **51** (Scheme 1.10).

As a typical example of the application of this methodology, the synthesis of (±)-19-norspironolactone **57** is shown in Scheme 1.11.[25] The synthetic strategy is characterized by the one-step creation of B, C, D, and E rings in a stereoselective manner via intramolecular [4 + 2] cycloaddition reaction of the *o*-quinodimethanes **56**, and then A-ring formation followed by functionalization of the C-7 position. These results show that all the investigated reactions proceed with high stereoselectivity, leading to the preferred formation of the *trans–anti* isomers. Thus it seems possible that the high stereoselectivity might reflect the severe steric interactions present in the *endo* transition

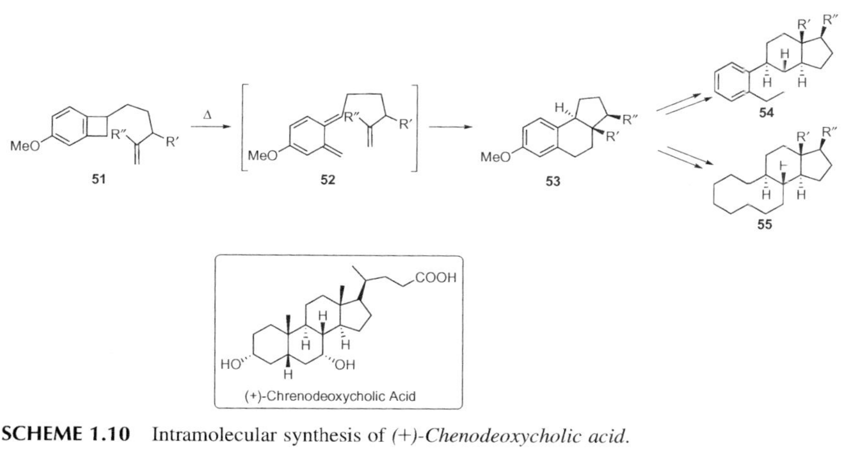

SCHEME 1.10 Intramolecular synthesis of *(+)-Chenodeoxycholic acid.*

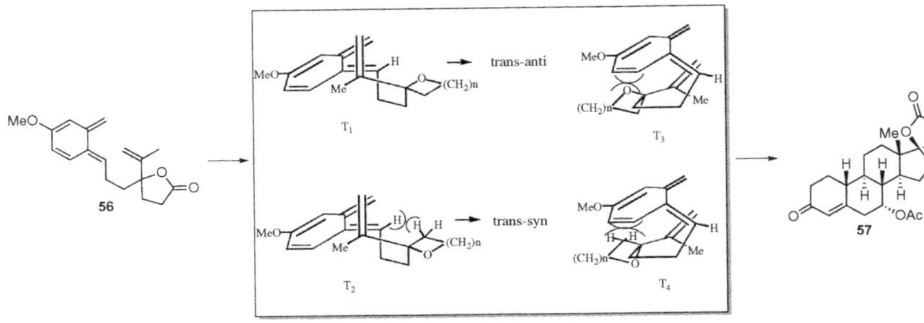

SCHEME 1.11 Transition Structures (TSs) for the intramolecular synthesis of (±)-19-norspironolactone **57**.

states T3 and T4 and the *exo* transition state T2 in comparison with the *exo* transition state T1.

Thus we could obtain information about the stereochemical outcome of the DA cycloaddition reactions of the olefinic *o*-quinodimethanes **56** that have asymmetrically substituted tertiary chiral centers.[26]

The facility of the thermally allowed conrotatory 4π-electrocyclic ring opening of cyclobutenes is known to depend on the electronic nature of the substituents on the cyclobutene ring.[27] In the course of the studies of the substituent effect on this reaction, the exceptionally facile ring opening reaction of cyclobutenes was observed when arylsulfinyl, arylsulfonyl, and diphenylphosphinyl substituents are located on the cyclobutene structures. All the reactions can be carried out in Triisopropylsilyl (THF) at −30°C for 10 min, using *n*-butyllithium as base, and were found to proceed in moderate to high yields.[28] The usefulness of small ring compounds for the syntheses of various pharmacologically important compounds and the novel reaction modes specific to these ring systems have been largely demonstrated. These findings may further serve as a means for the development of new reactions and synthetic methodologies.[29]

The application of DA cycloaddition reactions to the synthesis of antiviral compounds also finds in the hetero version of these [4 + 2] processes a wide panorama of valuable reactions. Here we have focused reader attention to some specific cases that are able to open a wider range of opportunities to access the target compounds.

It has been known for some time that aminosugar derivatives that inhibit glycoprotein processing have potential activity against HIV.[30] These naturally occurring aminodeoxysugars derive either from piperidines, pyrrolidines, pyrrolizines, or octahydroindolizines, and are deprived of the anomeric OH group. Compounds **63a** and **63b**, which are derivatives of (+)-amino erythrose, fall within the pyrrolidine group.

The synthesis of these compounds relies upon HDA cycloaddition reactions of nitrosocarbonyl intermediates.[31] In a previous publication

Tschamber et al. showed that nitrosodienophiles react face selectively with the convex α-side of type **59** azetidinodiazepines to give the direct and the inverse cycloadducts of type **60** in high yields (Scheme 1.12). These results agree with the previously described high face-selectivity of the DA process. The acylnitrosodienophile (i.e., benzyloxycarbonylnitroso dienophile) was prepared in situ from the corresponding hydroxamic acid **58**, according to some known procedures, and reacted at once with the diene component to afford **60**.[31] A primary amino group was introduced on **60** by azide elimination and then acetylation. Catalytic osmylation in the presence of the cooxidant *N*-methylmorpholine *N*-oxide (NMO) gave in excellent yield the expected *cis* diols of type **61**. These were transformed into the corresponding acetonide derivatives (i.e., **62**). One-pot hydrogenolyses (H_2/Pd/C) of the single N–O bond and of the benzyloxycarbonyl moiety triggered molecular rearrangements, which led ultimately to pyrrolidine derivatives **63a,b**. The formation of **63** results from a mechanistically straightforward multistep sequence: after hydrogenolysis of the O–Bn and of the N–O bonds, followed by decarboxylation, the hemiaminal functionality breaks up, leading to an aldehyde that condenses at once with the primary NH_2-C(1) amine. The ensuing Δ^1 pyrroline is then hydrogenated (Pd/C) to give pyrrolidine **63**. This multistep mechanism is akin to the one observed during the catalytic hydrogenolysis of a trihydroxytetrahydroxazine, which led also to a dihydroxypyrrolidine.

Compounds **63a** and **63b** were tested for their anti-HIV activity. These compounds were evaluated in two separate experiments in duplicate in a primary screen against HIV (strain GB 8) in JM cells (3 d assay). Activity was measured by syncytium formation and cytotoxicity in an 3-(4,5-dimethylthiazol-2-yl)-2,5-diphenyl-2*H*-tetrazolium (MTT) assay, using castanospermine as reference compound. In these assays **63a** and **63b** (hydrochlorides)

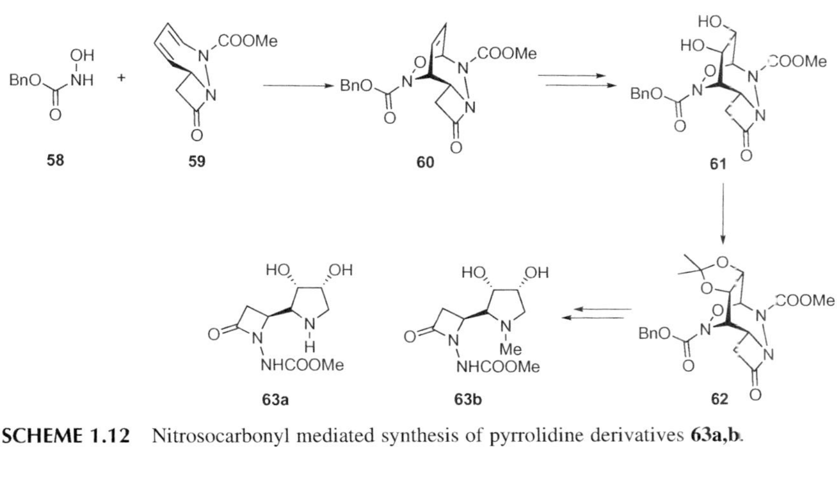

SCHEME 1.12 Nitrosocarbonyl mediated synthesis of pyrrolidine derivatives **63a,b**.

showed some antiviral activity, which was less pronounced than that of castanospermine.[32]

Ulosonic acids are a group of naturally occurring carbohydrates that are involved in many biologically important processes and have attracted considerable attention in the last few years.[33] The seven carbon-atom analog in this series, 3-deoxy-D-*arabino*-hept-2-ulosonic acid (DAH), is formed in plants by stereoselective condensation of phosphoenolpyruvate with D-erythrose 4-phosphate mediated by DAHP synthase, and it has been shown that DAH is a key intermediate in the biosynthesis of aromatic amino acids from glucose (Shikimate pathway). It is therefore not surprising that compounds of this type have become targets of many synthetic endeavors. In 1999, Wu et al. proposed an efficient methodology to the synthesis of the ethyl esters of 2,6-anhydro-3-deoxy-D-gluco-heptanoic acid **68a** and 2,6-anhydro-3-deoxy-D-allo-heptanoic acid **68b** using the chiral catalyst salen–Co(II) complex in the HDA reaction that led to significant improvement in both the yield and the selectivities.[34]

The key step in the synthesis involved the efficient elaboration of the silyl enol ether **65**, which was prepared from D-glyceraldehyde acetonide (Scheme 1.13). The synthesis is based on the HDA cycloaddition reaction between the diene **64** and the ethyl glyoxylate in the presence of salen–Co (II) complex to afford **65** in 65% yield. Hydroxylation of **65** with a catalytic amount of OsO_4 (10 mol%) and 2 equiv of NMO in acetone–water (3:1) at room temperature for 9 h provided a 4:1 mixture of **66** in 90% yield. The diastereoselectivity, however, was unacceptably low and separation of the products was very difficult. The asymmetric dihydroxylation on **65** was conducted by treatment with AD-mix-β[35] at 0°C or room temperature, affording only 10% conversion of the starting material, even after prolonged reaction

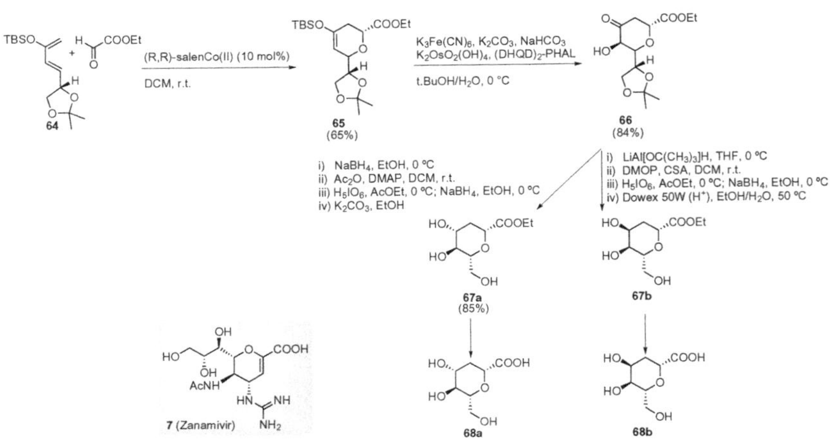

SCHEME 1.13 Synthetic strategy for Zanamivir **7** and related compounds.

time (7 d). Better conversion was later achieved by employing larger amounts of $K_2OsO_2(OH)_4$ (5 mol%); a single diastereoisomer **66** could be obtained in 84% yield. The excellent diastereoselectivity was attributed to the matched interactions between the substrate chirality and that of the chiral ligand.

Reduction of α-hydroxy ketone **66** with $NaBH_4$ at 0°C in ethanol provided the *trans*-diol in 94% yield. The hydroxyl groups were then protected as acetates, followed by selective hydrolysis of terminal *O*-isopropylidene acetals and the oxidative cleavage of the exposed glycol with periodic acid. The following by reduction of the resulting terminal aldehyde with $NaBH_4$ gave compound **67a** in 85% (two steps) isolated yield. Finally, deprotection of **67a** with K_2CO_3/EtOH afforded ethyl 2,6-anhydro-3-deoxy-D-*gluco*-heptanoate (**68a**). Treatment of **66** with hindered lithium tri-*t*-butoxy aluminum hydride in THF at 0°C for 10 h gave a 2:1 mixture of *cis*- and *trans*-isomers in 80% yield. Both the selectivity and the yield were improved (*cis/trans* = 3:1, 90% yield) when the reaction temperature was lowered to −15°C. Protection of the 4,5-dihydroxy as an isopropylidene acetal (91% yield), oxidation and isopropylidene removal by reaction with Dowex 50W (H^+) in ethanol/water (9:1) at 50°C afforded the ethyl 2,6-anhydro-3-deoxy-D-*allo*-heptanoate (**67b**) quantitatively.[36] These esters were useful precursors to the simple analogs of the antiinfluenza agent **GG167** (i.e., Zanamivir, **7**).[37] **GG167** (4-guanidino-2,4-dideoxy-2,3-dehydro-*N*-acetylneuraminic acid, **7**) is a known viral NA (sialidase) inhibitor that, following intranasal administration in ferrets, is at least 100−1000 times more effective than ribavirin and amantadine against influenza A and B viruses. It retains its activity even when treatments are delayed until 24 h postinfection and has no effect on the serum antibody response to infection.[38] The method for the stereoselective introduction of hydroxy groups at C-4 and C-5 illustrated in this work may be also applicable to syntheses of other members in this series of compounds. Some further applications of this strategy in the syntheses of Kdn, Neu5Ac, **GG167** analogs were investigated on the basis of the excellent HDA strategy.

Another promising NA inhibitor known as BCX-1812 (RWJ-270201) is currently under evaluation in clinical trials (Scheme 1.14).[39] This compound demonstrated notable selectivity and potency against NA for a wide range of influenza A and B viruses. It has been emphasized that the relative positions of the interacting functional groups are essential for the NA inhibitory potency, rather than the absolute positions of the central rings. The stereoselective synthesis of tetrasubstituted cyclopentane of BCX-1812 was described by Miller et al., emphasizing the utility of acylnitroso cycloaddition chemistry.[31] As the starting material to synthesize NA inhibitor, the racemic oxazanorbornene derivative (±)-**69** protected by a Boc group was synthesized (Scheme 1.14). Reductive N−O bond cleavage of (±)-**69** using $Mo(CO)_6$ and $NaBH_4$ gave the allylic alcohol (±)-**70**, which was converted

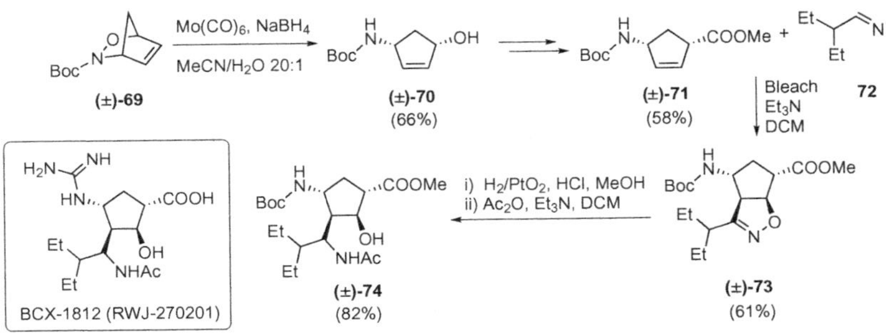

SCHEME 1.14 Nitrosocarbonyl mediated synthesis of BCX-1812 precursor **74**.

to its ethyl carbonate by treatment with ethyl chloroformate. The ethyl carbonate moiety was replaced by the nitromethyl group by reaction with $MeNO_2$ and a catalytic amount of Pd(0), to furnish the primary nitro compound that was transformed into the carboxylic acid by employing a mixture of $NaNO_2$ and AcOH in Dimethylformamide (DMF), followed by treatment with a catalytic amount of (TMS)-Cl in MeOH to give the corresponding methyl ester (±)-**71**. While subsequent elaboration described later was carried out on racemic (±)-**71**, either enantiomer could be prepared starting from the corresponding antipodes of (±)-**70**, available from enzymatic resolutions. The authors utilized a nitrile oxide [3 + 2] dipolar cycloaddition for incorporation of key additional framework components and functionality. The 1,4-*cis*-relationship of the rather bulky Boc-protected amine and methyl ester groups of (±)-**71** was anticipated to induce selective reaction of the nitrile oxide from the opposite face. However, while the regiochemistry of the reaction was less predictable, precedent indicated that it would proceed as desired.

Nitrile oxides can be derived from the corresponding oximes in situ, and it has been generally known that nitrile oxides bearing a wide range of substituents are capable of participating in [3 + 2] dipolar cycloaddition reactions (Scheme 1.14). As the precursor for [3 + 2] dipolar cycloaddition with compound (±)-**71**, oxime **72** containing diethyl units was prepared from the commercially available aldehyde and hydroxylamine (94% yield). The reaction of dipolarophile (±)-**71** and oxime **72**, which was transformed into the corresponding nitrile oxide dipole in situ in the presence of NaOCl and Et_3N, induced a [3 + 2] dipolar cycloaddition to provide bicyclic isoxazoline (±)-**73**. As confirmed by X-ray crystallographic analysis, the desired isomer (±)-**73** was the only isolated product despite the possibility for the generation of other unwanted stereo- and regioisomers. The isoxazoline ring of (±)-**73** was then subjected to hydrogenolysis in MeOH containing PtO_2 and an equivalent amount of HCl at 40 psi. Subsequent acetylation furnished compound (±)-**74** as precursor of the final product BCX-1812.

Upon the removal of the Boc protecting group, compound (±)-**74** was then converted to the corresponding protected guanidino derivative by treatment with 1,3-bis(*t*-Boc)-2-methyl-2-thiopseudourea in the presence of $HgCl_2$. Subsequent basic hydrolysis with 1 N NaOH afforded the corresponding acid in 95% yield. Cleavage of the Boc protecting groups using TFA in CH_2Cl_2 in the presence of Et_3SiH as *t*-butyl cation scavenger furnished the desired final product BCX-1812 in 78% yield.[40]

Moving from sugars or pseudo-sugars to alkaloids, the E-ring decarboxylated camptothecin (**75**) analogs, nothapodytines A and B, were isolated from *Nothapodytes foetida*.[41] Nothapodytine B (mappicine ketone) is an oxidized derivative of natural alkaloid mappicine (**76**), showing significant cytotoxicity in the human KB cell line (Fig. 1.4).[42]

Since nothapodytine B (mappicine ketone) has been identified as an antiviral agent with selective activities against herpes simplex virus (HSV)-1 and HSV-2, this family of alkaloids has received considerable attention by organic chemists. The limited supplies of mappicine ketone made it necessary to develop novel synthetic methodologies for its preparation. An example is offered by the novel synthetic approach to mappicine (**76**), employing the intramolecular HDA reaction as a key step. The proposed synthesis by Toyota, Komori, and Ihara relies upon the chemistry of 1-azabutadienes, although little was known about the practical applications of HDA reactions of 1-azabutadiene derivatives for the construction of polycyclic natural alkaloids.[43]

The starting compound for the synthesis of mappicine (**76**) is the quinoline-azide **77** (Scheme 1.15). The amide formation reaction was carried out to literature reported procedures: the reduction of **77** with wet triphenylphosphine followed by the condensation with fumaric acid monoethyl ester in the presence of BOP and Hunig's base gave the unsaturated amide **78** in 67% overall yield. Intramolecular HDA reaction of **78** led to the cycloadduct **79** (76%), which was treated with aqueous hydrogen bromide to provide the ethyl ester (93% yield) in a single step. The electron-withdrawing substituent group (ethyl ester in **79**) did not impede the cycloaddition; on the contrary, it

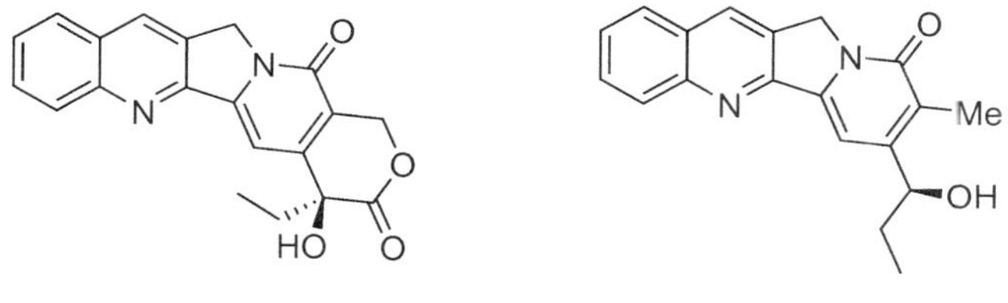

75 Camptothecin

76 Mappicine

FIGURE 1.4 Structures of *camptothecin* and *mappicine*.

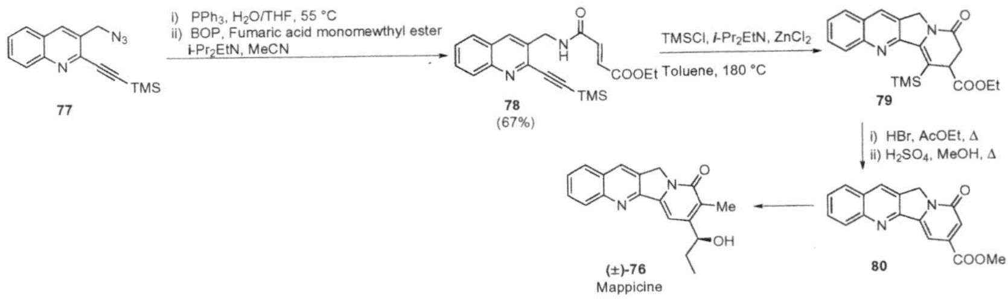

SCHEME 1.15 Synthesis of mappicine **76**.

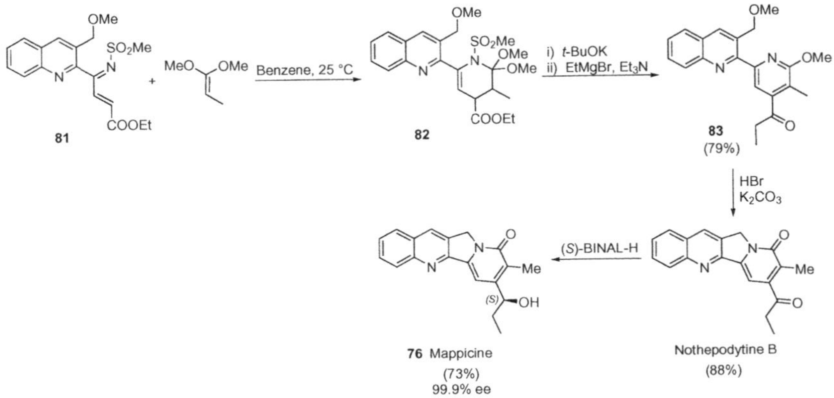

SCHEME 1.16 Enantioselective synthesis of mappicine **76**.

promoted the autoxidation of the corresponding cycloadduct. Finally, the ethyl ester was subjected to transesterification reaction with methanol in the presence of concentrated H_2SO_4 to afford the product **80**. Compound **80** had already been transformed into nothapodytine B and mappicine (**76**) by Kametani et al. by reduction of the ester moiety.[44]

In nearly the same years, American research groups proposed an alternative route to the same target. The key *N*-sulfonyl-1-aza-1,3-butadiene **81** was required for use in the $LUMO_{diene}$-controlled HDA reaction. Treatment of **81** with 1,1-dimethoxy-1-propene at room temperature (12 h, benzene) led to the formation of the sensitive [4 + 2] cycloadduct **82**. Notably, the deliberate incorporation of the noncomplementary C4 electron-withdrawing substituent resulted in an HDA cycloaddition that proceeded at 25°C, presumably by lowering the diene LUMO without altering the inherent [4 + 2] cycloaddition regioselectivity (Scheme 1.16).

Due to the expected sensitivity of the HDA adduct to hydrolysis, subsequent aromatization of the crude adduct (*t*-BuOK, THF, −35°C, 30 min) provided the methoxy-pyridine derivative in good yields (65%) without intermediate isolations. Presumably, aromatization proceeds by initial base-catalyzed elimination of methanesulfinic acid, which is facilitated by the C4–CO_2Me substitution, followed by elimination of methanol.[45] Addition of EtMgBr in the presence of a tertiary amine (EtMgBr, Et_3N, toluene, −10°C, 4 h, 79%) proceeded cleanly to give the corresponding ethyl ketone **83** without competitive tertiary alcohol formation by virtue of tertiary amine-promoted ketone enolization. The final step required the cyclization to form the C ring. This was accomplished in one operation by treatment of **83** with a saturated solution of HBr(g) in CF_3CH_2OH (80°C, 24 h) followed by the addition of K_2CO_3 (25°C, 1 h) to provide the *Nothepodytine B* (mappicine ketone) directly without workup and isolation of the intermediate. This approach worked beautifully to give the ketone in 88% overall yield, and the final product proved identical in all respects with the properties reported for authentic material. Reduction of *Nothepodytine B* (mappicine ketone) with $NaBH_4$ as first described by Kametani and later by Kingsbury and Comins provided (±)-mappicine (**76**).[44,46] Additionally, reduction of *Nothepodytine B* (mappicine ketone) with (*S*)-BINAL-H[47] provided (*S*)-(−)-mappicine (73%, 99.9% ee), which exhibited a CD spectrum identical with that described for naturally occurring material confirming the original absolute configuration assignment.[48] The cytotoxic evaluation (L1210 cell line) of **76** in both (+) and (−) forms are only weakly cytotoxic (IC_{50} = 23 and 13 μM, respectively).[49]

HIV protease inhibitors (PIs) continue to be an essential component of the arsenal in the fight against HIV infection and AIDS. The viral enzyme has the unusual ability to cleave peptide bonds between phenylalanine and proline, and inhibitors incorporating a hydroxyethylamine transition-state mimetic of Phe-Pro such as *Saquinavir* (inset of Scheme 1.17) are very potent and selective inhibitors of HIV protease. The reported synthesis does not allow ready substitution of the carbocyclic ring, a region that could potentially lead to enhanced potency and better pharmacokinetic properties of inhibitors. We have now developed a chiral DA route, which allows ready synthesis of a range of substituted carbocyclic mimetics of scissile Phe-Pro. The dienophile **84** did not react with the 2-(trimethylsilyloxy)-1,3-butadiene under thermal conditions, but reacted smoothly under Lewis acid catalysis to give, after work-up, a 4:1 mixture of diastereomeric trans-ketones in 83% yield from which the required isomer **85** was easily separable (Scheme 1.17).

Oxazolidinone **85** was thus converted to the acid by treatment with lithium hydroxide/hydrogen peroxide and thence to the required *tert*-butylamide **86** via the *N*-hydroxysuccinimide ester.

The equatorial and axial alcohol **87a** and **87b** could be obtained, respectively, by using L-selectride in excellent yields and selectivities. They also could be readily converted to a range of ethers by treatment with sodium hydride in DMF at 0°C and the corresponding alkyl halide. The axial alcohol **87b** could also be converted to the saturated carbocycle **88** by transformation to the mesylate, elimination, and reduction.[50] Following elaboration of the

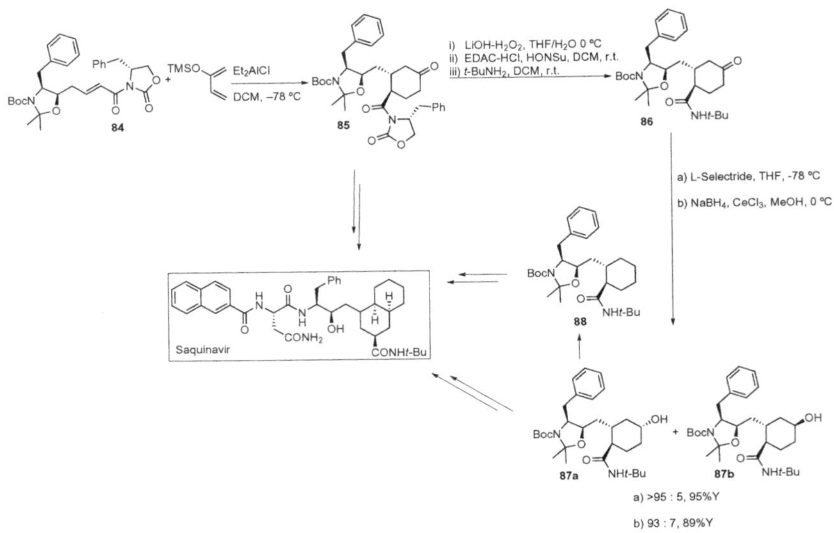

SCHEME 1.17 Synthetic strategy to Saquinavir.

carbocyclic ring, the Boc and acetonide protecting groups were easily removed using dichloromethane/TFA containing a few drops of water to give amines suitable for further elaboration to inhibitors of HIV protease.

(−)-Virantmycin (**96**) (Scheme 1.18) is an unusual chlorinated member of the large class of tetrahydroquinoline alkaloids. It was isolated from a strain of *Streptomyces nitrosporeus* by Omura et al.[51] and was found to be a potent DNA and RNA virus inhibitor as well as an antifungal agent.[52] Several total syntheses[53] of racemic (±)-Virantmycin, natural (−)-Virantmycin, and the unnatural (+)-antipode were proposed, stressing the importance and the promising pharmacological profile of the substance. Various synthetic approaches toward the construction of tetrahydroquinoline ring systems are available, including insertion, ring expansion, and ring contraction reactions. However, one of the most versatile and convergent strategies for the construction of tetrahydroquinoline alkaloids is the HDA reaction of *o*-azaxylylenes, which are formed by a 1,4-elimination from a suitable benzylic halide precursor and suitable dienophiles. One key fragment for our synthetic route toward Virantmycin is arylthiocarbamate **91** (Scheme 1.18). This substance has been known for more than 110 years, but synthetic methodologies for its reliable and high-yield formation are scarce. The classical method developed by Paal and Laudenheimer, heating 2-aminobenzyl alcohol (**89**) in carbon disulfide at reflux for several hours without additional solvent, leads to formation of the desired product **91**, via the intermediate **90**, in a maximum yield of 30%, which can be raised to about 40% if the reaction is performed in a sealed vial. A second possibility includes the reaction of 2-azidobenzyl

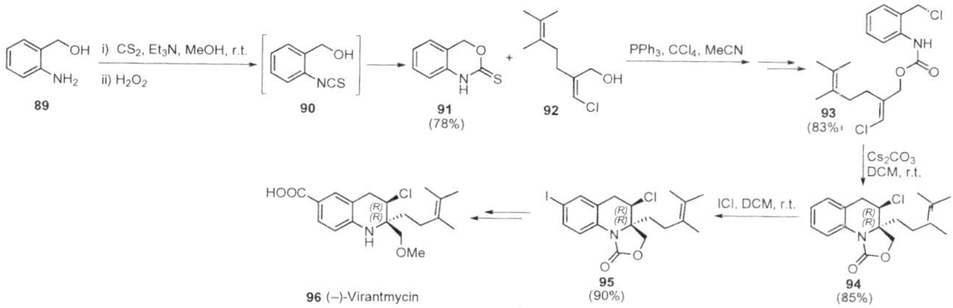

SCHEME 1.18 Synthetic strategy for the preparation of *(−)-Virantmycin* **96**.

alcohol in the presence of triphenylphosphane in a Staudinger-type reaction leading to an iminophosphorane. This intermediate can be reacted with carbon disulfide in a sealed vial at elevated temperature to yield the cyclized product **91** in a yield of 53%.

The allylic alcohol **92** was then allowed to react in an Appel-type reaction with cyclic thiocarbamate **91**. Using our standard conditions, the reaction proceeded smoothly, but purification of the product proved to be cumbersome, as the reaction byproduct triphenylphosphane sulfide cannot be easily removed by column chromatography. Exposure of the dissolved crude reaction product in Dichlromethane (DCM) and treatment with trifluoroacetic anhydride synthesize the desired carbamate **93** in 83% isolated yield due. Having completed the synthesis of advanced carbamate intermediate **93**, the stage was set for the key [4 + 2] cycloaddition: dry cesium carbonate (several hours at <0.1 Torr and 100°C) was added to a solution of **93** in DCM and stirring for 4 d at room temperature smoothly provided tricyclus **94** in a yield of 85%. The completion of the synthesis could be achieved by aromatic iodination using iodine monochloride in DCM, which provided the desired compound as a single regioisomer **95** in a yield of 90%. The synthesis of the advanced precursor **95** was completed in four consecutive steps starting from **91** in an overall yield of 49%. The remaining steps of the total synthesis toward (±)-Virantmycin were already performed.[54]

The identification of biologically active organic compounds is of utmost importance both for basic research and drug development. Besides the classical approach of identifying compounds by in vitro screens (i.e., by binding to an isolated protein), high-content cell-based screens have proven to be a powerful complementary approach.[55] The proposal to employ the frameworks of natural product classes as evolutionary selected and biologically prevalidated starting points in structural space for compound collection development represent an interesting research route, developing a tree-like classification of the natural product frameworks identified in nature. The biology-oriented synthesis of compound collections derived from these natural product frameworks provides a promising strategy in the search of new biologically active agents and can be efficiently carried out on polymer-bound substrates, as they allow rapid diversification without intermediate purification steps. Natural product-derived and -inspired compound collections have been synthesized on the solid support in a number of cases.[56] However, in order to meet the stereochemical demands of natural product frameworks, the use of enantioselective transformations on the solid phase presents a highly attractive though rarely addressed approach. Scheme 1.19 reports the enantioselective solid-phase synthesis of a natural product-inspired α,β-unsaturated δ-lactone collection and its investigation in cell-based screens, monitoring cell cycle progression, and viral entry into cells. The α,β-unsaturated δ-lactone motif belongs to the most frequently occurring scaffolds in nature and is the characteristic underlying structural element of antiproliferative agents, immunosuppressants, and inhibitors for different enzymes, in particular protein phosphatases. For the development of an asymmetric solid-phase synthesis of α,β-unsaturated δ-lactones, the enantioselective oxa HDA reaction between aldehydes and electron-rich 1-alkoxydienes **99** immobilized on the polymeric carrier is the key in stereodifferentiating transformation. 1-Alkoxydienes **99** were synthesized on the solid support starting from a Wang bromo-resin **97** and subjected to enantioselective cycloaddition with ethyl glyoxylate **100** as the reactive heterodienophile (Scheme 1.19).

The enantioselective HDA reaction with ethyl glyoxylate proceeded with 90%–95% ee in the presence of 50 mol% Ti-(*R*)-BINOL complex. The absolute and relative configurations of the cycloadducts were determined after release of the lactones from the resin and comparison of their specific rotations and characteristic Nuclear Magnetic Resonance (NMR) data with the corresponding values reported in literature. Through simple transformations, the esters **101** could be elaborated into the resin-bound final compounds of type **102**. After cleavage from the resin, compounds **103** were tested for their biological activity; the synthesized compound collection was subjected to the aforementioned cell-based screens, monitoring their

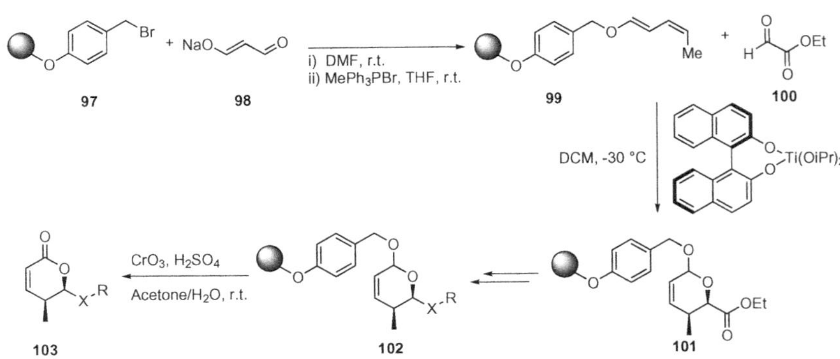

SCHEME 1.19 Solid phase synthesis of lactone **103**.

influence on cell cycle progression. In this screen, BSC-1 cells (from African green monkey) cultured in 384-well plates were treated for 8 h with 30 μM of the synthesized compounds. This cell line was chosen due to its proven advantage in assessing the influence of chemical compounds on the cell cycle.[57] Treated cells were fixed with formaldehyde and stained for chromatin and the actin- and microtubule cytoskeleton. Microscopic analysis revealed that several of the investigated compounds influenced the microtubule cytoskeleton in dividing and/or nondividing cells.[58]

In the previous paragraph we saw that Tamiflu (Oseltamivir phosphate, **19**) has been recognized as a key weapon in the combat of a new type of influenza, and its importance and demand are now increasing. Tamiflu was first developed by Gilead Science and is synthesized and marketed by Roche. Its production is performed starting from shikimic acid, a natural product, so that development of a new synthesis from nonnatural readily available chemical feedstock is required. The synthesis of Tamiflu was investigated by many groups in the world and the total synthesis was accomplished by Corey, Shibasaki, and others.[59] To construct such a polyfunctionalized cyclohexene unit, an HDA approach between pyrroles and acetylene was investigated by a Japanese research group. The synthesis started from the HDA reaction of *N*-Boc-pyrrole **104** (Scheme 1.20). A solution of bromoacetylene carboxylate (**105**) in *N*-Boc-pyrrole (**104**) was heated at 90°C for 39 h and the DA adduct, 7-*tert*-butyl 2-methyl 3-bromo-7-azabicyclo[2.2.1]hepta-2,5-diene-2,7-dicarboxylate (**106**), was isolated in 57% yield. Treatment of **106** with mCPBA gave 5,6-epoxide **107** in 46% yield. The oxidation took place in a stereoselective manner and *exo*-**107** was prepared as a single isomer. The use of oxone for the conversion resulted in the formation of a mixture of isomers. The authors succeeded in improving the yield of **107** up to 64% by repeating the reaction procedure twice. The reduction of **107** and basic hydrolysis yielded the corresponding carboxylic acid, that then opened the epoxide to give tricyclic lactone alcohol **108** in 89% yield in one

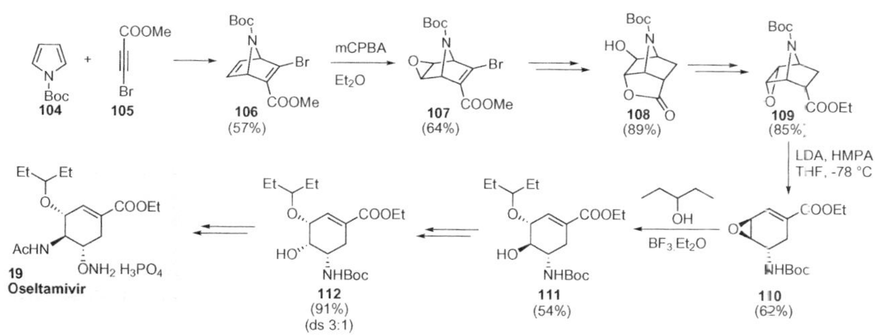

SCHEME 1.20 *N*-Boc pyrrole-based synthesis of Oseltamivir **19**.

step. Other synthetic steps were required to convert the lactone **108** into the *enco*-epoxide **109**, obtained in 85% yield.

The conversion of **109** into the cyclohexene-epoxide **110** was greatly enhanced when Hexa-methylphosphoramide (HMPA) was used as cosolvent of the reaction with Lithium Diisopropylamide (LDA) in THF at −78°C, yielding **110** up to 62% yield in the presence of 1.5 equivalents of HMPA.

The final stage of the synthesis was performed in the following way (Scheme 1.20). Installation of the 3-pentyloxy group was achieved by treatment with 3-pentanol in the presence of BF_3OEt_2 to obtain compound **111** in 54% yield. The epimerization at the hydroxyl group was then carried out by oxidation-reduction pathway. Thus the conversion to the corresponding ketone was achieved by treatment with the Dess–Martin periodinane (DMP). The reduction with $NaBH_4$ smoothly gave **112** in 91% yield. The diastereoselectivity was about 3:1. The conversion of **112** to Tamiflu **19** has already been reported and thus concludes the formal synthesis of Tamiflu from readily available organic feedstock. Overall yield was 5.4%, and all the steps proceeded in a sufficiently stereoselective manner.

The 2-azabicyclo[2.2.2]octanes (isoquinuclidines) are found widely in natural products such as iboga-type indole alkaloids, which have varied and interesting biological properties, and *vinca* alkaloids such as vinblastine and vincristin, which possess isoquinuclidines with the aspidosperma portion.[60] It has been indicated that ibogaine reduces cravings for alcohol and other drugs by means of its ability to boost the levels of a growth factor known as glial cell line–derived neurotrophic factor. In addition, it was also shown that the isoquinuclidines can be used as the synthetic intermediate for the synthesis of Oseltamivir phosphate (**19**). Furthermore, isoquinuclidines are also valuable intermediates in the synthesis of other alkaloids and in medicinal chemistry. It is therefore meaningful to establish an effective asymmetric synthetic methodology for chiral isoquinuclidines. A well-established route to the chiral ring system is through the asymmetric HDA reaction of 1,2-dihydropyridines with dienophiles. However, only a few examples of employing organometal catalysts or organocatalysts have been reported. Despite the obvious advantages of the catalytic enantioselective version using an organocatalyst, to the best of our knowledge, only one example employing a MacMillan catalyst has been reported for the organocatalytic asymmetric version of this reaction, which was used as the key reaction for the efficient practical total synthesis of Tamiflu by Fukuyama et al.[61] The effectiveness of acrolein derivative **114** was investigated using superior catalyst **115** (Scheme 1.21) in the presence of the aza-diene **113**. The reaction was carried out at 0°C in the presence of 10 mol% of **115** to give the DA adduct **116** in 83% yield, >99% ee.

This is the first example of an enantioselective DA reaction of 1,2-dihydropyridine with substituted dienophile using an organocatalyst. Reduction with $NaBH_4$ led to the alcohol **117**, which was finally converted into the desired target **19**.[62]

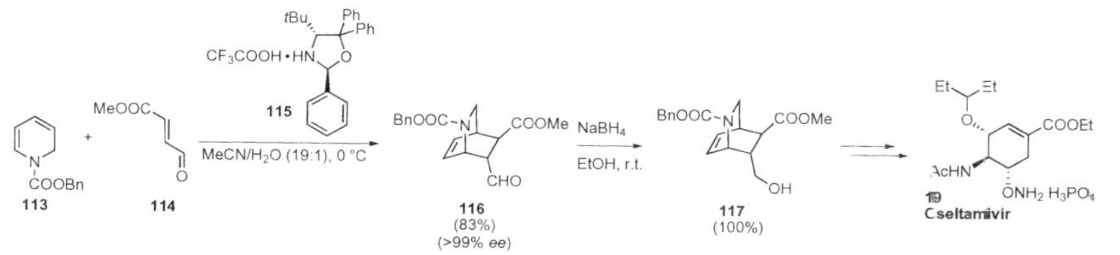

SCHEME 1.21 Enantioselective synthesis of Oseltamivir **19**.

We have already seen that Sialic acids are closely related to carboxylated sugars widespread in mammalian glycoproteins and glycolipids. They are typically found at the outermost end of glycan chains of all kind of cells and are involved in a variety of physiochemical and pathological processes. This class of sugars has structural and modulatory roles and serves as components of binding site for various pathogens and toxins; in addition, another important function they play is decorating microbial pathogens, allowing them to evade host immunity.[63] Influenza virus is subdivided into three serologically distinct types—A, B, and C—all of which recognize sialic acid and its derivatives as their functional receptors. Sialidase is the influenza virus surface enzyme recognizing the *N*-acetylneuraminic acid (Neu5Ac) moiety, which is typically associated as α-linked terminal saccharidic unit of mammalian glycoconjugates.[64] In the search for new sialidase-inhibitor antiinfluenza drugs, the synthesis of a new class of *exo* enitols obtained from Neu5Ac and their use as electron-rich dienophiles in chemo-, regio-, and stereoselective HDA reactions with α,α′-dioxothiones as electron-poor dienes, to form diasteromerically pure β-Neu5Ac derivatives, structurally attractive for the development of new inhibitors of sialidase, was developed by Italian and Greek research groups.[65]

The methyl ester **118** was converted at room temperature into the *tert*-butyldimethylsilyl triflate (TBDMSOTf) to afford the tris silyl derivative **119** in 81% yield (Scheme 1.22). The direct reduction of **119** by treatment with sodium borohydride in isopropanol and the subsequent oxidation to the lactone **120** was realized with sodium periodate in methanol/water as solvent (40% yield calculated over two steps). The desired *exo* enitol **121** was finally obtained by reaction of **120** with Tebbe reagent in pyridine. Compound **121** was isolated in 67% yield as single diasteroisomer.

The heterodienes selected to react with the dienophile **121** were obtained treating the commercially available β-ketoester **122** with phthalimidosolfenyl chloride (PhtNSCl) and pyridine under reported conditions.[65] The α,α′-dioxothione **124** was formed under mild conditions from the phtalimido derivative and trapped in situ by the *exo* enitol **121** in an inverse electron-demand DA reaction. The cycloaddition reaction was totally chemo- and regioselective and the cycloadduct **125** was obtained as single diasteroisomer. The ethyl ester **126**, isolated as pure α anomer, was hydrolyzed with

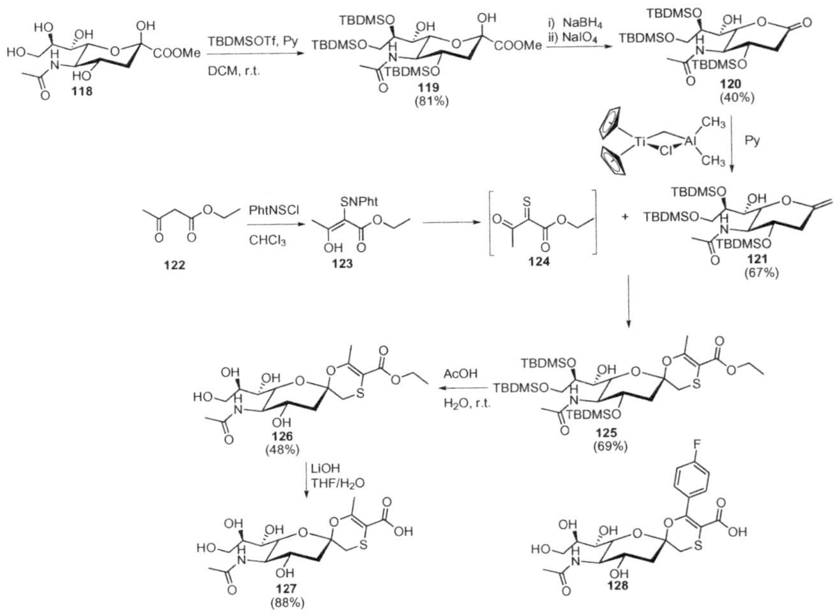

SCHEME 1.22 Synthesis of spiroketal **127**.

lithium hydroxide in THF/H_2O as solvent, to afford, in high yield, the completely deprotected spiroketal **127** as a diasteromerically pure compound.

To obtain a putative binding mode of the newly synthesized compounds within the NA active site, docking calculations were conducted for compounds **127** and the 4-fluorobenzene derivative **128**. Their ethyl ester precursors were not considered in this study, since they are expected to be readily hydrolyzed by endogenous esterases in vivo. The crystallographic structure of the complex between sialic acid (α-Neu5Ac) and the influenza virus sialidase was employed in order to make a direct comparison between **127**, **128** and the interactions of Neu5Ac with the active site residues (Fig. 1.5).

Two models for each compound were prepared, having the pyranose ring either in chair or boat configurations. Bound conformations retained all intermolecular interaction with the NA active site residues (Fig. 1.5A), displaying only minimal differences at the glycerol moiety. The estimated mean free energy of binding for Neu5Ac was −5.8 kcal/mol, corresponding to an inhibition constant of ~55 μM. Analysis of the docking results obtained for compounds **127** and **128** revealed a high number of distinct conformational clusters, many of which adopted a completely different binding mode in comparison with Neu5Ac. As a consequence, most of the hydrogen-bonding interactions displayed by Neu5Ac and the active site residues of NA are lost. On the other hand, docking of **127** and **128** in the same boat configuration as that of the bound Neu5Ac exhibited two conformational clusters that

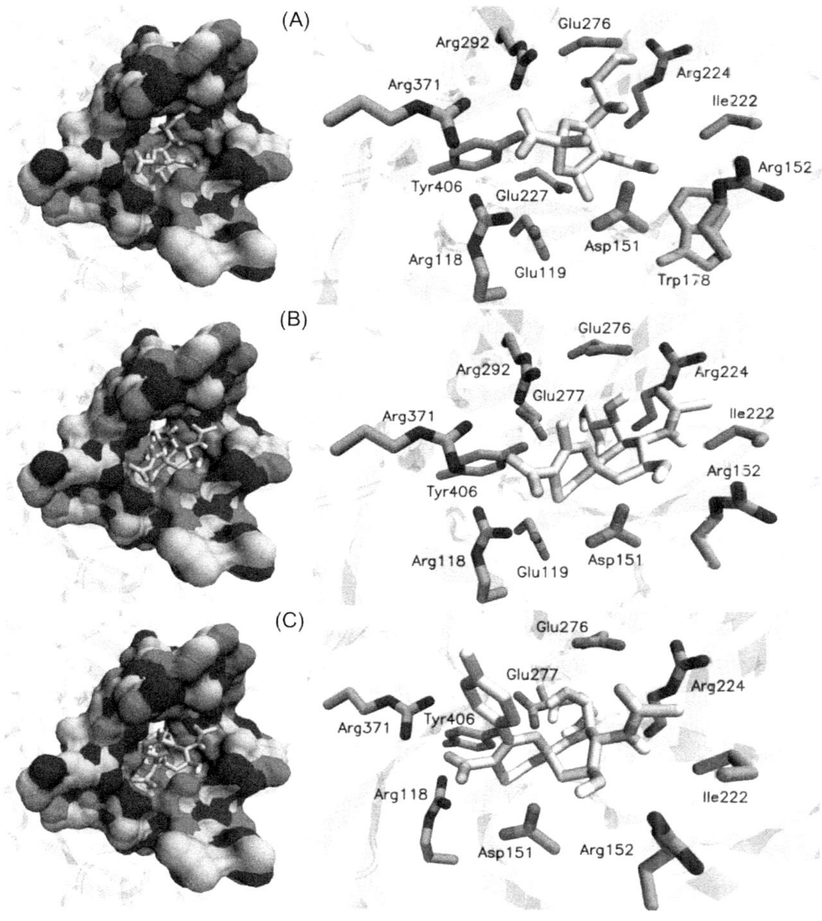

FIGURE 1.5 (A) Crystallographic structure of α-Neu5Ac complex with influenza virus sialidase from PDB ID 2BAT with the active site residues represented as surface (left) and sticks (right). (B) and (C) illustrate the predicted conformations of **127** and **128**, respectively, and the interacting active site residues of sialidase. Protein side-chain atoms are labeled. *From: Richichi, B.; Lunghi, C.; Papakyriakou, A.; Francesconi, O.; Nativi, C. Pure Appl. Chem. 2013, 85, 1803–1811.*

approximate the binding mode of Neu5Ac. For compound **127**, the highest energy conformation of the top ranked cluster (15%), with a mean free energy of binding of −5.0 kcal/mol, is shown in Fig. 1.5B. In this binding mode, the carboxylate group is slightly shifted with respect to that of sialic acid (<1 Å), thus maintaining the hydrogen-bonding network with Arg118, Arg292, and Arg371. However, for the proper accommodation of the oxathiine moiety inside the active site, the pyranose ring of **127** is rotated by

~45° and shifted toward residue Ile222 by ~3 Å, which results in loss of the hydrophobic contacts with Trp178. Compound **127** is predicted to interact with most of the residues displayed by Neu5Ac–sialidase complex, albeit with a lower affinity that is mainly due to a loss of ~1.4 kcal/mol in the electrostatic energy term. As for compound **127**, the *p*-fluorophenyl substituent gives rise to a larger shift of the oxathiine ring toward Arg118, thus losing contact with Arg292 (Fig. 1.5C). In this binding mode, the *p*-fluorophenyl group does not display contacts with any protein residue. In addition, the interaction between C4-OH of **128** and Glu119 is not present, and Glu276 exhibits only van der Waals contacts with the glycerol side-chain. For these reasons, compound **128** is predicted to bind sialidase with an even lower affinity (mean free energy of binding of −4.1 kcal/mol). Conclusively, the oxathiine ring of the spiroketal compounds might displace the pyranose ring from the crystallographic position displayed by sialic acid and its derivatives, though proper functionalization of the spiroketal scaffold could compensate for the interaction with key amino acid residues of NA. The new spiroketal scaffolds could represent a novel class of sialidase inhibitors.

The 5-aryl/heteryl/idene-4-thiazolidinone derivatives constitute an important series of biologically active compounds and important intermediates in modern medicinal and synthetic organic chemistry by possessing synthetic multifunctionality and bioactivity in diverse pharmacological areas.[66]

The synthetic strategy was based on 5-ethoxymethylidene-4-thioxo-2-thiazolidinone (5-ethoxymethylideneisorhodanine) as heterodiene in HDA reaction. Nevertheless, this type of HDA reaction is rarely used and only with a limited number of dienophiles, such as acrylic and maleic acid derivatives and β-nitrostyrene.[67]

HDA reaction of **129** with 1,4-naphthoquinone (Scheme 1.23) afforded **131** through (1) the intermediate dehydrogenation in the presence of hydroquinone excess, (2) the formation of an additional endocyclic double bond, and (3) EtOH elimination.

Furthermore, the HDA reaction of acetylene derivatives as dienophiles in [4 + 2]-cyclocondensation with **129** passed with ethanol elimination and simultaneous conjugated double-bond rearrangement. Thus reaction of **129** with propynoic acid afforded the corresponding 2-oxo-2H-thiopyrano [2,3-*d*]-1,3-thiazole derivative **133**. [4 + 2]-Cyclocondensation of **129** with aroylacrylic acids yielded regioselectively the 6-aryl-2-oxo-3,5-dihydro-2H-thiopyrano[2,3-*d*]-1,3-thiazol-6-yl methanones **135a,b**.

The synthesized compounds **131**, **133**, and **145a,b** were tested for the in vitro anticancer activity on 60 human tumor cell lines at concentration of 10 μM. Derivatives **145a,b**, besides low mean levels of tumor cells growth inhibition, have shown selectivity for leukemia cells. The compound **131** was characterized by high levels of growth inhibition of cancer cell lines SK-MEL-5 (melanoma), OVCAR-3 (ovarian cancer), and MDA-MB-435

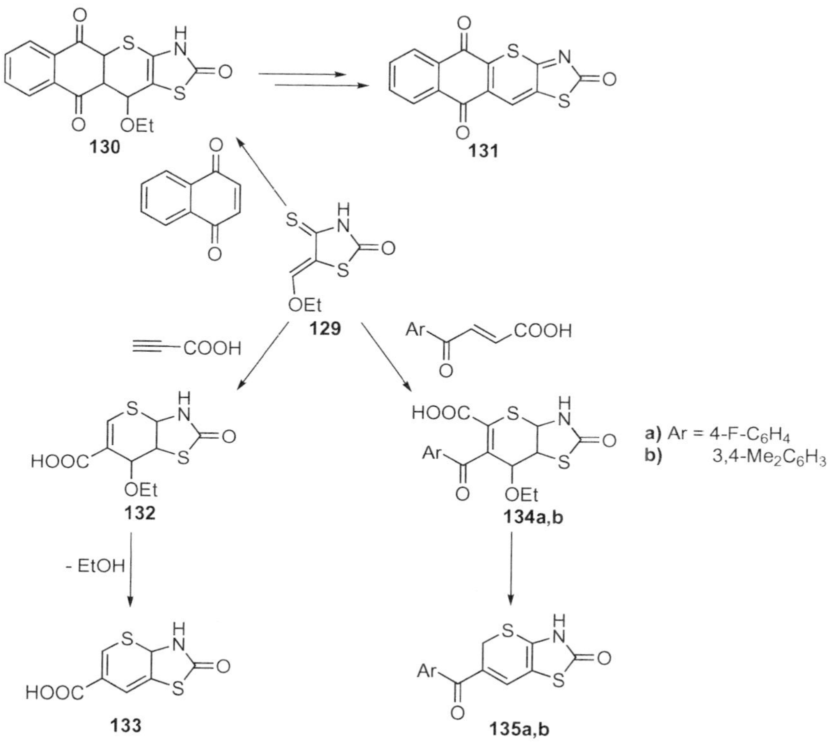

SCHEME 1.23 Synthetic strategy toward compounds **133** and **135a,b**.

(breast cancer). Moreover, compounds **131** and **133** were also tested for antiviral activity according to the antimicrobial acquisition and coordinating facility screening program. The studied compounds did not show significant activity against viruses flu A (H1N1), flu A (H3N2), flu A (H5N1), flu B, measles, parainfluenza virus, respiratory syncytial virus A and SARS, but compound **131** did possess moderate activity against coronavirus SARS.

Azabicyclic compounds are remarkable subunits of alkaloid natural products such as ibogaine, dioscorine, and coronaridine.[68] Among them isoquinuclidine ring systems are of eminent importance because of their potential as bioactive compounds or pharmaceuticals. There are also numerous natural compounds (terpenes and alkaloids) and drugs that contain purely carbocyclic scaffolds. Particularly, derivatives of several bridged carbobicyclic compounds with an exocyclic imine group were reported as being highly active against *Trypanosoma b. rhodesiense* and the resistant K1 strain of *Plasmodium falciparum*.[69]

Several approaches to azabicyclo compounds are known in literature but reactions that enable a convenient direct and metal-free route to complex bridged carbo- and heterobicycles using malononitrile and aldehyde as strikingly simple starting compounds have never been investigated. Because aldehydes are more electrophilic and simple starting materials, Scheme 1.24 reports the use of some aldehydes (1 equiv) with malononitrile 2 (1.5 equiv) in the presence of imidazole at 15 mol% loading.

The synthetic attention was directed to the artemisinin-derived aldehydes as substrates. Since the seminal discovery in 1972 of artemisinin (natural 1,2,4-trioxane sesquiterpene) from *Artemisia annua L.*, by Youyou Tu (Nobel Prize 2015),[70] diverse varieties of artemisinin-derived dimers and hybrids have been synthesized. Natural product hybrid molecules often possess strikingly improved or new biological activities compared with those of their parent compounds, as in the examples of different highly potent artemisinin-derived hybrids with antimalarial, antiviral, and anticancer activities.[71] Thus predicting the potential of these new transformations to structurally diverse and complex scaffolds of value to medicinal chemistry, artemisinin-derived hybrids were prepared. Scheme 1.24 shows two reactions of enantiomerically pure artemisinin-based aldehydes **136** and **138** with malononitrile in the presence of imidazole (15 mol%). Delightfully, new

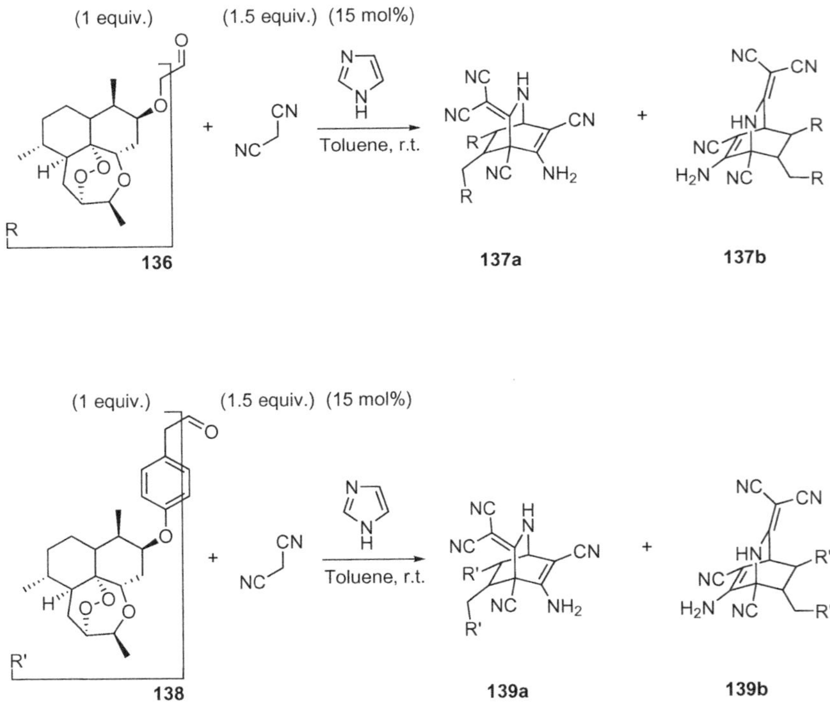

SCHEME 1.24 Malonodinitrile synthetic applications to azabicyclo compounds syntheses.

artemisinin-derived hybrids **137a,b** were synthesized in 52% overall yield, in 5:1 ratio. Hybrids **139a,b** were synthesized in 38% overall yield; a 3:1 ratio of corresponding products was formed.

Fig. 1.6 reports the complex mechanism suggested by the authors supported by Density Functional Theory (DFT) calculations as well as by some Mass Spectrometry (MS) investigations that allowed detection of some synthetic intermediates.

The obtained hybrids were investigated from the biological point of view. Remarkably, hybrids **139a,b** display superior potency against human cytomegalovirus (EC_{50} of 0.071 μm and 0.260 μm, respectively) compared with that of their parent compound artemisinin ($EC_{50} > 10$ μm) and are even more active than clinically used antiviral agent ganciclovir.[72]

Thiazolidinone-based molecules are attractive targets in the rational design of drug-like compounds, which possess antiinflammatory, antioxidant, antitumor, choleretic, diuretic, and other activities. In addition, 4-thiazolidinones are known as inhibitors of PRL-3 and JSP-1 phosphatases, HCV NS3 protease, antiapoptic proteins complex Bcl-XL-BH_3, TNFα-TNFRc-1 complex, Ras farnesyl transferase. Among thiazolidinone derivatives, 5-arylidene-4-thiazolidinones are of wide interest due to their diverse biological activity and clinical applications.[73] In previous studies thiopyrano [2,3-*d*]thiazoles were reported as potential antitripanosomal, antioxidant,

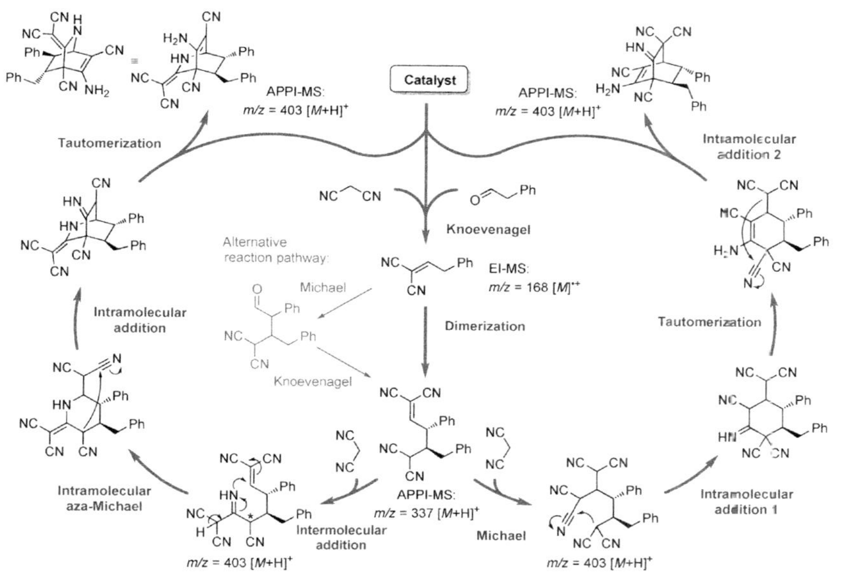

FIGURE 1.6 Suggested mechanism chart. *Modified from: Bock, C. M.; Parameshwarappa, G.; Bönisch, S.; Neiss, C.; Bauer, W.; Hampel, F.; Görling, A.; Tsogoeva, S. B. Chem. Eur. J.* ***2016****, 22, 5189–5197.*

anticancer, and antiinflammatory agents. A series of novel thiopyrano[2,3-*d*] thiazole-6-carbaldehydes were prepared and found active for antitumor and antiviral compounds. The synthesis of these compounds is based upon [4 + 2]-cyclization with acrolein via HDA reaction.

The synthesis of target thiopyrano[2,3-*d*]thiazoles followed the general pathway outlined in Scheme 1.25 for a specific compound. First, the starting 5-arylidene-4-thioxo-2-thiazolidinones **142** were obtained by Knoevenagel condensation of 4-thioxo-2-thiazolidinone **141** with several aldehydes in the presence of EDDA as the base catalyst. The reactivity of the sulfur atom at the 4-position of 5-ylidene-4-thioxo-2-thiazolidinones allows it to be used as a highly active heterodiene component in HDA reactions.

The synthesized compound **143** was evaluated for antiviral activity. Regarding Epstein–Barr virus, it was found that compound **143** showed excellent activity ($EC_{50} = 0.07$ μM, SI = 3279) by VCA ELISA test and possessed moderate activity against HSV-1 (SI = 31) and Varicella zoster virus (SI = 34) by viral Chlorinated Polyethylene (CPE) test.[74]

The design, synthesis, and evaluation of a new class of exceptionally potent HIV-1 PIs was reported by American and Japanese research groups in a study where the structure design strategies are based upon promoting extensive interactions in the active site of HIV-1 protease. The authors have examined and compared X-ray structural variations of critical active site interactions of numerous HIV-1 PIs bound to both wild-type and mutant proteases.[75] Therefore, a promising inhibitor design strategy would be the maximization of the active site interactions and particularly the promotion of a robust network of hydrogen-bonding interactions with backbone atoms of wild-type HIV-1 protease.[76] As depicted in Fig. 1.7, the design model to combat drug resistance involves the design of P2 and P2′ ligands that would form robust hydrogen bonds with polar groups in S2 and S2′ regions. Furthermore, the synthetic plan requires filling the hydrophobic S1 and S1′ subsites with hydrophobic P1 and P1′ ligands.

Among many possible classes of transition-state binders, the incorporation of an (*R*)-hydroxyethylamine sulfonamide isostere can be performed readily varying P1 and P1′ substituents to provide the nonpeptide drug-like

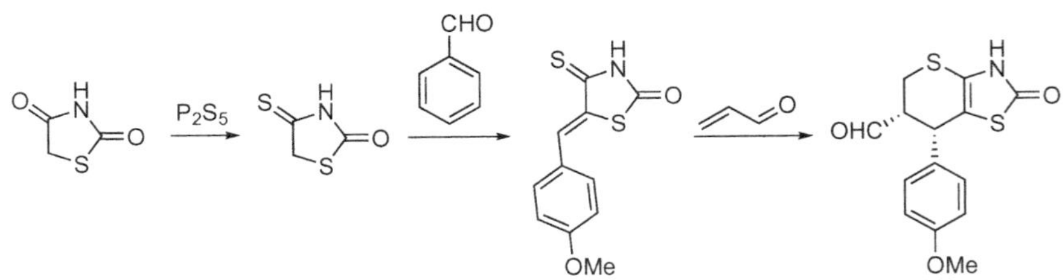

140 141 142 143

SCHEME 1.25 Thiazolidinone approach to compound **143**.

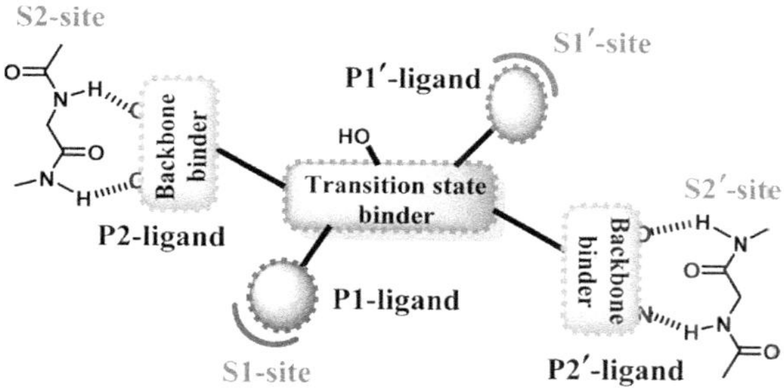

FIGURE 1.7 Proposed model for the design of PIs to combat drug resistance. The model suggests the formation of robust hydrogen bonds by the P2 and P2′ ligands simultaneously in both the S2 and S2′ subsites. The transition-state hydroxyl group binds to catalytic aspartates, and the P1 and P1′ ligands fill in the S1 and S1′ subsites. *PI*, Protease inhibitor. *From. Ghosh, A. K.; Venkateswara Rao, K.; Nyalapatla, P. R.; Osswald, H. L.; Martyr, C. D.; Aoki, M.; Hayashi, H.; Agniswamy, J.; Wang, Y.-F.; Bulut, H.; Das, D.; Weber, I. T.; Mitsuya, H. J. Med. Chem.* ***2017****, 60, 4267–4278.*

scaffold. Hence, the design, synthesis, and X-ray structural studies of a new class of PIs incorporating an unprecedented 6–5–5 ring-fused crown-like tetrahydropyranofura as the P2 ligand with the (*R*)-hydroxyethylsulfonamide isostere is proposed on the basis of a DA reaction of readily available ethyl vinyl boronate **144** and cyclopentadiene at −78°C, in the presence of the in situ generated catalyst **145**, providing the corresponding cycloadduct that was oxidized with H_2O_2 in the presence of $KHCO_3$, leading to the alcohol **146** in 80% yield over two steps in gram scale (Scheme 1.26).

Optical purity of alcohol **146** was determined to be 98% ee by chiral HPLC analysis. Alcohol **146** was converted to bicyclic acetal **147** in a three-step sequence, which was obtained in 62% yield. The lactol mixture was treated with TFA in DCM to afford, after removal of the TBS group with tetrabutylammonium fluoride, the alcohol **148** in 68% yields over two steps. This *exo*-alcohol **148** was converted to *endo*-alcohol by Dess–Martin oxidation followed by reduction of the resulting ketone with $NaBH_4$ in MeOH in 76% yield over two steps. Both *exo*- and *endo*-alcohols were converted to the corresponding mixed activated carbonate derivatives of type **149** in excellent yields. The final step to P2 ligands stands on the 4-methoxybenzenesulfonamide **150** coupling with the activated carbonates of the *endo*- and *exo*-(crown-like)-THF of type **149** in the presence of diisopropylethylamine to furnish inhibitors **151a** and **151b** in good yields.

The examination of the preliminary model of *endo*-(crown-like)-THF containing inhibitor **151b** and *exo*-(crown-like)-THF-derived inhibitor **151a** indicated that the acetal oxygens in the *endo*-derivative are suitably positioned to form hydrogen bonds with Asp30 and Asp29 amide NHs. Also, the tricyclic scaffold of inhibitor **151b** appeared to fill the hydrophobic pocket in the S2 site more effectively than inhibitor **151a**. Inhibitor **151b** containing

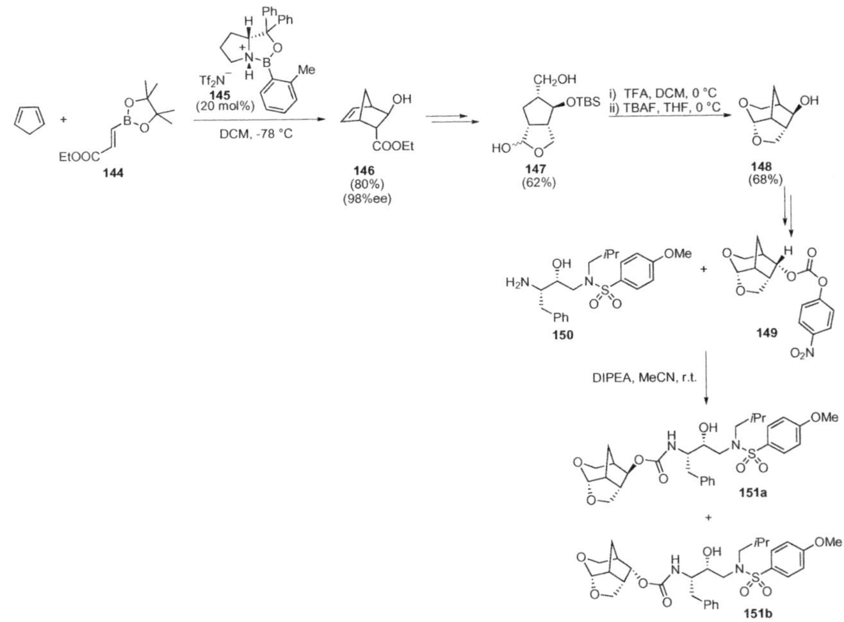

SCHEME 1.26 Enantioselective synthesis of compounds **151a,b**.

endo-crn-THF ligand and a 4-methoxybenzenesulfonamide as P2′ ligand exhibited an enzyme inhibitory K_i of 14 pM, compared to inhibitor **151a** containing *exo*-(crown-like)-THF ligand (K_i of 10.8 nM).

The X-ray crystal structure of the wild type HIV-1 protease cocrystallized with inhibitor **151b** was refined. It contains the protease dimer and the inhibitor bound to HIV-1 protease in two orientations related by a 180° rotation with 55%/45% relative occupancies (Fig. 1.8).[77]

The inhibitor is bound in the active site cavity by forming a series of hydrogen bonding interactions and numerous weaker CH· · ·O interactions with the main chain atoms of HIV-1 protease. The major conformation of the inhibitor forms hydrogen bonding interactions of its urethane NH with the carbonyl oxygen of Gly27. The inhibitor also forms tetracoordinated water-mediated interactions connecting the inhibitor carbonyl oxygen and sulfonamide oxygen with the amides of Ile50 and Ile50′ in the flaps. Furthermore, the *p*-methoxy group of the P2′-sulfonamide forms a hydrogen bond with the amide NH of Asp30′ as well as with its side chain carboxyl group.

We move now to the field of terpenes. In 1972 Hesp et al. reported the isolation and structure characterization of the tetracyclic diterpene *Aphidicolin* **157** (Scheme 1.27), produced by the fungus *Cephalosporium aphidicolia*.[78] *Aphidicolin* **157** shows marked activity against Herpes simplex

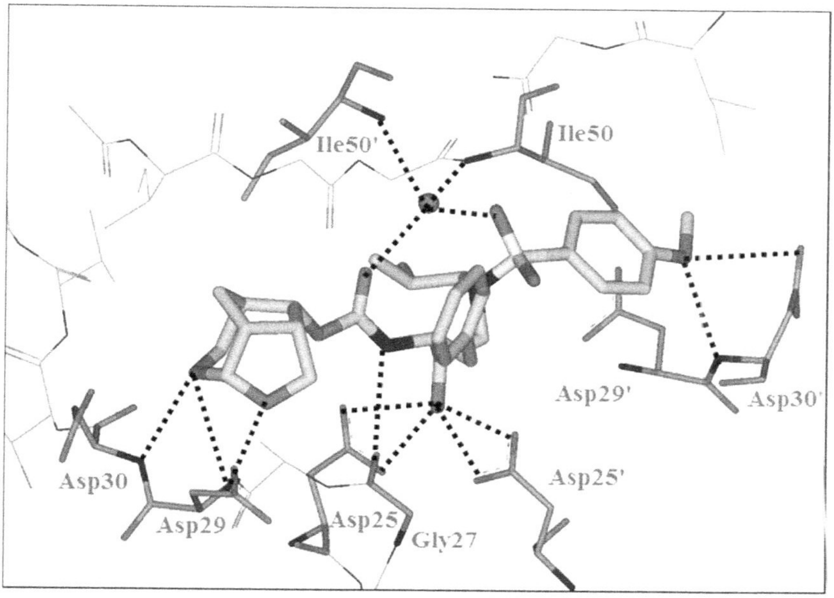

FIGURE 1.8 Inhibitor **151b**-bound X-ray structure of HIV-1 protease (PDB code 5ULT). The major orientation of the inhibitor is shown. The inhibitor carbon atoms are shown in gray, water molecules are black spheres, and the hydrogen bonds are indicated by dotted lines. *From: Ghosh, A. K.; Venkateswara Rao, K.; Nyalapatla, P. R.; Osswald, H. L.; Martyr, C. D.; Aoki, M.; Hayashi, H.; Agniswamy, J.; Wang, Y.-F.; Bulut, H.; Das, D.; Weber, I. T.; Mitsuya, H. J. Med. Chem.* ***2017****, 60, 4267–4278.*

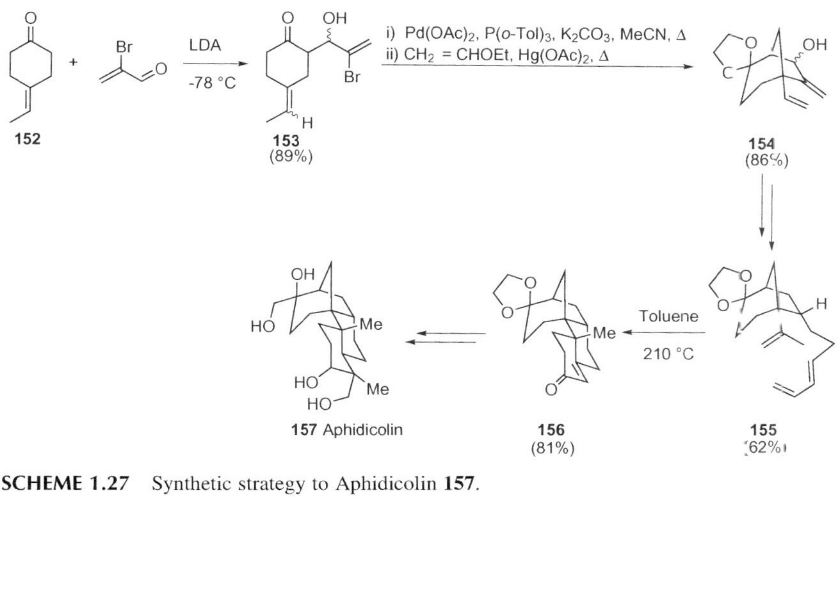

SCHEME 1.27 Synthetic strategy to Aphidicolin **157**.

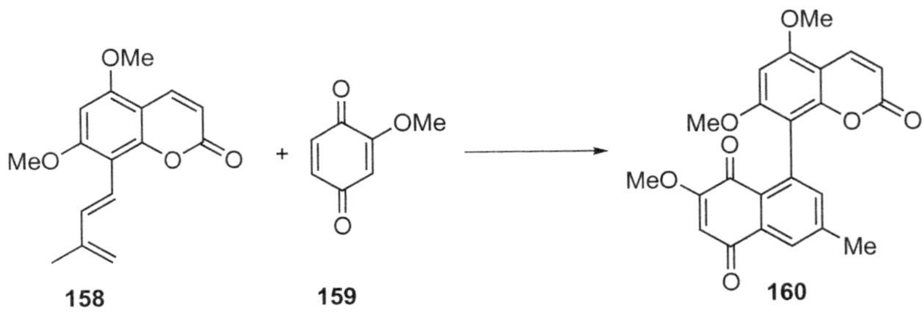

SCHEME 1.28 Synthesis of quinone **160**.

Type I virus both in vitro and in the rabbit eye. The synthetic approach relies upon the condensation of the cyclohexanone derivative **152** with 2-bromo-acrolein in the presence of LDA to afford **153** that is converted into the intermediate **154** through a critical Heck cyclization.

From **154**, via Claisen condensation and Wacker oxidation, the diene **155** is obtained in 62% yield, leading to the product of the intramolecular DA reaction **156**, in 81% yield. In few other steps, **156** is converted into the target compound **157**.[79]

It is well-known that a number of natural products contain a quinone moiety. They are antibiotics such as mitomycin C and anthracyclicnes and show remarkable biological activities. Toddacoumaquinone **160** is a unique coumarin-naphtoquinone dimer isolated from *Toddalia asiatica* and its synthesis was planned through a DA cycloaddition of the diene **158** with the quinone **159** (Scheme 1.28).

Compound **160** was tested for antiviral activities against HSV-1 and HSV-2 using acyclovir as control compound and showed a weak activity against both virus strains.[80]

The leaves of the rare littoral plant *Stemodia maritima* L. (family Scrophulariaceae) have long been used as a folk medicine in the Caribbean Islands for the treatment of venereal disease. Attracted by the reported medicinal properties, Manchand et al. carried out phytochemical investigations on this plant in 1973, isolating and characterizing two structurally unique tetracyclic diterpenes, stemodin and stemodinone **165** (Scheme 1.29).[81] Stemodane diterpenes, by virtue of their unique tetracyclic carbon skeleton, their alleged medicinal properties of *S. maritima*, and their close structural resemblance with aphidicolin, prompted a considerable number of synthetic investigations on this diterpene family.

The planned synthetic route proceeds via the dienol ether **162**, which was derived from **161** by the treatment of LDA in the presence of trimethylsilyl chloride (TMSCl). Thus, the crude ether **162**, without further purification, was converted to the bicyclic enedione **163** via DA reaction with 3-butyn-2-one, followed by desilylation of the resulting silyl enol ether. The bicyclic

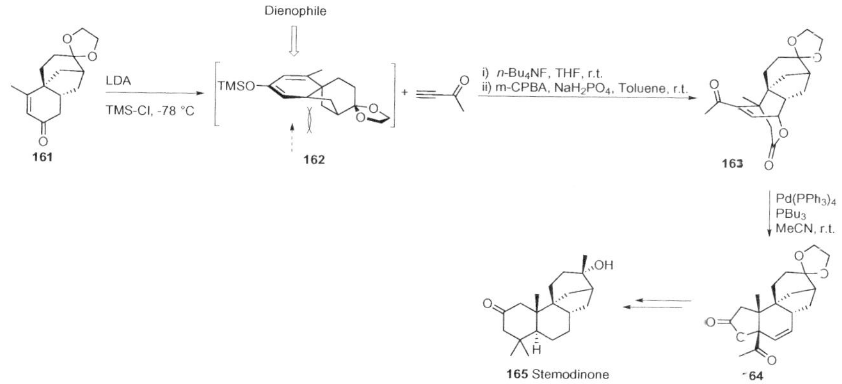

SCHEME 1.29 Synthesis of stemodinone **165**.

seven-membered lactone **163** was then easily isomerized to the *cis*-fused five-membered lactone **164**, having the desired stereochemical feature at C-10, by the Pd(0)-catalyzed lactone migration reaction. The final compound **165** was obtained in few more steps in the racemic form.[82]

Ascochlorin (**174**) is an antiviral antibiotic obtained from the filter cake of the fermented broth of *Ascochyta viciae* Libert. It has a strong inhibitory effect on viral growth in cultured cells.[83] The compound features are three stereogenic centers, a sesquiterpene unit, a 1,3-diene system, and an aromatic unit having an aldehyde, a methyl, a chloride, and two hydroxy substituents. Due to its high biological activity, the ascochlorin family has attracted the attention of synthetic chemists. Among the total syntheses of (±)-ascochlorin reported in the literature, an example of asymmetric DA reaction, followed by cuprate, and Julie olefination reactions as the key strategies in the total synthesis of this biologically active natural compound is reported. For the synthesis of a sesquiterpene unit, the tigloylsultam **166** was allowed to react with 1,3-butadiene to give the desired DA adduct **167** in 85% yield and in 96% de. The chiral auxiliary was easily removed by treatment of cycloadduct **167** with LiOH in the presence of H_2O_2, which gave cyclohexenecarboxylic acid **169** in 94% yield and the recovered sultam auxiliary **168** in 80% yield (Scheme 1.30).

Standard iodolactonization methodology was then applied to the crude carboxylic acid **169**. The reaction yielded the desired iodolactone **170** in 98% yield. Cleavage of the lactone unit with $LiAlH_4$ and oxidation of the allylic alcohol with MnO_2 afforded the desired unsaturated ketone. Reduction first, and oxidation of the resulting alcohol by Swern oxidation reaction conditions or by TPAP8 cleanly gave aldehyde **171**. Julie olefination with **172** afforded the desired compound **173**, precursor of **174**.[84]

In 1988, efforts to discover novel microbial metabolites led to the isolation of new classes of natural products, called pradimicins and

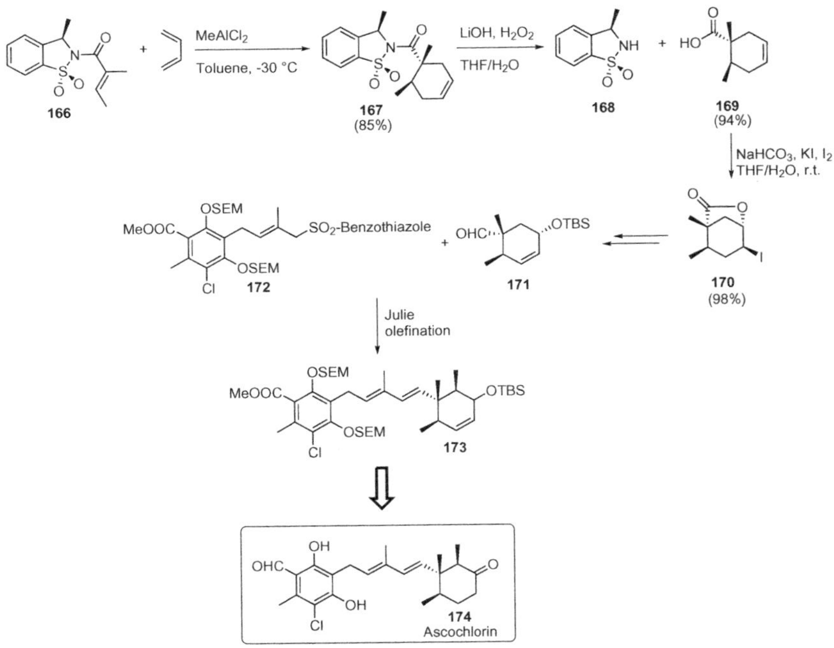

SCHEME 1.30 Synthetic strategy of Ascochlorin **174**.

benanomicins.[85] Pradimicins A, B, and C and were shown to be active against various fungi, yeasts, and viruses.[86] The unique biological activity that this class of compounds exhibits and the ongoing need for antifungal and antiviral treatments make it a valuable target in synthetic chemistry. The structures of the pradimicins and benanomicins consist of an amino acid, a disaccharide, and a benzo[*a*]naphthacenequinone core. There are approximately 20 congeners in this family to date, and their specific structures vary according to their amino acid and disaccharide components. Pradimicin A **181** (Scheme 1.31), one of the first and most abundant congeners isolated, was initially subjected to biological testing, but was claimed to have limited solubility in water. To overcome this issue, testing was also performed with some analogs, which have different and more water-soluble amino acid and disaccharide functionalities.

Pradimicins have shown activity against numerous pathogens, including systemic fungal infections and, potentially more importantly, against HIV. Scheme 1.31 shows an alternative route to Pradimicin A **181** that starts with protection of the alcohol **175** as TIPS derivative **176**, followed by deprotection of the triple bond to give the alkyne **177**. MW promoted DA cycloaddition reaction with Dimethylacetylenedicarboxylate (DMAD) afforded in 1 min the cycloaddutct **179**, subsequently oxidized with DDQ to **180**, the precursor of the final compound **181**.[87]

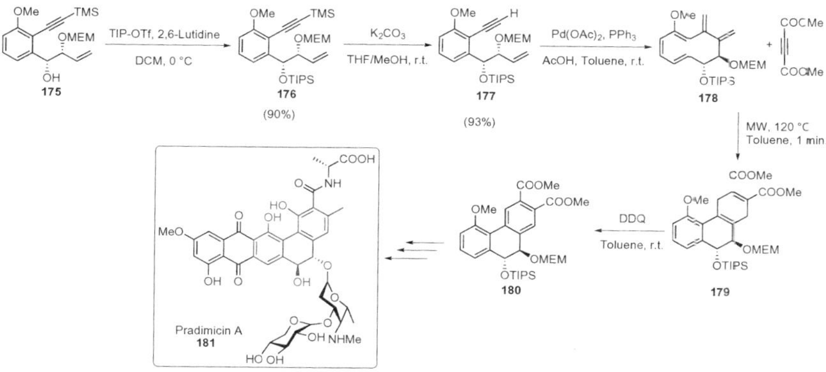

SCHEME 1.31 Synthetic strategy to Pradimicin A **181**.

Panduratin A (**188**) is the prototypical member of a family of related cyclohexenyl natural products and is isolated as a racemic mixture from *Boesenbergia rotunda*, a perennial herb found in Southeast Asia (Scheme 1.32).[88] The plant has been used in traditional medicine to treat conditions such as diarrhea, asthma, indigestion, dysentery, and many others. Studies have shown that these extracts have anti-HIV, antibacterial, antifungal, and antitumor activity. The synthesis proposed by an Australian research group relies upon the DA cycloaddition of methyl cinnamate (**182**) and (*E*)-ocimene (**183**), performed under high pressure conditions (19 kbar), to afford the regioisomeric panduratins **184a** and **184b** in a 1:2.9 ratio. Subsequent oxidation of alcohol, derived from **184a** by reduction with $LiAlH_4$, with the DMP provided aldehyde **185** as the branching point toward the final target. Addition of the organolithium reagent derived from bromobenzene **186** followed by DMP oxidation gave methoxymethyl ether (MOM)-protected ketone **187**. Removal of the MOM groups with acidic methanol cleanly afforded Panduratin A (**188**).[89]

The regiochemical outcome of the DA cycloaddition between methyl cinnamate (**182**) and (*E*)-ocimene (**183**) was in accord with an analysis based on computationally derived FMOs (Fig. 1.9).

Alignment of the larger diene HOMO ($2P_z$ atomic) coefficient with that of the corresponding dienophile LUMO coefficient predicts formation of ester **184b** as the major regioisomer.

Wickerol A (**194**) proved to be highly active against the type A/H1N1 influenza virus with an IC_{50} of 0.07 μg/mL. The natural product features a unique tetracyclic carbon skeleton that comprises two adjacent quaternary carbons and seven or eight stereocenters, respectively, two of which are quaternary as well. Although the 6–5–6–6 skeleton of the Wickerols contains only five- and six-membered primary rings, the molecule is remarkably strained due to numerous *syn* pentane interactions. Despite the formidable

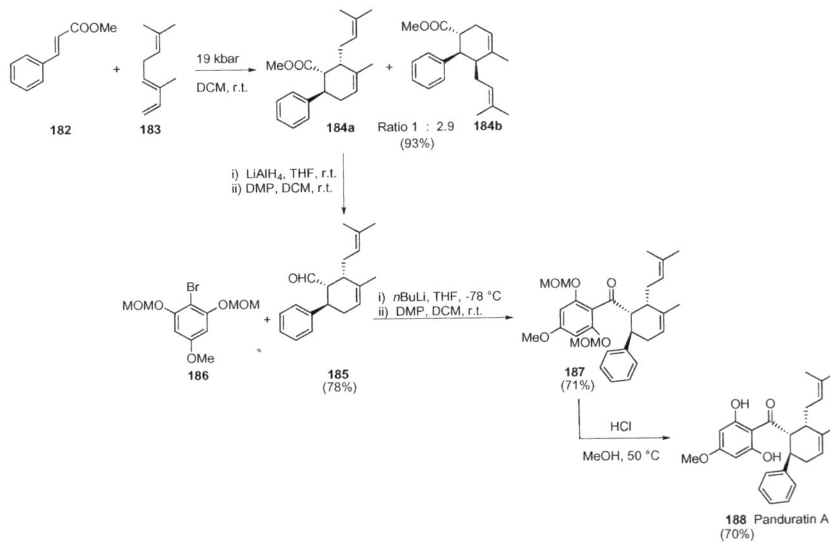

SCHEME 1.32 Synthesis of Panduratin A **188**.

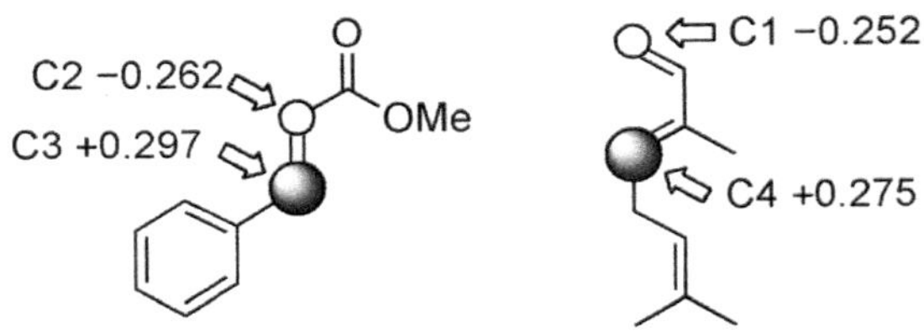

FIGURE 1.9 Atomic orbital coefficients of the 2p orbitals of methyl cinnamate (**182**), and (*E*)-ocimene (**183**) calculated at the B3LYP/6-31G(d) level of theory. *From: Pasfield, L. A.; de la Cruz, L.; Ho, J.; Coote, M. L.; Otting, G.; McLeod, M. D. Asian J. Org. Chem.* ***2013****, 2, 60–63.*

challenge posed by the Wickerols, only a single synthetic study was disclosed.[90] In 2017 an asymmetric synthesis of Wickerol A (**194**) was proposed as a testament to the power of catalysis in cycloadditions and the many pleasant and unpleasant surprises that we encounter when working with complex, sterically congested molecules. Key intermediate **191**, which contains all carbons of the Wickerols except two, could be accessed with a DA addition of known enone **189** to diene **190**. The synthesis commenced with the DA cycloaddition of the achiral enone **189** to the chiral diene **190** (Scheme 1.33). The intermolecular cycloaddition represented a major challenge because both partners, though electronically matched, are sterically very hindered. Initially, thermal or high pressure conditions were explored, but unsuccessfully.

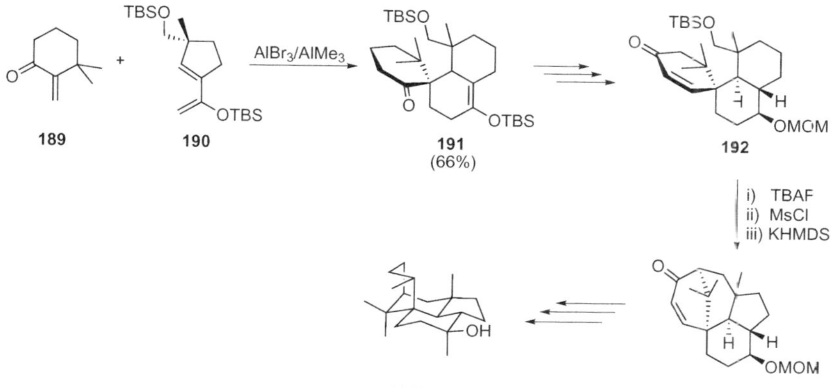

SCHEME 1.33 Synthetic strategy to Wickerol A **194**.

Lewis acid activation was found to be the solution of the problem. Jung's conditions, employing a mixture of $AlBr_3/AlMe_3$ as catalyst, gave the desired *endo*-product **191** as single diastereomer.[91] Both the quaternary and the neopentyl stereocenters were set with the correct simple and induced diastereoselectivity. Presumably, the enone approaches the diene via an *endo*-transition state and the methyl group provides less steric hindrance than the siloxy methyl moiety. Having set the correct configuration, the subsequent modification conducted to the compound **192**, which led to the ring system of Wickerol A (**194**) complete with the introduction of the missing methyl groups. The methyl group on the D-ring could be added with complete diastereoselectivity by exposure of **193** to methyl lithiocyanocuprate in the presence of $BF_3 \cdot Et_2O$. Given the steric hindrance at the neopentyl β-position and the introduction of additional *syn*-pentane strain, it is remarkable that this reaction proceeded at all. Wickerol A (**194**) was then obtained through a sequence of reaction required for the insertion of another methyl group.[92]

A research group of Belgian authors demonstrated that both D- and L-cyclohexenylguanine are highly potent and selective antiherpes virus (HSV-1, HSV-2, VZV, CMV) agents. Their activity profile is similar to that of the known antiviral drugs acyclovir and ganciclovir, and they represent the most potent antiviral nucleosides containing a six-membered carbohydrate mimic that have ever been reported. Remarkably, the antiviral activity profiles of both enantiomers are very similar. The enantioselective synthesis of the D- and L-enantiomers of these compounds starting from (*R*)-carvone was already described.[93] However, the synthesis is long and time-consuming, and it is not suited for the preparation of large amounts of the final products that are needed for full biological evaluation. This fact is one of the major drawbacks of this type of synthetic approach: the need of high amounts of products because of the length of the synthetic route causing low final chemical yields. The Belgian authors set up a novel synthetic route to overcome these problems based on the DA reaction between the dienophile **195** and the Danishefsky's diene **196** in the presence of hydroquinone at 180°C for 1.5 h (Scheme 1.34).

Adduct **197** was obtained in 62% yield as a 4:1 mixture of the *endo*- and *exo*-adducts. The highly functionalized adduct **197** is sensitive to acid and moisture and was immediately reduced with excess $LiAlH_4$, giving the triol **198** in 66% yield. This reaction involves the reduction of two ester groups and concomitant rearrangement of the silyloxyenol ether to the enone intermediate, which is then further reduced to **198**. Reaction of triol **198** with benzaldehyde dimethyl acetal in the presence of para-Toluensulfoni acid (PTSA) at room temperature for 1 d afforded the benzylidene acetal **199** (70% yield). Introduction of the purine base moiety was performed by the reaction with 2-amino-6-chloropurine under Mitsunobu reaction conditions to generate the chloropurine **200** with inversion of the configuration at C1. The N7-isomer (14%) of **200** was also isolated. Finally, acidic hydrolysis of **200** gave the racemic (±)-**201** in 27% overall yield starting from **199**.[94]

The example just reported is a typical case of carbocyclic nucleoside synthesis. Carbanucleosides are modified nucleosides in which a methylene group has replaced the oxygen atom of the furanose ring.[95] Due to the absence of a glycosidic linkage, carbanucleosides are chemically more stable and not subjected to the phosphorylases that cleave the *N*-glycosidic linkage in conventional nucleosides. Many compounds of this class with

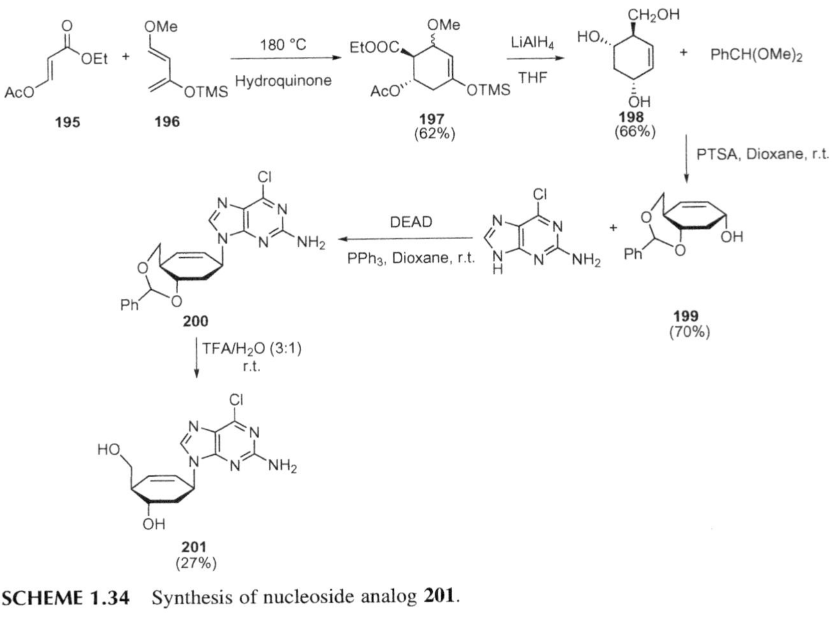

SCHEME 1.34 Synthesis of nucleoside analog **201**.

potent and selective biological activity were identified, including carbovir and abacavir, which are currently used therapeutically as anti-HIV agents.[96] An important aspect of the anti-HIV therapy is the suppression of viral replication in the brain; in this regard an enhanced lipophilicity of potential drugs is likely to be advantageous. The modification of some structures by fusion of a carbocyclic or heterocyclic rings to the carbasugar moiety increases the lipophilicity (and hence the access to the central nervous system, a major reservoir of HIV),[97] maintaining at the same time the rigidity imposed by the C2′—C3′ double bond. The synthesis of carbocyclic nucleosides containing a fused isoxazoline ring and lacking a methylene group in the side chain of the carbocyclic unit was suggested as a new class of analogs.[98] Nucleosides lacking a methylene group in the side chain have been reported and in some cases display reduced cytotoxicity. The synthetic route that was exploited is based on the HDA cycloaddition of dienes with transient acyl nitroso moieties,[31] affording privileged structures for the synthesis of carbocyclic nucleosides. The HDA cycloadduct *N*-benzoyl-2,3-oxazanorborn-5-ene **202** was obtained from freshly distilled cyclopentadiene that quantitatively trapped the nitrosocarbonyl benzene generated in situ in good yields through the mild oxidation of benzonitrile oxide Ph-CNO (BNO) with a slight excess of NMO according to the reported procedure (Scheme 1.35).[31]

The HDA cycloadduct **202** was found to be an excellent dipolarophile toward BNO addition, affording the two regioisomeric cycloadducts **203a,b**, which were easily isolated in quantitative yields by chromatographic separation in a 3:2 regioisomeric ratio. A proper tuning of the subsequent synthetic steps had to be done for the scale-up necessary for the synthesis of nucleosides in the due amount for the biological tests. The ease detachment of the benzoyl group from the cycloadducts **203a,b** was achieved through NaOH/MeOH treatment at room temperature for 24 h, affording the fairly stable hydroxylamines. The reaction was conducted by adding a slight excess of NaOH, added portionwise in 2—3 h, to a methanol solution of cycloadducts **203a,b**, stirred until complete disappearance of the starting materials. The hydrogenolytic cleavage of the N—O bond of the hydroxylamines was performed by the reaction with Pd/C 10% in ethyl acetate at room temperature for 3 h.

Hence, the aminols **204a,b** were obtained and submitted to the transformations into the desired nor-nucleosides through the linear construction of the purine and pyrimidine rings. The synthesis of the purine-nucleosides requires the two steps. The first one is the condensation of the regioisomeric aminols **204a,b** with the 5-amino-4,6-dichloropyrimidine to give the pyrimidine derivatives of type **205a,b**. This condensation is indeed the most difficult step in many reported syntheses, which attained only low yields (around 50%). Better yields of **205a,b** were achieved by increasing the reaction temperature gradually up to the boiling point of the solvent and working in sealed tubes. The pyrimidine derivatives **205a,b** were obtained in 79% yields

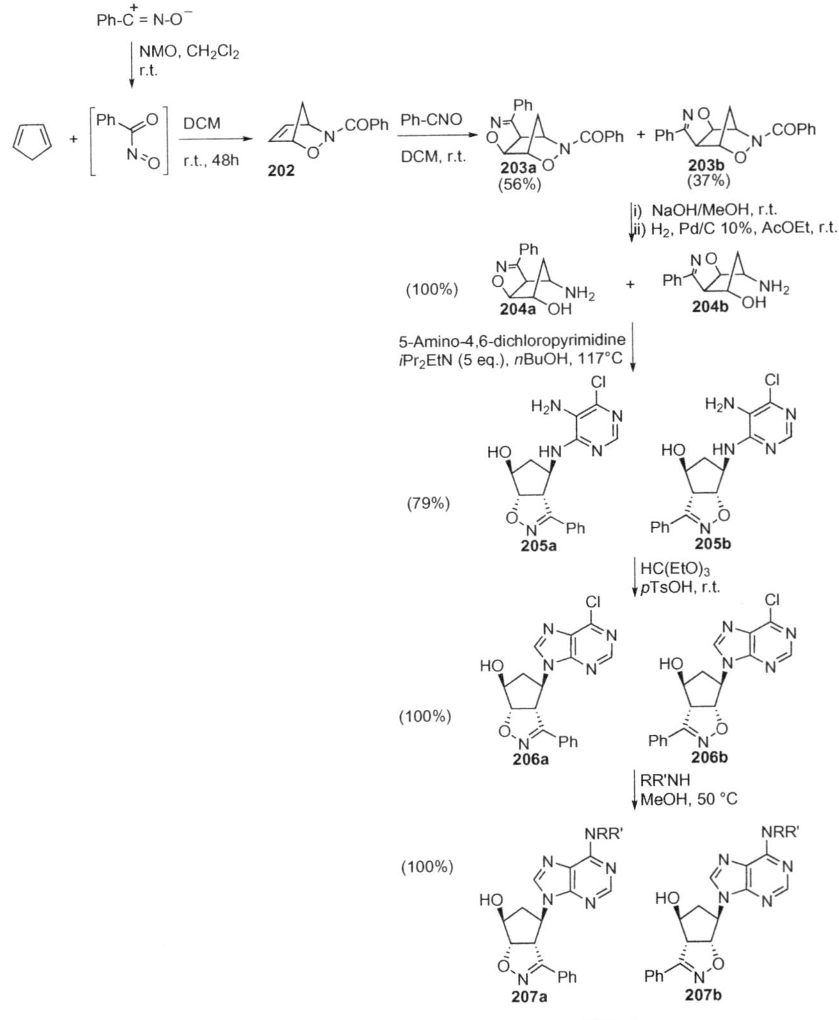

SCHEME 1.35 Nitrosocarbonyl approach to nucleoside analogs **207a,b**.

for both regioisomeric compounds. The conversion of the pyrimidine derivatives **205a,b** into the chloropurine compounds **206a,b**, was achieved by treatment with ethyl orthoformate and a catalytic amount of pTsOH. The final chlorine replacement method with selected amines was conducted upon heating MeOH solutions of **205a,b** at 50°C in the presence of an excess NH_3 or some primary or secondary amines affording the adenine derivatives **207a,b** in more than 92% yields, which have been submitted to the biological tests. Compounds **207a,b** were evaluated for their inhibitory activity against a wide variety of viruses, including HSV-1, vaccinia virus, vesicular stomatitis virus, parainfluenza-3 virus, reovirus-1, sindbis virus, coxsackie virus B4, Punta Toro virus, and respiratory syncytial virus.

All synthesized compounds showed no specific antiviral effects, however some differences between the regioisomeric nucleosides can be noted. Regioisomers of type **b**, involving a phenyl group distal to the heterocyclic base, display better response than regioisomers of type **a**, which have proximal phenyl and heterocyclic base. Moreover, all compounds were evaluated for antiviral activity against HIV-1 (strain IIIB) and HIV-2 (strain Rob) in MT-4 cell culture and none of these compounds showed inhibitory activity at concentration up to 400 μg/mL. The modest antiviral activity showed from these compounds could be likely linked to the lack of substrate activity for cellular and/or viral nucleoside kinases, or alternatively, the lack of recognition of the compounds by the viral DNA or RNA polymerases.[99]

The same strategy was applied for the synthesis of isoxazolino-carbocyclic nor-nucleosides incorporating an anthracene moiety through nitrosocarbonyl intermediates chemistry.[31] A variety of analogs were attained, starting from the already reported stereodefined heterocyclic aminols through the linear construction of purine heterocyclic rings. The synthesis hinges on the *exo* selective 1,3-dipolar cycloaddition of the stable anthracenenitrile oxide to the *N*-benzoyl-2,3-oxazanorborn-5-ene **202** and simple elaborations of the regiorisomeric cycloadducts **208a,b** (Scheme 1.36). The synthetic route of Scheme 1.35 was here applied again and the chloropurine compounds **209a,b** were prepared.[100]

A selection of nucleoside primary and secondary amines derivatives **210a,b** were initially tested for their inhibitory activity against a variety of viruses, including hepatitis B and C, human papillomavirus (HPV), as well as influenza viruses of type A and B. Modest antiviral activities were observed in hepatitis assays while the activities in the cases of influenza viruses were almost negligible. Good antiviral activity was found for compound of type **210b** (R = H; R′ = Et), with no cellular toxicity at the dose tested in the case of HPV. It was reported that the sulfonyl derivative **211** is one of the most active compounds against HPV proposed in literature (Fig. 1.10) and displays its antiviral activity by an allosteric mechanism: it is a hyperbolic competitive inhibitor of the ATPase activity of HPV6 E1 and also inhibits its helicase activity.[101]

Fig. 1.10 shows the sulfonyl structure **211** (orange shading) optimized through DFT calculations along with that of compound **210b** (R = H; R′ = Et) and the overlap between them obtained through Root Means Squared (RMS) fitting of the aromatic moieties. As we can see, there is a nice overlap of the anthracene and naphthalene aromatic rings while the second phenyl ring of the diphenyl moiety crosses the purine ring directing the sulfonyl group in the same position where the ethyl group is located. Moreover, the isoxazoline ring nicely overlaps the amide group. The larger

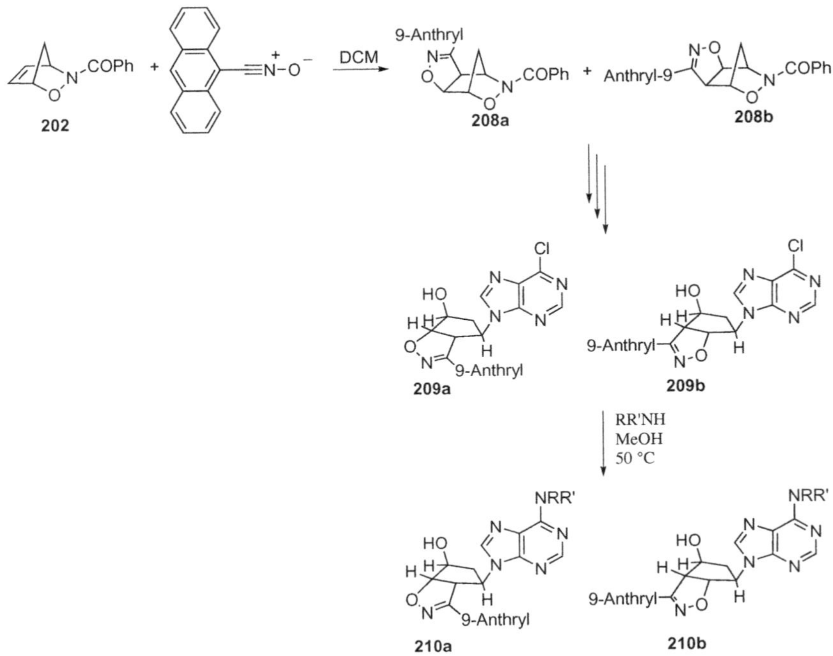

SCHEME 1.36 Synthetic strategy anthracene-substituted nucleoside analogs **210a,b**.

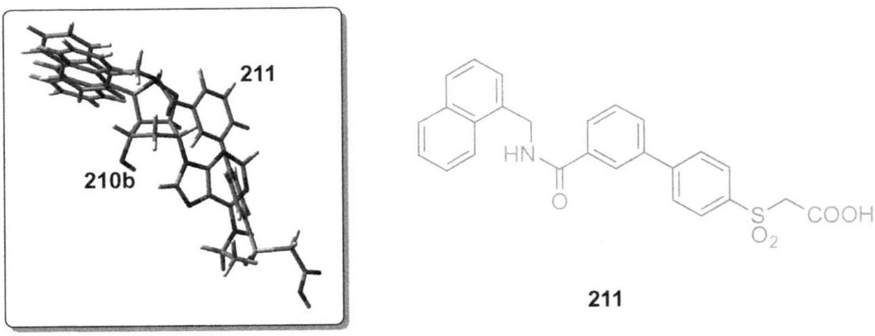

FIGURE 1.10 Structure of sulfonyl derivative **211** and structure overlap with compound **210b**.

activity shown by those compounds, which bear a naphthyl residue, could suggest a possible mechanism based on DNA-intercalation. Some isoxazolidinyl polycyclic aromatic hydrocarbons have been proposed as DNA-intercalating antitumor agents and molecular modeling studies on these structures confirm the degree of binding when a polycyclic aromatic residue is linked to an isoxazoline heterocyclic ring.[102]

These investigations were completed by the synthesis of isoxazolino-carbocyclic anthracene nor-nucleosides, prepared through nitrosocarbonyl chemistry, containing uracil residues, tested for their inhibitory activity against some viruses, such as HSV-1 and HSV-2, Zoster virus, and hepatitis B and C. The activities were almost negligible in most of the cases. Again, a remarkable antiviral activity was found for a specific regioisomer **212** with no cellular toxicity at 1−100 μM dose concentration in the case of HPV.[103] Regarding HPV, its complex replication machinery begins with the specific binding of E2 to the viral DNA and the recruitment of protein E1, which rapidly oligomerizes to a hexamer (around the nucleic acid) displaying its helicase activity. With this mechanism in mind, the binding affinity of the uracil derivative **212** was evaluated to the HPV11 E2 protein (Fig. 1.11). Flexible

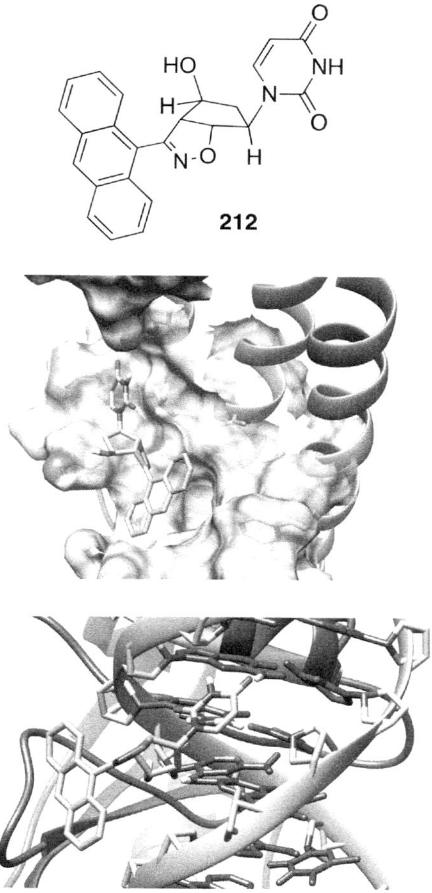

(A)

(B)

FIGURE 1.11 Structure of uracil derivative **212**. *From: Memeo, M. G.; Lapolla, F.; Maga, G.; Quadrelli, P. Tetrahedron Lett.* ***2015****, 56, 1986–1990.*

molecular docking of the uracil derivative **212** on this receptor gives a docking score that is comparable and even higher than the ones obtained through the self-docking of the crystallographic inhibitor. As Fig. 1.11 shows, **212** is anchored at the receptor pocket involved in the protein–protein interaction (PPI) through the isoxazoline ring and the histidine His32 residue.[104]

The anthracene and the uracil substituted cyclopentanol moieties are perfectly complementary to the pocket involved in the PPI, and, in particular, the anthracene substituent shades a hydrophobic pocket involved in the specific E2–E1 interaction. A second round of molecular docking calculations was conducted on the specific sequence of DNA recognized by viral E2. Strong indications were obtained that **212** can specifically bind the DNA minor groove. The anticonfiguration between the uracil and anthracene substituents allows for the **212** backbone adaptations on the minor DNA groove. The interactions of the uracil ring as well as the cyclopentane hydroxyl group with the viral DNA are established with the heterobases of the nucleic acid, reasonably inhibiting in this way the DNA replication machinery. The last docking calculations were carried out on the viral helicase protein E1. In this case low docking scores along with lack of specific interactions suggests the low probability that E1 represents the biological target.

Upon changing the nitrile oxide structure, the regioisomeric cycloadducts **213a,b** of the bromonitrile oxide to the *N*-benzoyl-2,3-oxazanorborn-5-ene **202** were easily prepared and elaborated into a novel class of uracil nor-nucleoside derivatives (Scheme 1.37). In the key synthetic step represented by the reductive N–O bond cleavage, an unusual double ring opening afforded the aminol intermediates **214a,b** containing a β-hydroxynitrile structure. By adapting known protocols, the aminols entered the linear construction of uracil rings.[105]

The novel nucleosides **216a,b** were found structurally similar to a potent antiviral compound, brivudin (BVDU), and molecular modeling and docking selected one of the two regioisomeric structures as a promising candidate for antiviral tests, due to the nice level of binding with the thymidine kinase (TK), the enzyme involved in virus replication. In order to verify the stability of the poses and to better simulate the induced-fit effect of the ligands on the enzyme, optimization and MD simulations were performed on the complexes of **216a,b** within TK. The superimposition of **216a,b** with BVDU is shown in Fig. 1.12.[106]

In the case of compound **216a**, the stabilizing pairwise interaction of the nitrogen basis with Gln125 is maintained, some interactions are lost, and some new interactions are observed. Adaptation of **216a** into the TK binding site seems to be driven by the proximity of the CN group to the side chain of Tyr101, which causes a change in the hydrogen bond network found for the cyclopentane moiety. Actually, the OH at C3′ remains in contact with the side chain of Tyr101, while the OH at C4′ establishes an interaction with Glu225. Moreover, the lone pair of the latter hydroxyl group accepts a

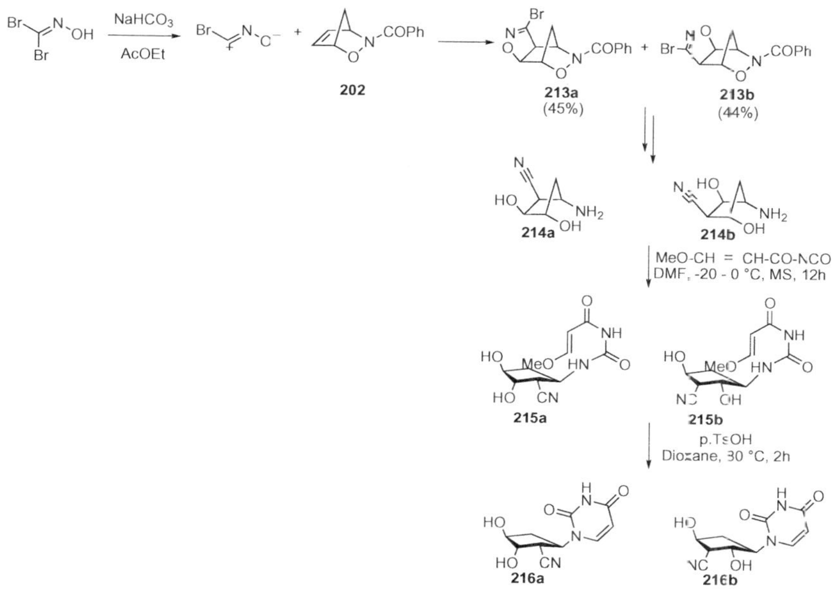

SCHEME 1.37 Bromonitrile oxide strategy to nucleoside analogs **216a,b**.

hydrogen bond from the charged side chain of Arg222, a residue in turn blocked by salt bridges with Glu225 and Glu83. Conversely, in the case of compound **216b**, only one of the two hydrogen bonds of the nucleobase with Gln125 is maintained. Moreover, the OH groups at C2′ and C4′ of the cyclopentane ring interacts with Glu225 and Glu83, respectively. Furthermore, it is worthy pointing out that both compounds **216a,b**, compared with BVDU, lack the bromovinyl chain with the consequent loss of the related hydrophobic interactions that might exert an anchoring effect in the pocket of the enzyme shaped by Trp88, Tyr132, Arg163, and Ala167.

On pursuing the studies on antiviral compounds and in the search of novel small molecules with potential biological activity, some nucleoside analogs were prepared and found to be moderately active against human herpes and varicella viruses or strongly active against HPV as well as influenza A virus H1N1 by taking advantage of the synthetic protocol relied upon the nitrosocarbonyl chemistry.[31]

Pyridine and quinoline hydroximoyl chlorides **219a–d** were prepared according to the literature procedures (Scheme 1.38). The selected aldehydes **217a–d** were converted in high yields into the corresponding oximes **218a–d** following the classical methods and from the latter, the desired hydroximoyl chlorides were obtained upon chlorination with chlorine gas.[107]

The in situ generation of the nitrile oxides **220a–d** is required in the present cases since none of the nitrile oxides at hand displayed any stability

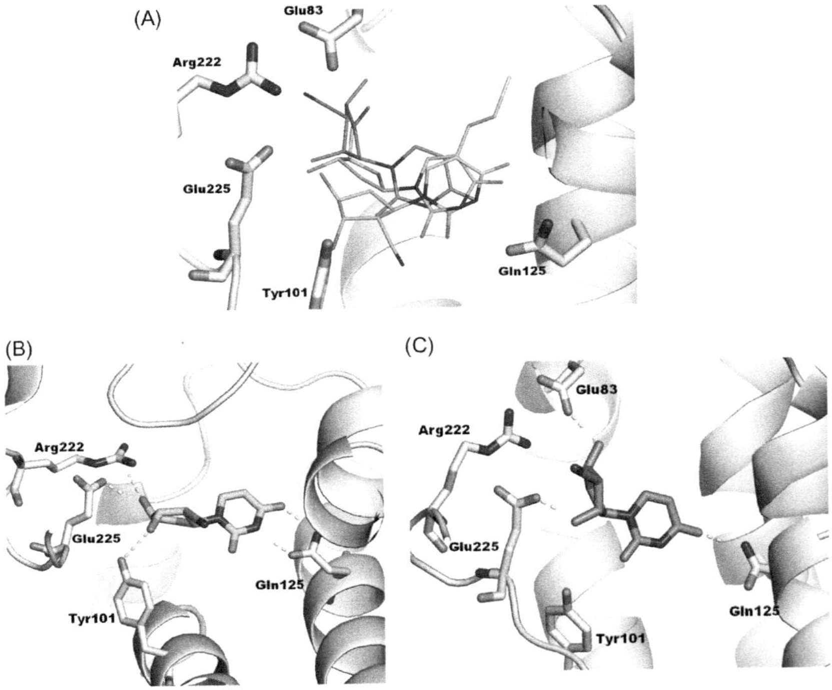

FIGURE 1.12 (A) Overlay of *BVDU* with **216a** and **216b** after 5 ns of MD simulations, starting from the best docking solutions. The residues of the active site shown belong to the complex with *BVDU*. (B) Predicted binding mode of compound **216a** after MD simulation. (C) Predicted binding mode of compound **216b** after MD simulation. Hydrogen bonds are shown as dashed lines. *BVDU*, Brivudin. *From: Savion, M.; Memeo, M. G.; Bovio, B.; Grazioso, G.; Legnani, L.; Quadrelli, P. Tetrahedron.* ***2012****, 68, 1845–1852.*

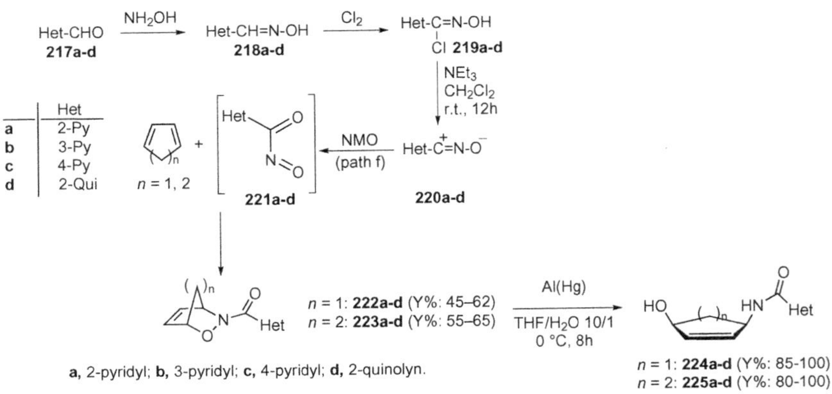

SCHEME 1.38 Nitrosoxcarbonyl strategy to nucleoside analogs **224** and **225**.

at room temperature, possibly in the solid state, for a long time. The mild oxidation of these 1,3-dipoles with NMO is conducted one-pot in the presence of the required trapping diene (freshly distilled cyclopentadiene or 1,3-cyclohexadiene) to afford the nitrosocarbonyl HDA cycloadducts **222a–d** and **223a–d**. The derivatives **224a–d** and **225a–d**, obtained through N–O bond reductive cleavage, were tested against a variety of viruses but positive results were obtained only in the case of the respiratory influenza A H1N1 virus. The majority of the compounds were found to be inactive, but in a single case **225d** the EC_{50} values are just 10 times those of the reference compound while the SI_{50} values must be increased by the same factor.

One more time, HPV finds in the nitrosocarbonyl chemistry a valuable synthetic tool to achieve a remarkably active target against this virus. A fleeting heterocyclic nitrosocarbonyl **230** derived from the corresponding nitrile oxide is at work in a short-cut synthesis of 4-bromo-*N*-[(1R*,4S*)-4-hydroxy-2-cyclohexen-1-yl]-2-thiazolecarboxamide **232** (Scheme 1.39). The synthetic strategy is based on HDA cycloaddition of the in situ generated nitrile oxide **229** to 1,3-cyclohexadiene followed by mild reductive cleavage of the N–O bond in the cycloadduct **231**.

The new 2-thiazolecarboxamide derivative **232** is obtained in good yields and the results of the in vitro antiviral tests indicated that the product was active against HPV virus. Moreover, some structural evidence shines some

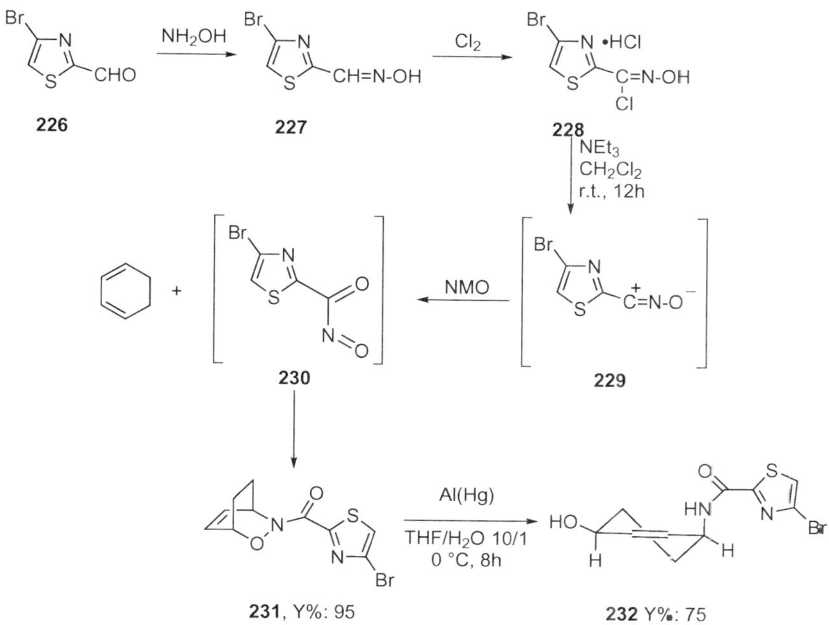

SCHEME 1.39 Bromothiazole nucleoside analog synthesis **232**.

light on future perspectives on the application of pericyclic reactions or the synthesis of biological active molecules. The findings demonstrated that the synthetic methodology works well for the preparation of heterocyclic substituted novel compounds that displayed interesting and promising activity against viruses and in particular against HPV. The results clearly showed that compound **232** was inefficient against herpes, hepatitis, and influenza viruses. An exception emerged against HPV-11. Compound **232** showed an EC_{50} value <1 along with a CC_{50} value >100. The secondary screening demonstrated that compound **232** was found less effective in raft cultures than in primary assay in HEK 293 cells. The $EC_{50/90}$ values are higher than those of the reference compound U0126, as well as the CC_{50} value. The $SI_{50/90}$ values however are close to the reference.[108]

The identical strategy but using a different starting material is at work in another example of HPV active molecule, a new adenine derivative, prepared through the chemistry of nitrosocarbonyl intermediates.[31] The synthesis relies upon the functionalization of adenine by inserting an ester moiety, to be transformed into the corresponding hydroxamic acid **233** (Scheme 1.40). The oxidation conducted in situ in the presence of cyclopentadiene afforded the HDA cycloadduct **235**, further elaborated to give the target nucleoside analog **236**.

This single product displayed a selective high activity against HPV. The EC_{50} and EC_{90} values are remarkably lower than the values collected for Cidofovir used as a control drug. The EC_{50} is 39 times lower while the CC_{50} value is 97.11 μM.[109]

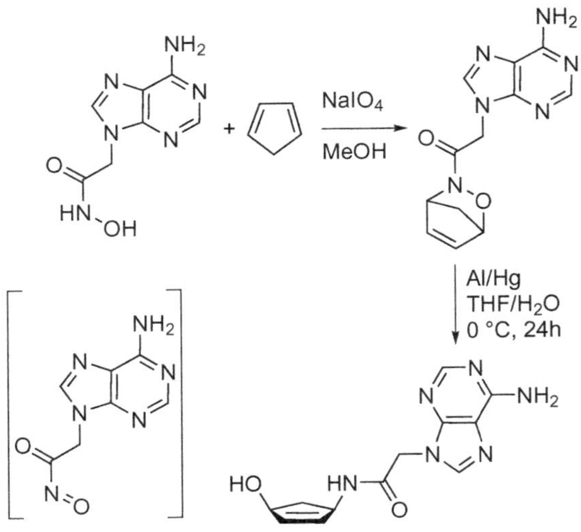

SCHEME 1.40 Nitrosocarbonyl strategy to nucleoside analog.

A number of synthetic carbocyclic nucleosides, where the furan ring is replaced by a carbocyclic ring, was found to display important therapeutic properties. The US FDA has in fact approved Abacavir for HIV-1 treatment and Entecavir for chronic HBV infections (Fig. 1.13). Highly oxygenated troponoids have been identified as lead compounds for HBV infections. To date, structure-function studies on these molecules have been limited due to a scarcity of synthetic methods for their preparation. A successful adaptation of a [5 + 2] oxidopyrilium cycloaddition/ring-opening synthesis was applied to prepare HBV antagonists such as the 2,3,7-trihydroxy-5-methylcyclohepta-2,4,6-trien-1-one shown in Fig. 1.13.

Analogously, compounds containing tricyclic spacer displayed interesting antiviral activities. An interesting synthetic approach to these systems is shown in Scheme 1.41: furan and ethyl fumarate were allowed to react to afford the DA cycloadduct **237** that is reduced with $LiAlH_4$ to give the diol **238**. Treatment of the diol with benzyl azidoformate allowed for the intramolecular cyclization to give, after deprotection of the amino group, the aminol **239**, which was condensed with the commercially available 4,6-dichlro-5-amino-pyrimidine to give the adduct **240**. Standard acid-catalyzed orthoformate cyclization led to the 6-chloropurine nucleoside **241** that was found active against Coxsackie virus (EC_{50} 1.86 and CC_{50} 50 μM).[110]

The same authors followed a similar strategy to obtain the DA cycloadduct **242** (Scheme 1.42) from furan and acrolein, even in very low yields. Upon functionalization of the C=C double bond, the *exo* regioisomeric

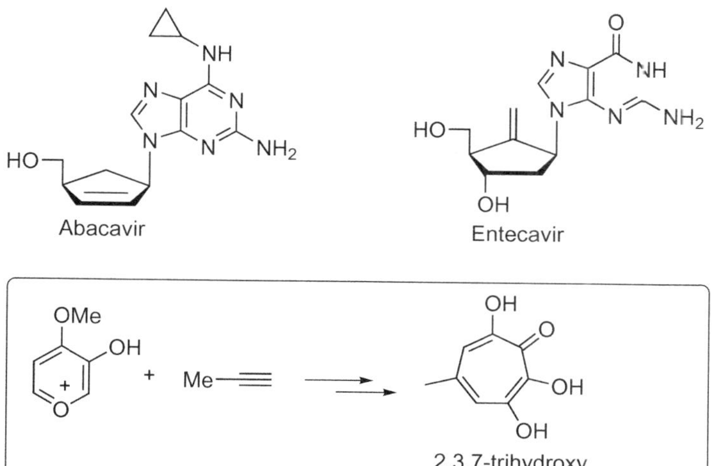

FIGURE 1.13 Structures of abacavir and entecavir; [5 + 2] cycloaddition/ring opening reaction. *Adapted from: Hirsch, D. R.; et al.* Org. *Biomol. Chem.* ***2018***, 16, *62–69.*

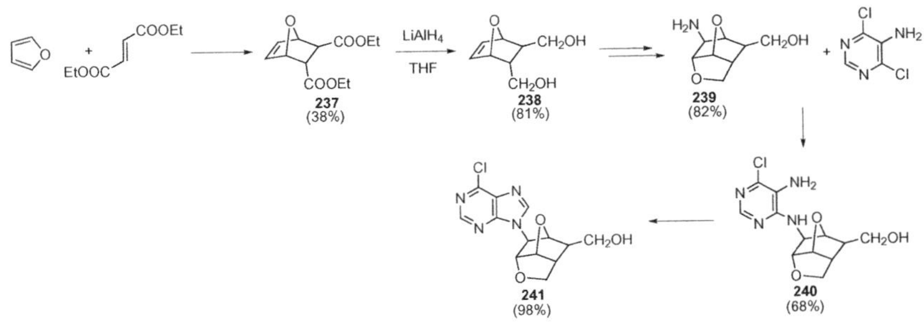

SCHEME 1.41 Synthetic strategy to nucleosidic compound **241**.

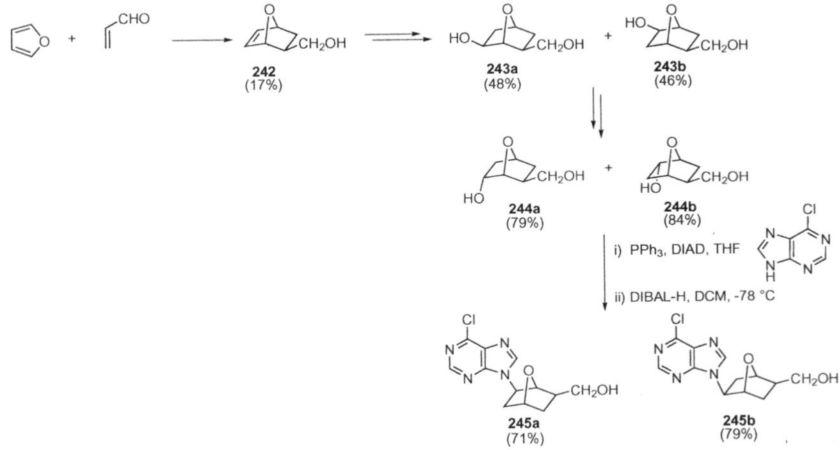

SCHEME 1.42 Synthetic strategy to compounds **245a,b**.

diols **243a,b** were obtained and converted into the *endo* ones. The stereochemical inversion was needed to apply the Mitsunobu protocol for the insertion of the 6-chloropurine ring, leading to the nucleosides **245a,b** in very good yields.

Both the products were tested against the Coxsackie virus displaying the following activities: **245a**, EC_{50} 1.40 and CC_{50} >50 μM; **245b**, EC_{50} 2.21 and CC_{50} >50 μM.[111]

This group of Czech and Belgian researchers is actively involved in the synthesis of novel nucleosides. The preparation of novel 1′-homonucleoside derivatives locked in a West conformation by 1′,4′-bridge, consisting of annulated benzene or naphthalene ring, was illustrated. The crucial step of the synthesis was DA reaction of an appropriate aryne with a suitable furane derivative. Antiviral properties of novel compounds were studied and slight activity against HCV was detected in several compounds, although largely correlated with cytotoxicity.[112]

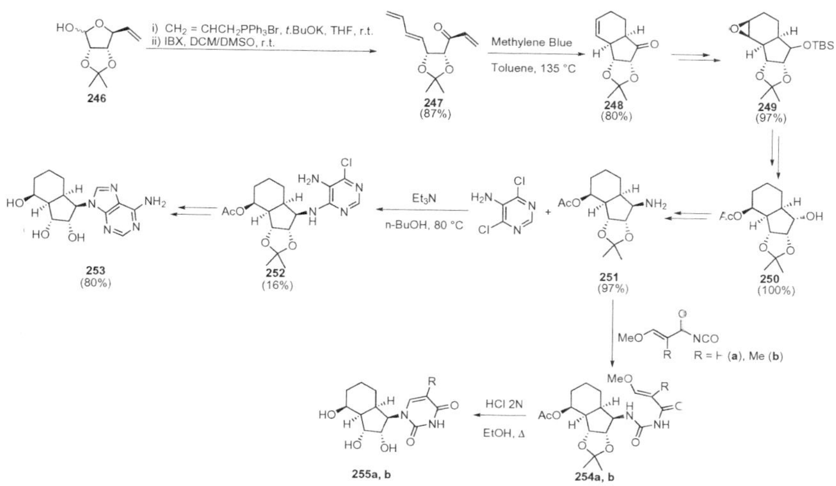

SCHEME 1.43 Synthesis of uracil derivative **255a,b**.

Three nucleosides **253**, **255a,b** showed screening results suggestive of having weak in vitro antiviral activity against Coxsackie B3 virus. Adenosine **253** (concentration = 1 μg/mL) showed 4.9% of inhibition and thymidine **255b** (concentration = 1 μg/mL) showed 5.5% of inhibition and higher concentration (concentration = 100 μg/mL) led to highly toxic uridine **255a** (concentration = 1 μg/mL), which gave the best result, showing more than 12.3% of inhibition. Their synthesis is shown in Scheme 1.43; the starting point was the lactol **246**, as a good candidate for the Wittig reaction with allyltriphenylphosphonium bromide to introduce the diene functionality in **247**.[113]

After screening with different bases, potassium *tert*-butoxide (KOtBu) in THF gave the best yield of triene of 88% with a *Z* to *E* ratio of 4:1. The mixture of trienes was subjected to oxidation with IBX as an oxidant and furnished trienone **247** in 87% yield. Trienone **247** was highly unstable and polymerized quickly after standing in vacuum. Trienone **247** underwent the Intra Molecular Diels-Alder (IMDA) reaction in toluene, in the presence of methylene blue, in a sealed tube to give hydrindenone **248** in good yield. Hydrindenone **248** was first reduced with $NaBH_4$ to the corresponding alcohols in a 1:4 ratio. One of them could be oxidized back and protected to give a silyl ether that was epoxidized to give **249**. After the reduction of the epoxide, the newly formed hydroxyl group was acetylated and the silyl protection removed to afford **250**, further converted through azide chemistry into the amine **251**. The linear approach was then adopted to synthesize the target molecules. Reaction of amine **251** with 5-amino-4,6-dichloropyrimidine furnished pyrimidine **252** in miserable yield. Standard orthoformate cyclization afforded the adenine derivative **253**. Conversely, condensation with freshly prepared isocyanates (R = H, Me) obtained the ureas **254a,b**, and were subsequently cyclized to the uracil derivative **255a** and the timine derivative **255b**.

Antivirals and 1,3-dipolar cycloaddition reactions

1,3-Dipoles represent a valuable class of reacting intermediates frequently used in the synthesis of antiviral compounds. Examples of application of the chemistry of these in situ generated reacting species have been already accounted in previous pages, where some nitrile oxides were at work in the synthesis of some biological active molecules.[98–100,103,105,107–109] Their roles were for both the introduction of suitable residues necessary to display the correct antiviral activity and that of starting material for the generation of other reacting species, such as the nitrosocarbonyl intermediates,[31] which represented the key intermediates for the synthesis of the target compounds.

This section reports some remarkable examples of the use of some of the most frequently used 1,3-dipoles in the synthesis of antiviral compounds. Not all the 1,3-dipoles will be reviewed, just those successfully employed for the preparation of biologically active molecules; in vitro data will be also reported here.

One of the most frequently used 1,3-dipoles is azide. The versatility of this 1,3-dipole is well known since its use in "click chemistry" as well as for the preparation of triazoles, valuable heterocycles in the synthesis of biologically active compounds. Organic azides belong to the propargyl-allenyl category of dipoles and since the discovery of triazole formation from phenyl azide and dimethyl acetylenedicarboxylate in 1893, they found large success in the construction of heterocyclic frameworks and core structures of natural and bioactive compounds.[114]

A simple case is shown in Scheme 1.44 where treatment of 4-acetoxybutylbromide with sodium azide in anhydrous DMF gave the 4-acetoxybutylazide in 96% yield. The 6-(methylthio)-9-(prop-2-yn-1-yl)-9*H*-purine was reacted with the azido-compound via 1,3-dipolar cycloaddition reaction, in anhydrous toluene under reflux, to afford a mixture of two possible regioisomers.

The regiosiomer **256** was obtained in 89%. It is well known from the literature[115] that the addition of azides to unsymmetrical acetylenes is determined by steric and electronic factors. In general, such addition tends to give mainly the isomers with electron-withdrawing groups at the 4-position and electron-releasing groups at 5-position. Thus, after separation of the regioisomers, treatment with methanol saturated with ammonia in a sealed reacting vessel at 100°C led to the carboacyclic nucleoside **257**. The carboacyclic nucleoside **258** was synthesized in 73% yield via treatment of **257** with CH_3ONa/CH_3OH at room temperature. Other compounds were prepared, but

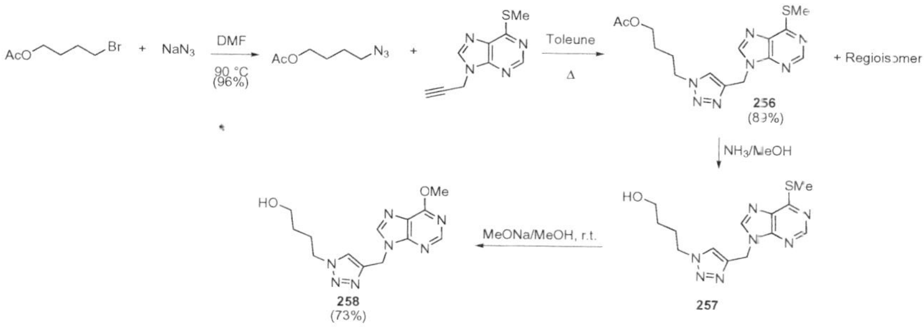

SCHEME 1.44 Azido-strategy to compound **258**.

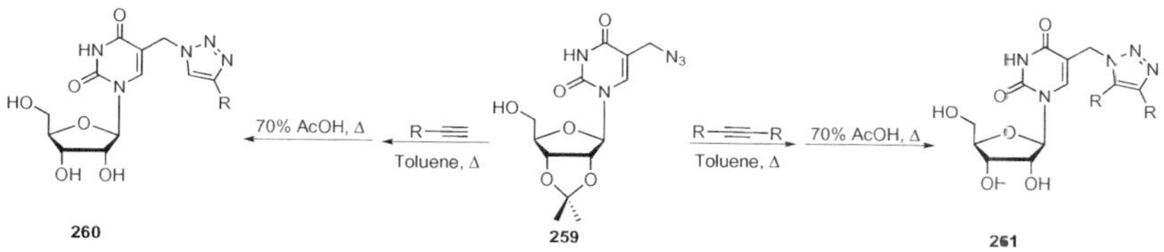

SCHEME 1.45 Azidomethyluridine synthetic applications to nucleoside analogues.

only compounds **157** and **258** have a low activity, in vitro, against parainfluenza virus type 3 and reovirus type 1, respectively ($MIC_{50} = 240\ \mu g/mL$, $MCC > 400\ \mu g/mL$). Neither anti-HIV-1(III_B) nor anti-HIV-2(ROD) activity was observed in MT-4 cells for the products at hand.[116]

Some 5-(1,2,3-triazol-1-ylmethyl)uridine derivatives of type **260,261** were synthesized via the 1,3-dipolar cycloaddition of a 5-azidomethyluridine derivative **259** with substituted mono- and di-substituted acetylenes (Scheme 1.45).

The antiviral activities of these compounds against hepatitis A virus (HAV, MBB cell culture-adapted strain) and HSV-1 were tested. For the antiviral activity against HAV-27 it is noted that at both concentrations tested, 10 and $20\ \mu g/10^5$ cells, compounds of type **260** revealed the highest antiviral activity in this series, and compounds of type **261** revealed high activity at $10\ \mu g/10^5$ cells using amantadine (C*) as a control. For the antiviral activity against HSV-1 the results revealed that compounds of type **260** showed the highest effect at $10\ \mu g/10^5$ cells, while compounds of type **261** showed moderate activity.[117]

The synthesis and antiviral evaluation of a series of C5-(1,4- and 1,5-disubstituted-1,2,3-triazolo)-nucleoside derivatives of type **264** was described by Agrofoglio et al. The key steps of these syntheses are the regioselective Huisgen's 1,3-dipolar cycloaddition reactions, using either copper-catalyzed azide-alkyne cycloaddition (CuAAC) or ruthenium-catalyzed azide-alkyne cycloaddition (RuAAC) under microwave activation starting from a triple bond functionalized and protected nucleoside structure **262** and a series of substituted azides **263** (Scheme 1.46).

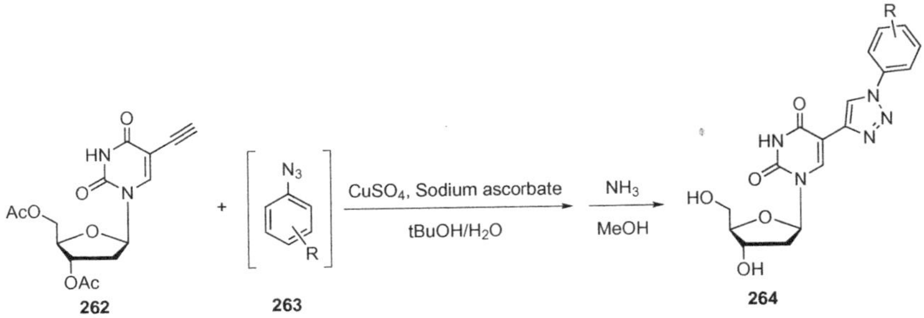

SCHEME 1.46 Synthesis of analog **264**.

Some compounds among the synthesized series possess activity against HSV-1 and HSV-2, varicella-zoster virus, human cytomegalovirus, and vaccinia virus. Their cytostatic activities were determined against murine leukemia cells, human T-lymphocyte cells, and cervix carcinoma cells. Compounds were also evaluated on a wide panel of RNA viruses, including Vesicular stomatitis virus, influenza viruses type A (H1N1 and H3N2) and B in MDCK cell cultures, parainfluenza-3 virus, reovirus-1, Sindbis virus, and Punta Toro virus in Vero cell cultures and Vesicular stomatitis, Coxsackie B4, and respiratory syncytial virus, however with no specific antiviral effect. Some cellular activity was observed, which most likely implies that the compounds are able to penetrate into cells. Some of the synthesized compounds showed activities against several DNA viruses in the micromolar range. Whereas the 1,4-regioisomers exhibited antiviral activity, it is interesting to note that the (1,5)-regioisomers were found inactive against this wide panel of viruses and display no toxicity. In order to design compounds with a significantly increased potency, it will be necessary to study their phosphorylation by human and viral kinases; this process would be greatly aided by a cocrystal structure of the best compounds bound in the enzyme active sites. Also, it will be of interest to investigate the properties of their monophosphorylated derivatives against thymidylate synthase and their triphosphorylated derivatives against the viral and cellular DNA polymerases to reveal the mechanism of biological action.[118] Copper-catalyzed 1,3-dipolar cycloaddition reaction between azidoristocetin aglycon and various propargyl glycosides were used to prepare new sugar derivatives of ristocetin. Some of these compounds were found to be active against Gram-positive bacteria and showed favorable antiviral activity against the H1N1 subtype of influenza A virus.[119]

O^6-(Benzotriazol-1H-yl)guanosine and its 2′-deoxy analog can be readily converted to the O^6-allyl derivatives of type **265** that upon diazotization with *t*-BuONO and TMS-N_3 yield the C-2 azido derivatives of type **266** (Scheme 1.47).

American and Belgian researchers have previously analyzed the solvent dependent azide · tetrazole equilibrium of C-6 azidopurine nucleosides, and

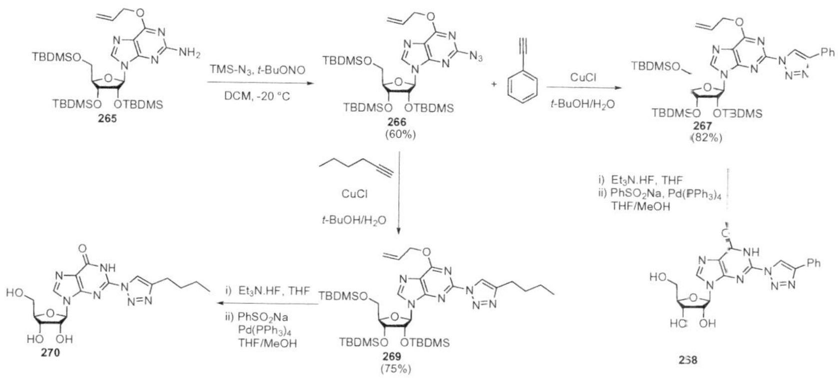

SCHEME 1.47 Synthesis of nucleosides **268** and **270**.

in contrast to these, the O^6-allyl C-2 azido nucleosides appear to exist predominantly in the azido form, relatively independent of solvent polarity.[120] Consistent with the presence of the azido functionality, each neat C-2 azide displayed a prominent IR band at 2126-2130 cm^{-1}. A screen of conditions for the ligation of the azido nucleosides with alkynes showed that CuCl in t-BuOH/H_2O is optimal, yielding C-2 1,2,3-triazolyl nucleosides **267** in 70%–82% yields. Removal of the silyl groups with $Et_3N \cdot 3HF$ followed by deallylation with $PhSO_2Na/Pd(PPh_3)_4$ gave the C-2 triazolylinosine nucleosides **268-270**. Products were desilylated for biological assays. The two C-2 triazolyl adenosine analogs demonstrated pronounced antiproliferative activity in human ovarian and colorectal carcinoma cell cultures. When evaluated for antiviral activity against a broad spectrum of DNA and RNA viruses, some of the C-2 triazolylinosine derivatives showed modest inhibitory activity against cytomegalovirus. Compound **270**, however, displayed a somewhat more pronounced activity as anti-CMV (EC_{50} 73 μM).[121]

The efficient synthesis of a new series of polyhydroxylated dibenzyl ω-(1*H*-1,2,3-triazol-1-yl)alkylphosphonates as acyclic nucleotide analogs of type **273** was described by Belgian and Polish researchers starting from dibenzyl ω-azido-(polyhydroxy)alkylphosphonates **271** and selected alkynes **272** under microwave irradiation (Scheme 1.48).

Selected *O,O*-dibenzylphosphonate acyclonucleotides were transformed into the respective phosphonic acids. All compounds were evaluated in vitro for activity against a broad variety of DNA and RNA viruses and for cytostatic activity against murine leukemia L1210, human T-lymphocyte CEM, and human cervix carcinoma HeLa cells. Compound **273(Tymine)** exhibited antiviral activity against Influenza A H3N2 subtype ($EC_{50} = 20$ μM-visual CPE score; $EC_{50} = 18$ μM-MTS method; MCC > 100 μM, CC50 > 100 μM) in MDCK cell cultures, and **273(Uracil-Bz)** was active against vesicular stomatitis virus and respiratory syncytial virus in HeLa cells ($EC_{50} = 9$ and

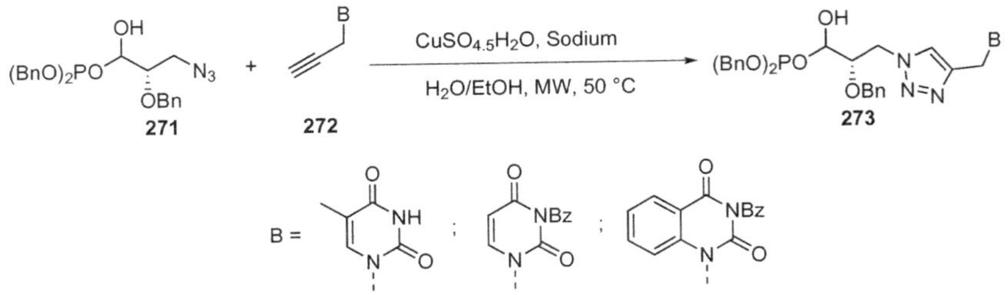

SCHEME 1.48 Synthesis of analogs **273**.

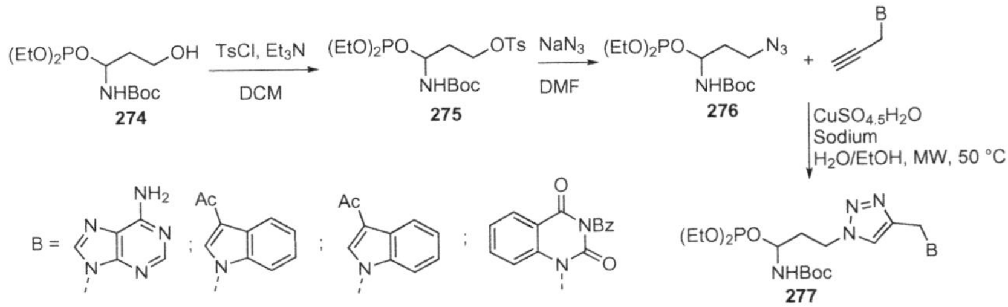

SCHEME 1.49 Synthesis of derivatives **277**.

12 μM, respectively). Moreover, compound **273(BenzoUracil-Bz)** showed activity against both HSV-1 and HSV-2 in HEL cell cultures (EC_{50} = 2.9 and 4 μM, respectively) and feline herpes virus in Crandell–Rees feline kidney (CRFK) cells (EC_{50} = 4 μM), but at the same time it exhibited cytotoxicity toward uninfected cells (MCCP4 lM). Several other compounds have been found to inhibit proliferation of L1210, CEM as well as HeLa cells with IC_{50} in the 4–50 μM range.[122]

A new series of 1-amino-3-(1*H*-1,2,3-triazol-1-yl)propylphosphonates (*R*)- and (*S*)-**277** were obtained from enantiomerically pure (*R*)- and (*S*)-1-*tert*-butoxycarbonyl (Boc)-amino-3-azidopropylphosphonates **276** and *N*-propargylated nucleobases in good yields (Scheme 1.49).

The synthesis stands on the tosylation of the alcohol **274** and azido functionalization of **275**. The cycloaddition to the *N*-propargylated nucleobases afforded the protected compounds **277**.

All 1,2,3-triazolylphosphonates (*R*)- and (*S*)-**277** were evaluated for their activities against a broad range of DNA and RNA viruses. Compound (*R*)-**277** (B = 3-acetylindole) was moderately active against vesicular stomatitis virus in HeLa cell cultures (EC_{50} = 45 μM). In addition, (*S*)-**277** (B = adenine), (*R*)-**277** (B = *N*3-Bz-benzuracil), (*R*)-**277** (B = 3-acetylindole), and (*R*)-**277** (B = 5,6-dimethylbenzimidazole) were cytotoxic toward CRFK cells (CC_{50} = 2.9, 45, 72, and 96 μM, respectively). The cited compounds **277** were also slightly cytostatic to different tumor cell lines.[123]

The same authors of the previous work prepared a new series of 4-substituted [(1,2,3-triazol-1-yl)acetamido]methylphosphonates of type **280** as acyclic nucleotide analogs, synthesized from diethyl (2-chloroacetamido) methylphosphonate **278** via azidation followed by 1,3-dipolar cycloaddition with selected alkynes derived from natural nucleobases or their mimetics (Scheme 1.50).

All compounds were tested for their antiviral activities against DNA and RNA viruses as well as for cytostatic activity or cytotoxicity. Among all tested compounds, [(1,2,3-triazol-1-yl)acetamido]-methylphosphonate **280** bearing a methyl-imidazole substituent showed activity against the vesicular stomatitis virus (EC_{50} = 45 μM) in HeLa cell cultures.[124]

Nucleoside analogs are commonly used in HIV/AIDS therapy and as a first line therapeutics for many tumor types. After entering through the membrane, nucleoside analogs are phosphorylated by intracellular nucleoside kinases into active 5′-mono-, di-, and triphosphates. The drug resistance to nucleoside analogs refers to nucleoside transporter deficiency, reduced nucleoside kinase activity, overexpression of multidrug resistance proteins, or modifications in apoptotic pathways.[125] A conceivable strategy to avoid resistance is to use a prephosphorylated nucleoside directly as a drug that can thus bypass intracellular phosphorylation. However, nucleoside triphosphates (NTPs) are very poorly internalized by cells. Attempts to get around this problem by masking the phosphate groups by turning them into phosphate esters have not been successful until recently.[126] The fully substituted di- and triphosphate esters become unstable. Difficulties in the development of lipophilic NTP prodrugs arise due to hydrolytical instability of the pyrophosphate bond, while in the natural NTPs this bond is kinetically stable due to negative charges that slow down cleavage by nucleophiles.

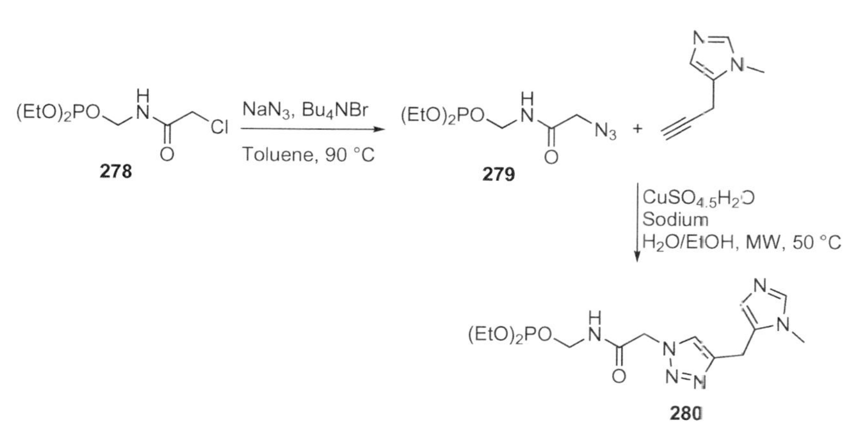

SCHEME 1.50 Synthesis of compound **280**.

An alternative to the lipophilic prodrug approach could be in using drug delivery systems such as liposomes, nanogels, nanoparticles, and so on, which could potentially solve the problem of poor cell absorption of NTPs. SiO_2 nanoparticles represent a simple and versatile method for the preparation of SiO_2-dNTP conjugates. There are several examples when nucleoside analogs were originally approved as antiviral agents and then it has been shown that they display promising anticancer properties. Thus conjugates of antiviral phosphorylated nucleoside analogs may also possess potential antitumor activity. Some Russian researchers prepared triphosphates of anti-HIV nucleoside drugs lamivudine and zalcitabine with γ-alkynyl phosphoramidate group, obtained their conjugates with SiO_2 nanoparticles (SiO_2-dNTP) via Cu(I)-catalyzed azide-alkyne cycloaddition, and assessed their antiviral and cytotoxic properties. Scheme 1.51 shows the synthetic route.

The pivotal step is represented by the insertion of a triple bond on the phosphorylated derivative **282** to give **283** that undergoes 1,3-dipolar cycloaddition with the azide **284** to afford the final compound **285**. The conjugates of phosphorylated lamivudine and dideoxycytidine (zalcitabine) showed higher potency than the parent nucleosides. The conjugate of phosphorylated azidothymidine (AZT) was less active against HIV-1 than the parent nucleoside probably because of the replacement of its 3′-azido group by

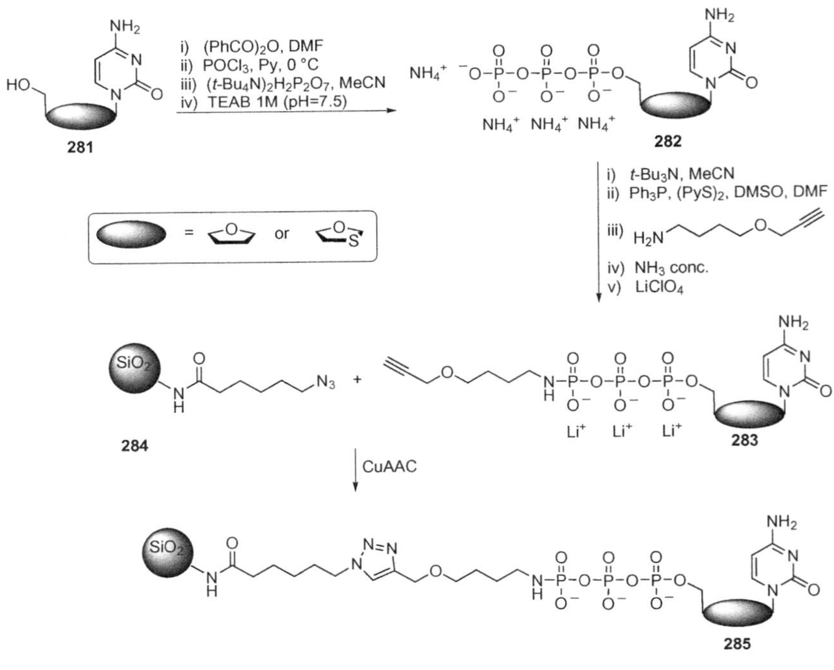

SCHEME 1.51 SiO_2 nanoparticles synthesis.

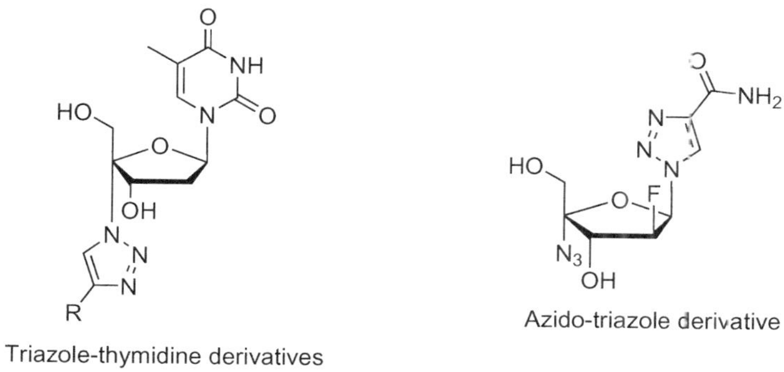

FIGURE 1.14 Structures of triazole-thymidine and azido-triazole derivatives.

1,2,3-triazole ring. These results show an opportunity for using SiO_2 nanoparticles as a transport for delivering phosphorylated nucleosides to cells in order to increase their efficiency as antiviral and anticancer drugs.[127] Nonobligate chain terminating nucleosides with a linear substituent (azido or ethynyl group) at the 4′ position represent an important class of compounds in antiviral discovery, particularly against HCV and HIV. 3′- AZT-derived 1,2,3-triazoles can be potent inhibitors of HIV-1. To gauge the medicinal chemistry impact of functionalizing the 4′-linear substituent and possibly generate novel antiviral nucleoside scaffolds, azide-alkyne cycloaddition reactions with 4′-AZT were investigated. Fig. 1.14 shows the structures of families of thymidine and triazole derivatives.

Antiviral screening identified a few triazole analogs moderately active against HIV-1 (18%–62% inhibition at 10 mM) and/or influenza A virus (15%–50% inhibition at 10 mM), and none active against West Nile virus or HCV. These results suggest that the linear 4′ azido group of ADRT may be essential for target binding and that its chemical manipulation could largely compromise antiviral potency.[128] Azido-triazole derivatives were found to be potent anti-HBV agents.[127b]

The use of multivalent carbohydrate compounds to block cell-surface lectin receptors is a promising strategy to inhibit the entry of pathogens into cells and could lead to the discovery of novel antiviral agents. One of the main problems with this approach, however, is that it is difficult to make compounds of an adequate size and multivalency to mimic natural systems such as viruses. Hexakis adducts of [60]fullerene are useful building blocks in this regard because they maintain a globular shape at the same time as allowing control over the size and multivalency. Water-soluble tridecafullerenes decorated with 120 peripheral carbohydrate subunits, so-called superballs, were prepared through the collaboration of several European research groups and could be synthesized efficiently from hexakis adducts of [60]fullerene in one step by using copper-catalyzed azide-alkyne cycloaddition click chemistry (Fig. 1.15).

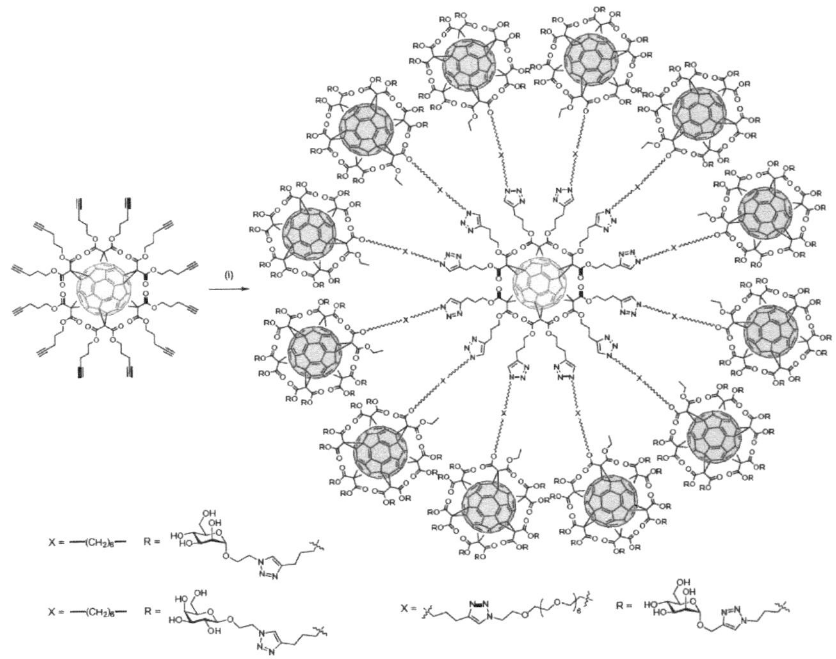

FIGURE 1.15 Syntheses of the tridecafullerenes using CuAAC click chemistry. The core fullerene (endowed with 12 alkyne groups) is joined to the peripheral fullerene-based compounds by click chemistry. Reagents and conditions: (1) $CuBr \cdot S(CH_3)_2$, sodium ascorbate, Cu(0), DMSO, 25°C, 48 h, 73%–79%; (2) $CuSO_4 \cdot 5H_2O$, sodium ascorbate, THF/H_2O, 80°C (MW), 2 h, 76%. *CuAAC*, Copper-catalyzed azide-alkyne cycloaddition. *Modified from: Munoz, A.; Sigwalt, D.; Iliescas, B. M.; Luczkowiak, J.; Rodriguez-Perez, L.; Nierengarten, I.; Holler, M.; Remy, J.-S.; Buffet, K.; Vincent, S. P.; Rojo, J.; Delgado, R.; Nierengarten, J.-F.; Martin, N. Nat. Chem.* ***2016****, 8, 50–57.*

Infection assays show that these superballs are potent inhibitors of cell infection by an artificial Ebola virus with half-maximum inhibitory concentrations in the subnanomolar range.[129]

In the previous example we have seen the application of a CuAAC methodology for the synthesis of antiviral compounds. There is considerable attention directed at chemically modifying nucleic acids with robust functional groups in order to alter their properties. Since the breakthrough of CuAAC, there have been several reports describing the synthesis and properties of novel triazole-modified nucleic acid derivatives for potential downstream DNA- and RNA-based applications. A review will focus on highlighting representative novel nucleic acid molecular structures that have been synthesized via the click azide-alkyne cycloaddition. Many of these derivatives show compatibility for various applications that involve enzymatic transformation, nucleic acid hybridization, molecular tagging, and purification and gene silencing. The details of these applications can be found in these manuscripts where the chemistry of triazoles seems to be really promising for future developments of antiviral and anticancer compounds.[130,131]

In the field of macromolecular structures, distinct and well-defined glycopolymers for deciphering the biological roles of natural bioactive polysaccharides were prepared from Fucose monomers, chemically synthesized and decorated with specific sulfation patterns including unsulfate, monosulfate, disulfate, and trisulfate groups (Fig. 1.16).

The six fucoidan-mimetic glycopolymers were successfully fabricated through Microwave (MW)-assisted ring-opening metathesis polymerization in an emulsion system. Three glycopolymers associated with 2-*O*-sulfation exhibited better inhibitory effects on the H1N1 virus, while glycopolymers with monosulfate groups were more effective against the H3N2 virus. These findings would promote the development of novel antiinfluenza A virus drugs based on natural fucoidans.[132]

Another class of 1,3-dipoles that has been extensively used in the synthesis of antiviral compounds is constituted of nitrones. Nitrones are arguably the most useful tools because of their ability to generate nitrogen- and oxygen-based functionality and to introduce chiral centers stereoselectively. They were first prepared by Beckmann in 1890 and several methods are known for their preparation, apt to fit the synthetic requirements to be used in the planned syntheses. They constitute a valuable entry to isoxazolidines; these heterocyclic rings are a remarkable alternative to sugar moieties in nucleosides and they mimic the spacer of natural compounds. The facility of preparation of these 1,3-dipoles and the variety of substituents that can be

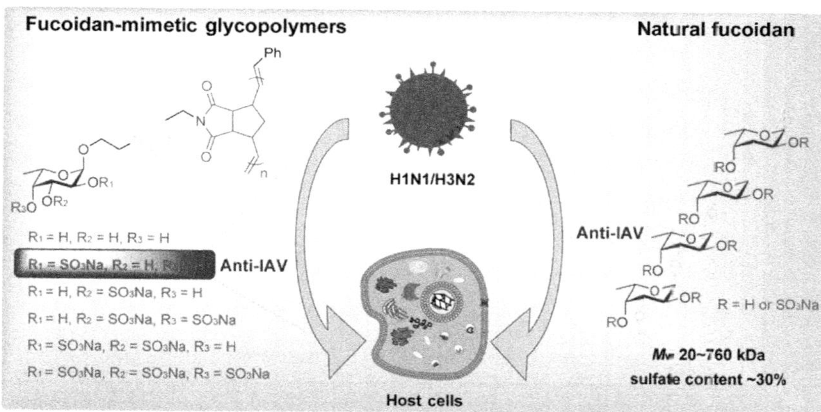

FIGURE 1.16 Synthesis of fucoidan-mimetic glycopolymers. *From: Fan, F.; Cai, C.; Wang, W.; Gao, L.; Li, J.; Li, J.; Gu, F.; Sun, T.; Li, J.; Li, C.; Yu, G. ACS Macro Lett.* ***2018****, 7, 330–335.*

placed on the 1,3-dipolar structures determine the success of these reactive species in the synthesis of new biologically active compounds.

Among the numerous examples reported in literature, it is worth mentioning the use of unconventional nitrones bearing phosphonate substituents leading to a new class of truncated phosphonated oxazanucleosides. The phosphonate nitrone **286** undergoes 1,3-dipolar cycloaddition reaction with the ethyl 2-acetoxyacrylate in THF under reflux to afford the isoxazolidine **287** as minor component of the diasteroisomeric mixture (80% yield). Upon adapting the Vorbrüggen protocol, uracil, 5-fluorouracil, and thymine were inserted on the isoxazolidine skeleton to give the final compounds **288a-c** (Scheme 1.52).

The biological tests indicate that these anomers **288a-c** inhibit the reverse transcriptase of Avian Moloney virus (AMV), human T-lymphotropic virus (HTLV-1), and HIV. In particular, the 5-fluorouracil derivative **288c** is the more promising compound, acting on Abelson Murine Leukemia virus and HIV at the concentration of 1 and 10 nM, respectively. The level of the inhibitory activity of this product toward HTLV-1 and HIV was 10-fold higher than that of tenofovir and similar to that of AZT. Moreover, this compound does not show any cytotoxicity according to 3-(4,5-dimethylthiazol-2-yl)-5-(3-carboxymethoxyphenyl)-2-(4-sulfophenyl)-2H-tetrazolium (MTS) assay.

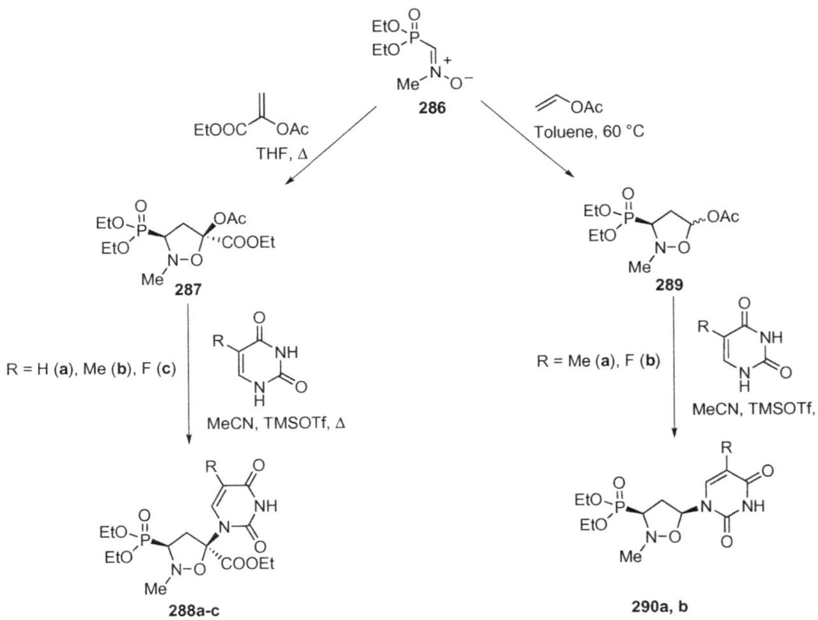

SCHEME 1.52 Nitrone synthesis of isoxazolidines.

The 1,3-dipolar cycloaddition to vinyl acetate led also to the diasteromers of type **289** in 62% yield. Insertion of thymine and 5-fluorouracil led to the nucleosides **290a,b** that were found to inhibit completely the reverse transcriptase of the AMV and HIV at concentration 1 nM, at a level comparable with that of tenofovir (1 nM) and 10-fold lower than AZT (10 nM). Moreover, MTS assays indicate a very low toxicity (CC_{50} > 500 μM) in comparison with AZT (CC_{50} 12.14 μM). These and many other applications of phosphonate nitrones have been reviewed[133] and the processes studied from the mechanistic point of view; the cycloadditions, in fact, showed a complete regioselectivity and a nearly exclusive *cis* stereoselectivity. M062X/6-31G(d,p) calculations were conducted to rationalize the regio- and the stereochemical outcome.[134]

A novel series of 5-arylcarbamoyl-2-methylisoxazolidin-3-yl-3-phosphonates of type **293A,B** were synthesized via the 1,3-dipolar cycloaddition reaction of the *N*-methyl-*C*-(diethoxyphosphoryl)nitrone **286** with *N*-substituted naphthalimide acrylamides **292A,B** prepared from the amine **291A,B** upon functionalization with acryloyl chloride in presence of triethylamine (Scheme 1.53).

All *cis*- and *trans*-isoxazolidine phosphonates obtained were assessed for antiviral activity against a broad range of DNA and RNA viruses. Isoxazolidines *trans*-**293A** and *trans*-**293B** exhibited the highest activity (EC_{50} = 8.9 μM) toward cytomegalovirus. Compounds *cis*- and *trans*-**293A** as well as *cis*- and *trans*-**293B** were found potent against HSV and vaccinia viruses (EC_{50} in the 45–58 μM range). Antiproliferative evaluation of all obtained isoxazolidines revealed promising activities toward tested cancer cell lines with IC_{50} in the 1.1–19 μM range.[135]

Another class of 1,3-dipole that found attention in the synthesis of nucleoside analogs is that of nitrile oxides. Although less used in the synthesis of nucleoside because of their high instability and difficulties for their preparation, some cases are reported in which the choice of the proper substituent can solve selectivity problems and facilitate the access to biologically active molecules.

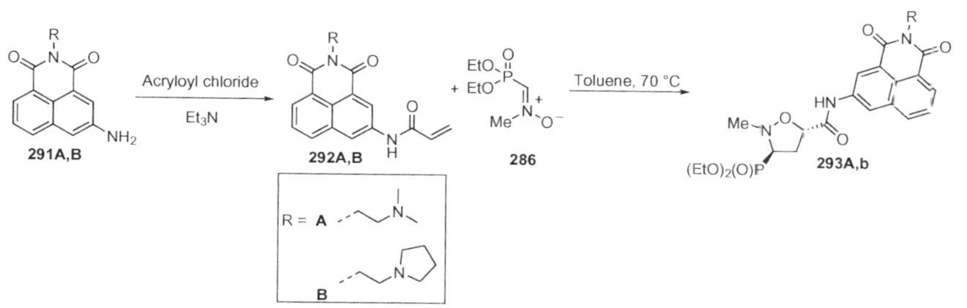

SCHEME 1.53 Nitrone synthetic strategy to compound **293**.

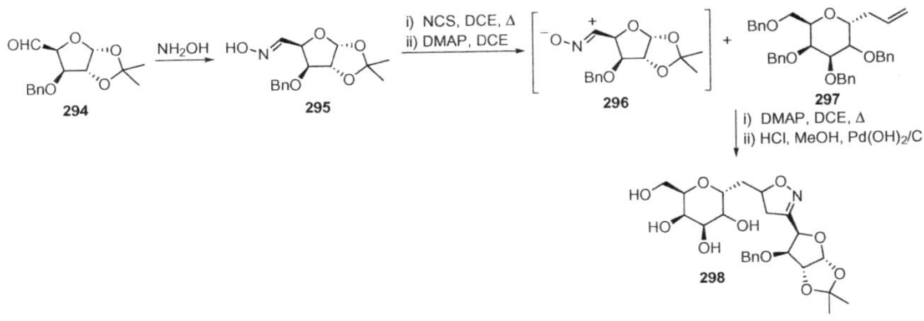

SCHEME 1.54 Synthetic strategy to compound **298**.

A series of novel isoxazoline-linked pseudodisaccharide derivatives **298** were regiospecifically synthesized by the 1,3-dipolar cycloaddition reaction of α-allyl-*C*-glycopyranoside **297** and sugar-derived nitrile oxide **296** with good yields. The nitrile oxide was prepared starting from the aldehyde **284** upon conversion into the corresponding oxime **295** and chlorination with *N*-chlorosuccinimide (NCS) followed by in situ generation of the 1,3-dipole in the presence of the suitable dipolarophile **297** (Scheme 1.54).

The structure of the compound **298** were characterized by NMR spectroscopy and MS spectrometry and confirmed by the X-ray crystallographic analysis. The glycosidase inhibitory activities of **298** and other compounds were determined with hydrolytic reactions of glycosidases using acarbose as a control. However, the compound **298** showed a selectively moderate inhibition to β-glucosidase; the inhibition of **298** was 20% at the concentration of 2.7 μmol/mL. HIV-RT and antitumor activity were also preliminarily evaluated. Some of these types of cycloadducts exhibited potent inhibitory activity to HIV-RT.[136]

The introduction of a second heteroatom into the furanose ring would result in several new classes of nucleoside analogs. The presence of multiheteroatoms in the five-membered ring is desired to retain the required conformation for the recognition between the nucleoside and the target enzyme (reverse transcriptase). The use of oxygen or sulfur atoms has produced several diheteroatom-substituted analogs and some of the dioxolanyl or oxathiolanyl nucleosides are potent virus inhibitors.[137] It has been synthetically challenging to introduce both oxygen and nitrogen into the five-membered ring while keeping a glycosidic bond. The N–O compounds such as dihydroisoxazolidine display a mutual induction effect of two heteroatoms exerting several desired properties. First, the nitrogen atom becomes less basic to prevent its protonation; therefore, the synthesized compounds are expected to be stable at physiological pH. Second, the oxygen atom may also become less nucleophilic, making the glycosidic bond more resistant to acidic hydrolysis. The use of nitrile oxides in the preparation of antiviral derivatives via 1,3-dipolar cycloaddition reactions represents a valuable method to achieve these targets.[138]

[2 + 2] Cycloaddition reaction for antiviral compounds

Nucleosides are structurally defined compounds where the spacer often plays a key role in recognition and in the biological activity outcome. Cycloaddition reactions offer a variety of methods for the synthesis of nucleoside and in particular regarding the spacer structure. Within the frame of cyclic spacers and given the most frequently encountered five- and six-membered ring structures, the cyclobutane moiety received considerable attention due to the broad spectrum of antiviral and anticancer compounds that were discovered. Moreover, cyclobutane amino acids and dipeptides were found in nature, some of them with interesting properties as antibiotics and antivirals.

The cyclobutane structural unit can be easily prepared through the photochemical [2 + 2] cycloaddition reaction or on the use of suitable precursors bearing the preformed cyclobutane ring. Absolute chirality can be raised from chiral precursors obtained from natural sources or by the use of chiral catalysts. Functionalization of the carbocyclic ring is also a challenging task, crucial for the synthesis of polyfunctionalized nucleosides and amino acids.

In the search for therapeutic agents against HSV and HIV, much attention was focused on the preparation of analogs of natural nucleosides as antimetabolites. Such analogs were modified in the carbohydrate portion of the molecule, the spacer, and in the base portion. Examination of molecular models suggested that a cyclobutane ring might serve as a surrogate for the oxetan ring of oxetanocin A. This observation led to the synthesis of the racemic carbocyclic analogs of oxetanocin A, compounds **302** and **303** that exhibited broad-spectrum activity against HSV and HIV (Scheme 1.55). They were considered promising agents for AIDS treatment.

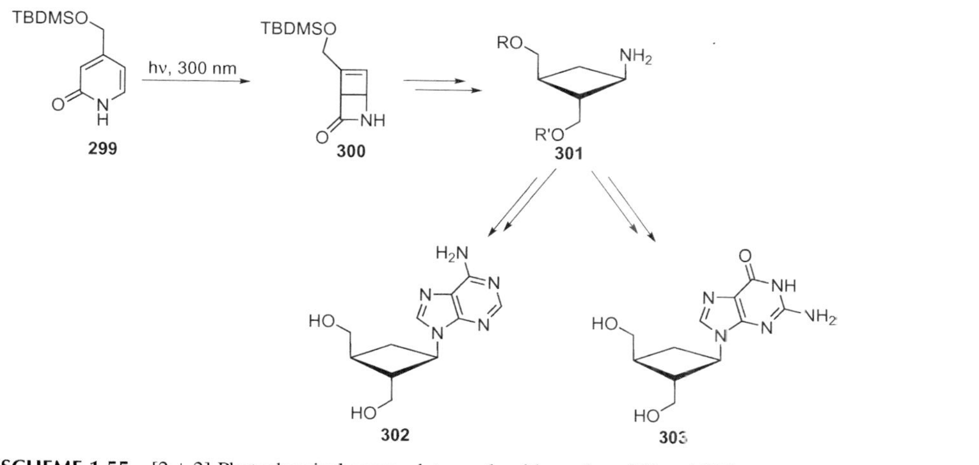

SCHEME 1.55 [2 + 2] Photochemical approach to nucleoside analogs **302** and **303**.

For the synthesis of these analogs, many preparative methods for optically active cyclobutane derivatives were developed. Some reported syntheses resort to a racemic or enantiomerically pure substituted cyclobutanone as the key intermediate. Furthermore, the nucleoside strategy may rely on a convergent synthesis or a linear approach. In Scheme 1.55, the 4-(((*tert*-butyldimethylsilyl) oxy)methyl)pyridin-2(1*H*)-one **299** was irradiated at 300 nm and the intermediate **300** was obtained as a result of a [2 + 2] photochemical ring closure. Synthetic elaboration of **300** led to the protected cyclobutane-amine **301** that is the starting point for the linear construction of adenine and guanine, respectively, to afford the compounds **302** and **303**. This represents just a simple example of the way these compounds can be easily obtained; many other approaches are reported in literature, not only regarding the preparation of antiviral compounds but also for the synthesis of amino acids, peptides, and other biologically active molecules.[139]

Acknowledgment

Financial support by the University of Pavia and MIUR (PRIN 2011, CUP: F11J12000210001) is gratefully acknowledged. We also thank "VIPCAT – Value Added Innovative Protocols for Catalytic Transformations" project (CUP: E46D17000110009) for valuable financial support.

References

1. Domingo, L. R.; Ríos-Gutiérrez, M.; Pérez, P. *Molecules* **2016,** *21*, 748–770.
2. Domingo, L. R. *Molecules* **2016,** *21*, 1319–1334.
3. (a) Nishiwaki, N., Ed. *Methods and Applications of Cycloaddition Reactions in Organic Syntheses;* John Wiley & Sons, Inc., 2014. (b) Maretina, I. A.; Ionin, B. I., Eds. *Alkynes in Cycloadditions;* John Wiley & Sons, Ltd, 2013; p 310. (c) Synthesis of Heterocycles via Cycloadditions II. In *Topics in Heterocyclic Chemistry;* Hassner, A., Ed.; Vol. 13; Springer-Verlag Berlin Heidelberg, 2008; p 212. (d) Synthesis of Heterocycles via Cycloadditions I. In *Topics in Heterocyclic Chemistry;* Hassner, A., Ed.; Vol. 12; Springer-Verlag Berlin Heidelberg, 2008; p 226. (e) Kobayashi, S.; Jorgensen, K. A., Eds. *Cycloaddition Reactions in Organic Synthesis;* Wiley-VCH, 2002; p 332. (f) Synthetic Applications of 1,3-Dipolar Cycloaddition Chemistry Toward Heterocycles and Natural Products. In *Chemistry of Heterocyclic Compounds;* Padwa, A.; Pearson, W. H., Eds.; Vol. 59; Wiley, J. and Sons: Chichester: UK, 2002; p 940. (g) Harmata, M., Ed. *Advances in Cycloaddition*, Vol. 6; JAI Press Inc., 1999; p 249. (h) Lautens, M., Ed. *Advances in Cycloaddition*, Vol. 4; JAI Press Inc., 1997; p 210. (i) Curran, D. P., Ed. *Advances in Cycloaddition*, Vol. 3; JAI Press Inc., 1993; p 210. (j) Curran, D. P., Ed. *Advances in Cycloaddition*, Vol. 2; JAI Press Inc., 1990; p 219. (k) Zefirov, N. S.; Kazimirchik, I. V.; Lukin, K. A., Eds. *Cycloaddition of Dichlorocarbene to Olefins; Izd.;* Nauka, 1985; p 152. (l) Padwa, A., Ed. *1,3-Dipolar CycloadditionChemistry*, Vol. 2; John Wiley and Sons, 1984; p 704. (m) Padwa, A., Ed. *1,3-Dipolar Cycloaddition Chemistry*, Vol. 1; John Wiley and Sons., 1984; p 817. (n) 1, 2-Cycloaddition Reactions; In *The Formation of Three and Four-Membered Heterocycles;* Muller, L. L.; Hamer, J., Eds.; Interscience Publishers, 1967;

p 362. (o) Ulrich, H., Ed. *Organic Chemistry: A Series of Monographs*, Vol. 9; Academic Press, 1967; p 364. (p) 1,4-Cycloaddition Reactions: The Diels-Alder Reaction in Heterocyclic Syntheses. In *Organic Chemistry: A Series of Monographs;* Hamer, J., Ed.; Vol. 8; Academic Press, 1966; p 530. (q) Topics in Current Chemistry Collections; In *Cycloadditions in Bioorthogonal Chemistry;* Vrabel, M.; Carrell, T. Eds.; Springer, 2016; p 107.

4. Rezza, G. *Virus. Scienze della Vita*; Treccani: Roma, 2007; pp 611–618. For a cutting-edge view on antiviral compounds, see De Clerq, E. *Med. Chem. Rev.* **2013**, *33*, 1249–1277.
5. (a) Colman, P. M. In *The Influenza Viruses: Influenza Virus Neuraminiaase, Enzyme and Antigen;* Krug, R. M., Ed.; Plenum: New York, 1989; pp 175–218. (b) Colman, P. M. *Protein Sci.* **1994,** *3*, 1687–1696. (c) See also Kovac, P., Ed. Carbohydrate Chemistry; CRC Press: Boca Raton, FL, 2012; Vol. 1, pp 239–258, Chp 26–28.
6. Zhang, L.; Williams, M. A.; Mendel, D. B.; Escarpe, P. A.; Chen, X.; Wang, K.-Y. Graves, B. J.; Lawton, G.; Kim, C. U. *Bioorg. Med. Chem. Lett.* **1999,** *9*, 1751–1756.
7. Kerrigan, S. A.; Smith, P. W.; Stoodley, R. J. *Tetrahedron Lett.* **2001,** *42*, 4709–4712.
8. Kerrigan, S. A.; Pritchard, R. G.; Smith, P. W.; Stoodley, R. J. *Tetrahedron Lett.* **2001,** *42*, 7687–7690.
9. von Itzstein, M.; Wu, W.-Y.; Kok, G. B.; Pegg, M. S.; Dyason, J. C.; Jin, B.; Phan, T. V.; Smythe, M. L.; White, H. F.; Oliver, S. W.; Colman, P. M.; Varghese, J. N.; Ryan, D. M.; Woods, J. M.; Bethell, R. C.; Hotham, V. J.; Cameron, J. M.; Penn, C. R. *Nature* **1993,** *363*, 418–423.
10. (a) Kim, C. U.; Lew, W.; Williams, M. A.; Liu, H.; Zhang, L.; Swaminathan, S.; Bischofberger, N.; Chen, M. S.; Mendel, D. B.; Tai, C. Y.; Laver, W. G.; Stevens, R. C. *J. Am. Chem. Soc.* **1997,** *119*, 681–690. (b) For a recent review, see Farina, V.; Brown, J. D. *Angew. Chem. Int. Ed.* **2006,** *45*, 7330–7334.
11. (a) Abrecht, S.; Harrington, P.; Iding, H.; Karpf, M.; Trussardi, R.; Wirz, B.; Zutter, U. *Chimia* **2004,** *58*, 621–629. (b) Abrecht, S.; Federspiel, M. C.; Estermann, H.; Fischer, R.; Karpf, M.; Mair, H.-J.; Oberhauser, T.; Rimmler, G.; Trussardi, R.; Zutter, U. *Chimia* **2007,** *61*, 93–99.
12. (a) Yeung, Y.-Y.; Hong, S.; Corey, E. J. *J. Am. Chem. Soc.* **2006,** *128*, 6310–6311. (b) Ryu, D. H.; Corey, E. J. *J. Am. Chem. Soc.* **2003,** *125*, 6388–6390.
13. (a) Yamatsugu, K.; Kamijo, S.; Suto, Y.; Kanai, M.; Shibasaki, M. *Tetrahedron Lett.* **2007,** *48*, 1403–1406. (b) Mita, T.; Fukuda, N.; Roca, F. X.; Kanai, M.; Shibasaki, M. *Org. Lett.* **2007,** *9*, 259–262.
14. Satoh, N.; Akiba, T.; Yokoshima, S.; Fukuyama, T. *Angew. Chem. Int. Ed.* **2007,** *46*, 5734–5736.
15. (a) Ahrendt, K. A.; Borths, C. J.; MacMillan, D. W. C. *J. Am. Chem. Soc.* **2000,** *122*, 4243–4244. (b) Lelais, G.; Mac-Millan, D. W. C. *Aldrichimica Acta* **2006,** *39*, 79–87.
16. Moriarty, R. M.; Chany, C. J., II; Vaid, R. K.; Prakash, O.; Tuladhar, S. M. *J. Org. Chem.* **1993,** *58*, 2478–2482.
17. Shibasaki, M.; Kanai, M. *Eur. J. Org. Chem.* **2008,** 1839–1850.
18. Gong, J.; Xu, W. *Curr. Med. Chem.* **2008,** *15*, 3145–3159.
19. Matsuya, Y.; Qin, H.; Nagaoka, M.; Nemoto, H. *Heterocycles* **2004,** *62* 207–211.
20. Mosmann, T. *J. Immunol. Methods* **1983,** *65*, 55–63.
21. Matsuya, Y.; Sasaki, K.; Ochiai, H.; Nemoto, H. *Biorg. Med. Chem.* **2007,** *15*, 424–432.
22. (a) Shie, J.-J.; Fang, J.-M. *J. Chin. Chem. Soc.* **2014,** *61*, 127–141. (b) Grondal, C.; Jeanty, M.; Enders, D. *Nat. Chem.* **2010,** *2*, 167–178.

23. (a) Kametani, T.; Nemoto, H.; Ishikawa, H.; Shiroyama, K.; Fukumoto, K. *J. Am. Chem. Soc.* **1976,** *98*, 3378–3379. (b) Kametani, T.; Nemoto, H.; Ishikawa, H.; Shiroyama, K.; Matsumoto, H.; Fukumoto, K. *J. Am. Chem. Soc.* **1977,** *99*, 3461–3466. (c) Kametani, T.; Matsumoto, H.; Nemoto, H.; Fukumoto, K. *J. Am. Chem. Soc.* **1978,** *100*, 6218–6220.
24. Kametani, T.; Suzuki, K.; Nemoto, H. *J. Am. Chem. Soc.* **1981,** *103*, 2890–2891.
25. Nemoto, H.; Fujita, S.; Nagai, M.; Fukumoto, K.; Kametani, T. *J. Am. Chem. Soc.* **1988,** *110*, 2931–2938.
26. Nickisch, K.; Bittler, D.; Casals-Stenzel, J.; Laurent, H.; Nickolson, R.; Nishino, Y.; Petzoldt, K.; Wiechert, R. J. *J. Med. Chem.* **1985,** *28*, 546–550.
27. Jefford, C. W.; Bernardinelli, G.; Wang, Y.; Spellmeyer, D. C.; Buda, A.; Houk, K. N. *J. Am. Chem. Soc.* **1992,** *114*, 1157–1165.
28. (a) Kametani, T.; Tsubuki, M.; Nemoto, H.; Suzuki, K. *J. Am. Chem. Soc.* **1981,** *103*, 1256–1258. (b) Nemoto, H.; Suzuki, K.; Tsubuki, M.; Minemura, K.; Fukumoto, K.; Kametani, T.; Furuyama, H. *Tetrahedron* **1983,** *39*, 1123–1132.
29. Nemoto, H. *Chem. Pharm. Bull.* **2007,** *55*, 961–974.
30. Karplas, A.; Fleet, G. W. J.; Dwek, R. A.; Petursson, S.; Namgoong, S. K.; Ramsden, N. G.; Jacob, G. S.; Rademacher, T. W. *Proc. Natl. Acad. Sci. U.S.A.* **1988,** *85*, 9229–9233.
31. For a review on nitrosocarbonyl intermediates see Memeo, M. G.; Quadrelli, P. *Chem. Rev.* **2017,** *117*, 2108–2200.
32. Tschamber, T.; Craig, C. J.; Muller, M.; Streith, J. *Tetrahedron* **1996,** *52*, 6201–6214.
33. (a) Corfield, A. P.; Schauer, R. In *Sialic Acids: Chemistry, Metabolism and Function in Cell Biology Monographs;* Schauer?R., Ed.; Vol. 10; Springer Verlag: Wien: New York, 1982; p 5. (b) Schauer, R. *Adv. Carbohydr. Chem. Biochem.* **1982,** *40*, 131. (c) Ng, S. S.; Dain, J. In *Biological Roles of Sialic Acid;* Rosenberg, R. A., Schengrund, C. L., Eds.; Plenum Press: New York, 1976; p 59. (d) Jeanloz, R. W.; Codington, J. F. In *Biological Roles of Sialic Acid;* Rosenberg, R. A., Schengrund, C. L., Eds.; Plenum Press: New York, 1976; p 201.
34. Hu, Y.-J.; Huang, X.-D.; Yao, Z.-J.; Wu, Y.-L. *J. Org. Chem.* **1998,** *63*, 2456–2461.
35. Hashiyama, T.; Morikava, K.; Sharpless, K. B. *J. Org. Chem.* **1992,** *57*, 5067–5068.
36. Li, L.-S.; Wu, Y.; Wu, Y.-L. *J. Carbohydr. Res.* **1999,** *18*, 1067–1077.
37. Smith, P. W.; Sollis, S. L.; Howes, P. D.; Cherry, P. C.; Cobley, K. N.; Taylor, H.; Whittington, A. R.; Scicinski, J.; Bethell, R. C.; Taylor, N.; Skarzynski, T.; Cleasby, A.; Singh, O.; Wonacott, A.; Varghese, J.; Colman, P. *Bioorg. Med. Chem. Lett.* **1996,** *6*, 2931–2936.
38. Ryan, D. M.; Ticehurst, J.; Dempsey, M. H. *Antimicrob. Agents Chemother.* **1995,** *39*, 2563–2564.
39. (a) Sorbera, L. A.; Graul, A.; Castaner, J. *Drugs Future* **2000,** *25*, 249–251. (b) Chand, P.; Kotian, P. L.; Dehghani, A.; El-Kattan, Y.; Lin, T.-H.; Hutchison, T. L.; Babu, Y. S.; Bantia, S.; Elliott, A. J.; Montgomery, J. A. *J. Med. Chem.* **2001,** *44*, 4379–4392.
40. Mineno, T.; Miller, M. J. *J. Org. Chem.* **2003,** *68*, 6591–6596.
41. Wu, T. S.; Chan, Y. Y.; Leu, Y. L.; Chern, C. Y.; Chen, C. F. *Phytochemistry* **1996,** *42*, 907–908.
42. Wu, T. S.; Leu, Y. L.; Hsu, H. C.; Ou, L. F.; Chen, C. C.; Chen, C. F.; Ou, J. C.; Wu, Y. C. *Phytochemistry* **1995,** *39*, 383–385.
43. Toyota, M.; Komori, C.; Ihara, M. *J. Org. Chem.* **2000,** *65*, 7110–7113.
44. Kametani, T.; Takeda, H.; Nemoto, H.; Fukumoto, K. *J. Chem. Soc. Perkin Trans. 1* **1975,** 1825–1828.

45. Boger, D. L.; Hüter, O.; Mbiya, K.; Zhang, M. *J. Am. Chem. Soc.* **1995,** *117*, 11839–11849.
46. (a) Kametani, T.; Takeda, H.; Nemoto, H.; Fukumoto, K. *Heterocycles* **1975,** *3*, 167–170. (b) Comins, D. L.; Saha, J. K. *J. Org. Chem.* **1996,** *61*, 9623–9624. (c) Josien, H.; Curran, D. P. *Tetrahedron* **1997,** *53*, 8881–8886.
47. Noyori, R.; Tomino, I.; Tanimoto, Y.; Nishizawa, M. *J. Am. Chem. Soc.* **1984,** *106*, 6709–6716.
48. Govindachari, T. R.; Ravindranath, K. R.; Viswanathan, N. *J. Chem. Soc Perkin Trans. 1* **1974,** 1215–1217.
49. Boger, D. L.; Hong, J. *J. Am. Chem. Soc.* **1998,** *120*, 1218–1222.
50. Crackett, P.; Demont, E.; Eatherton, A.; Frampton, C. X. S.; Gilbert, J.; Kahn, I.; Redshaw, M. S.; Watson, W. *Synlett* **2004,** 679–683.
51. Omura, S.; Nakagawa, A.; Hashimoto, H.; Oiwa, R.; Iwai, Y.; Hirano, A.; Shibukawa, N.; Kojima, Y. *J. Antibiot.* **1980,** *33*, 1395–1396.
52. Nakagawa, A.; Iwai, Y.; Hashimoto, H.; Miyazaki, N.; Oiwa, R.; Takahashi, Y.; Hirano A.; Shibukawa, N.; Kojima, Y.; Omura, S. *J. Antibiot.* **1981,** *34*, 1408–1415.
53. (a) Hill, M. L.; Raphael, R. A. *Tetrahedron Lett.* **1986,** *27*, 1293–1296. (b) Hill, M. L., Raphael, R. A. *Tetrahedron Lett.* **1990,** *46*, 4587–4594. (c) Morimoto, Y.; Matsuda, F., Shirahama, H. *Synlett* **1991,** *3*, 201–203. (d) Morimoto, Y.; Matsuda, F.; Shirahama, H. *Tetrahedron* **1996,** *52*, 10609–10630. (e) Steinhagen, H.; Corey, E. J. *Org. Lett.* **1999,** *1*, 823–824.
54. Keck, D.; Vanderheiden, S.; Bräse, S. *Eur. J. Org. Chem.* **2006,** *21*, 4916–4923.
55. Stockwell, B. R. *Nature* **2004,** *432*, 846–854.
56. Berg, T. *Angew. Chem. Int. Ed.* **2005,** *44*, 5008–5011.
57. Mayer, T. U.; Kapoor, T. M.; Haggarty, S. J.; King, R. W.; Schreiber, S. L.; Mitchtison, T. J. *Science* **1999,** *286*, 971–974.
58. Leßmann, T.; Leuenberger, M. G.; Menninger, S.; Lopez-Canet, M.; Müller, O.; Hümmer, S.; Bormann, J.; Korn, K.; Fava, E.; Zerial, M.; Mayer, T. U.; Waldmann, H. *Chem. Biol.* **2007,** *14*, 443–451.
59. Kamimura, A.; Nakano, T. *J. Org. Chem.* **2010,** *75*, 3133–3136 and references therein.
60. (a) Buchi, G.; Coffen, D. L.; Kocsis, K.; Sonnet, P. E.; Ziegler, F. E. *J. Am. Chem. Soc.* **1966,** *88*, 3099–3109. (b) Marazano, C.; LeGoff, M.; Fourrey, J.; Das, B. C. *J. Chem. Soc. Chem. Commun.* **1981,** 389–391. (c) Raucher, S.; Bray, B. L. *J. Org. Chem.* **1987,** *109*, 442–446. (d) Reding, M. T.; Fukuyama, T. *Org. Lett.* **1999,** *1*, 973–976.
61. (a) Satoh, N.; Akiba, T.; Yokoshima, S.; Fukuyama, T. *Angew. Chem. Int. Ed.* **2007,** *46*, 5734–5736. (b) Satoh, N.; Akiba, T.; Yokoshima, S.; Fukuyama, T. *Tetrahedron* **2009,** *65*, 3239–3245.
62. Nakano, N.; Osone, K.; Takeshita, M.; Kwon, E.; Seki, C.; Matsuyama, H.; Takanoa, N.; Koharia, Y. *Chem. Commun.* **2010,** *46*, 4827–4829.
63. Vimr, E. R.; Kalivoda, K. A.; Deszo, E. L.; Steenbergen, S. M. *Microbiol. Mol. Biol. Rev.* **2004,** *68*, 132–153.
64. Wagner, R.; Matrosovich, M.; Klenk, H. D. *Rev. Med. Virol.* **2002,** *12*, 159–166.
65. Richichi, B.; Lunghi, C.; Papakyriakou, A.; Francesconi, O.; Nativi, C. *Pure Appl. Chem.* **2013,** *85*, 1803–1811.
66. (a) Singh, S. P.; Parmar, S. S.; Raman, K.; Stenberg, V. I. *Chem. Rev.* **1981,** *81*, 175–203. (b) Theocharis, S.; Margeli, A.; Kouraklis, G. *Curr. Med. Chem. Anticancer Agents* **2003,** *3*, 239–251. (c) Theocharis, S.; Margeli, A.; Vielh, P.; Kouraklis, G. *Cancer Treat. Rev.* **2004,** *30*, 545–554. (d) Sing, W. T.; Lee, C. L.; Yeo, S. L.; Lim, S. P.; Sim, M. M.

Bioorg. Med. Chem. Lett. **2001,** *11*, 91–94. (e) Rao, A.; Balzarini, J.; Carbone, A.; Chimirri, A.; De Clercq, E.; Monforte, A. M.; Monforte, P.; Pannecouque, C.; Zappalà, M. *Farmaco* **2004,** *59*, 33–39. (f) Murphy, G. J.; Holder, J. C. *Trends Pharmacol. Sci.* **2000,** *21*, 469–474. (g) Bailey, C. J. *Trends Pharmacol. Sci.* **2000,** *21*, 259–265.

67. Atamanyuk, D.; Zimenkovsky, B.; Atamanyuk, V.; Lesyk, R. *Synth. Commun.* **2014,** *44*, 237–244.
68. (a) Cappendijk, S. L. T.; Dzoljic, M. R. *Eur. J. Pharmacol.* **1993,** *241*, 261–265. (b) Page, C. B.; Pinder, A. R. *J. Chem. Soc.* **1964,** 4811–4816. (c) Delorenzi, J. C.; Attias, M.; Gattass, C. R.; Andrade, M.; Rezende, C.; da Cunha Pinto, A.; Henriques, A. T.; Bou-Habib, D. C.; Saraiva, E. M. B. *Antimicrob. Agents Chemother.* **2001,** *45*, 1349–1354.
69. (a) Seebacher, W.; Schlapper, C.; Brun, R.; Kaiser, M.; Saf, R.; Weis, R. *Eur. J. Pharm. Sci.* **2005,** *24*, 281–289. (b) Seebacher, W.; Schlapper, C.; Brun, R.; Kaiser, M.; Saf, R.; Weis, R. *Eur. J. Med. Chem.* **2006,** *41*, 970–977.
70. Tu, Y. *Nat. Med.* **2011,** *17*, 1217–1220.
71. (a) Reiter, C.; Fröhlich, T.; Zeino, M.; Marschall, M.; Bahsi, H.; Leidenberger, M.; Friedrich, O.; Kappes, B.; Hampel, F.; Efferth, T.; Tsogoeva, S. B. *Eur. J. Med. Chem.* **2015,** *97*, 164–172. (b) Reiter, C.; Fröhlich, T.; Gruber, L.; Hutterer, C.; Marschall, M.; Voigtländer, C.; Friedrich, O.; Kappes, B.; Efferth, T.; Tsogoeva, S. B. *Bioorg. Med. Chem.* **2015,** *23*, 5452–5458.
72. Bock, C. M.; Parameshwarappa, G.; Bönisch, S.; Neiss, C.; Bauer, W.; Hampel, F.; Görling, A.; Tsogoeva, S. B. *Chem. Eur. J.* **2016,** *22*, 5189–5197.
73. (a) Geronikaki, A. A.; Lagunin, A. A.; Hadjipovlou-Litina, D. I.; Eleftheriou, P. T.; Filimonov, D. A.; Poroikov, V. V.; Alam, I.; Saxena, A. K. *J. Med. Chem.* **2008,** *51*, 1601–1609. (b) Shin, M. H.; Ke, F. Y. *Bioorg. Med. Chem. Lett.* **2004,** *12*, 4633–4643. (c) Lesyk, R. B.; Zimenkovsky, B. S. *Curr. Org. Chem.* **2004,** *8*, 1547–1577. (d) Ahn, J. H.; Kim, S. J.; Park, W. S.; Cho, S. Y.; Ha, J. D.; Kim, S. S.; Kang, S. K.; Jeong, D. G.; Jung, S.-K.; Lee, S.-H.; Kim, H. M.; Park, S. K.; Lee, K. H.; Lee, C. W.; Ryu, S. E.; Choi, J.-K. *Bioorg. Med. Chem. Lett.* **2006,** *16*, 2996–2999. (e) Liu, W.; Bulgaru, A.; Haigentz, M.; Stein, C. A.; Perez-Soler, R.; Mani, S. *Curr. Med. Chem. Anticancer Agents* **2003,** *3*, 217–223. (f) Lee, J.; Kim, J.; Koh, J. S.; Chung, H. H.; Kim, K. H. *Bioorg. Med. Chem. Lett.* **2006,** *16*, 1954–1956.
74. Lozynskyi, A.; Golota, S.; Zimenkovsky, B.; Atamanyuk, D.; Gzella, A.; Lesyk, R. *Phosphorus, Sulfur Silicon* **2016,** *191*, 1245–1249.
75. Ghosh, A. K.; Chapsal, B. D.; Weber, I. T.; Mitsuya, H. *Acc. Chem. Res.* **2008,** *41*, 78–86.
76. Ghosh, A. K.; Sridhar, P. R.; Leshchenko, S.; Hussain, A. K.; Li, J.; Kovalevsky, A. Y.; Walters, D. E.; Wedekind, J.; Grum-Tokars, V.; Das, D.; Koh, Y.; Maeda, K.; Gatanaga, H.; Weber, I. T.; Mitsuya, H. *J. Med. Chem.* **2006,** *49*, 5252–5261.
77. Ghosh, A. K.; Venkateswara Rao, K.; Nyalapatla, P. R.; Osswald, H. L.; Martyr, C. D.; Aoki, M.; Hayashi, H.; Agniswamy, J.; Wang, Y.-F.; Bulut, H.; Das, D.; Weber, I. T.; Mitsuya, H. *J. Med. Chem.* **2017,** *60*, 4267–4278.
78. (a) Bundret, K. M.; Dalziel, W.; Hesp, B.; Jarvis, J. A. J.; Neidle, S. *J. Chem. Soc. Chem. Commun.* **1972,** 1027–1028. (b) Dalziel, W.; Hesp, B.; Stevenson, K. M.; Jarvis, J. A. *J. Chem. Soc. Perkin Trans. 1* **1973,** 2841–2851.
79. (a) Toyota, M.; Nishikawa, Y.; Fukumoto, K. *Tetrahedron Lett.* **1994,** *35*, 6495–6498. (b) Toyota, M.; Nishikawa, Y.; Seishi, T.; Fukumoto, K. *Tetrahedron* **1994,** *50*, 10183–10192.
80. (a) Ishikawa, T.; Kotake, K.-I.; Ishii, H. *Chem. Pharm. Bull.* **1995,** *4*, 1039–1041. (b) Desimoni, G.; Faita, G.; Quadrelli, P. *Chem. Rev.* **2013,** *113*, 5924–5988.

81. Manchand, P. S.; White, J. D.; Wright, H.; Clardy, J. *J. Am. Chem Soc.* **1973,** *95*, 2705–2706.
82. Tanaka, T.; Murakami, K.; Kanda, A.; Patra, D.; Yamamoto, S.; Satoh, N.; Kim, S.-W.; Rahman, S. M. A.; Ohno, H.; Iwata, C. *J. Org. Chem.* **2001,** *66*, 7107–7112.
83. (a) Tamura, G.; Suziki, S.; Takatsuki, K. A.; Arima, K. *J. Antibiot.* **1968,** *21*, 539–544. (b) Takatsuki, A.; Tamura, G.; Arima, K. *Appl. Microbiol.* **1969,** *17*, 825–829.
84. Dogan, O.; Oppolzer, W. *Turk. J. Chem.* **2001,** *25*, 273–281.
85. Tsunakawa, M.; Nishio, M.; Ohkuma, H.; Tsuno, T.; Konishi, M.; Naito, T.; Oki, T.; Kawaguchi, H. *J. Org. Chem.* **1989,** *54*, 2532–2536.
86. Kakushima, M.; Sawada, Y.; Nishio, M.; Tsuno, T.; Oki, T. *J. Org. Chem.* **1989,** *54*, 2536–2539.
87. Zilke, L.; Hall, D. G. *Eur. J. Org. Chem.* **2012,** *22*, 4153–4163.
88. Win, N. N.; Awale, S.; Esumi, H.; Tezuka, Y.; Kadota, S. *Chem. Pharm. Bull.* **2008,** *56*, 491–496.
89. Pasfield, L. A.; de la Cruz, L.; Ho, J.; Coote, M. L.; Otting, G.; McLeod, M. D. *Asian J. Org. Chem.* **2013,** *2*, 60–63.
90. Ye, D. K. J.; Richard, J.-A. *Tetrahedron Lett.* **2014,** *55*, 2183–2186.
91. (a) Lee, E.; Song, H. Y.; Kang, J. W.; Kim, D.-S.; Jung, C.-K.; Joo, J. M. *J. Am. Chem. Soc.* **2002,** *124*, 384–385. (b) Jung, M. E.; Ho, D.; Chu, H. V. *Org. Lett.* **2005,** *7*, 1649–1651. (c) Jung, M. E.; Guzaev, M. *J. Org. Chem.* **2013,** *78*, 7518–7526.
92. Liu, S.-A.; Trauner, D. *J. Am. Chem. Soc.* **2017,** *139*, 9491–9494.
93. (a) Wang, J.; Froeyen, M.; Hendrix, C.; Andrei, G.; Snoeck, R.; De Clercq, E.; Herdewijn, P. *J. Med. Chem.* **2000,** *43*, 736–745. (b) Wang, J.; Herdewijn, P. *J. Org. Chem.* **1999,** *64*, 7820–7827. (c) Wang, J.; Busson, R.; Blaton, N.; Rozenski, J.; Herdewijn, P. *J. Org. Chem.* **1998,** *63*, 3051–3058.
94. Wang, J.; Morral, J.; Hendrix, C.; Herdewijn, P. *J. Org. Chem.* **2001,** *66*, 8478–8482.
95. (a) Agrofoglio, L.; Challand, S. R. *Acyclic, Carbocyclic and l-Nucleosides;* Kluver Academic Publishers: Dordrecht, The Netherlands, 1998. (b) Crimmins, M. T. *Tetrahedron* **1998,** *54*, 9229–9272. (c) Ueda, T. Chapter 1 In *Chemistry of Nucleosides and Nucleotides;* Townsend?L. B., Ed.; Vol. 1; Plenum Press: New York, 1988.
96. Vince, R.; Hua, M. *J. Med. Chem.* **1990,** *33*, 17–21.
97. Belmonte, L.; Barré, P.; de Bracco, M. M. E.; Ruibal-Ares, B. H. *Curr. Med. Chem.* **2003,** *10*, 303–312.
98. Quadrelli, P.; Scrocchi, R.; Caramella, P.; Rescifina, A.; Piperno, A. *Tetrahedron* **2004,** *60*, 3643–3651.
99. Quadrelli, P.; Martinez Vazquez, N.; Scrocchi, R.; Corsaro, A.; Pistarà, V. *Sci. World J.* **2014,** *2014*, 492178 Article ID 492178.
100. Moggio, Y.; Legnani, L.; Bovio, B.; Memeo, M. G.; Quadrelli, P. *Tetrahedron* **2012,** *68*, 1384–1392.
101. (a) White, P. W.; Faucher, A.-M.; Massariol, M.-J.; Welchner, E.; Rancourt, J.; Cartier, M.; Archambault, J. *Antimicrob. Agents Chemother.* **2005,** *49*, 4834–4842. (b) Faucher, A.-M.; White, P. W.; Brochu, C.; Grand-Maitre, C.; Rancourt, J.; Fazal, G. *J. Med. Chem.* **2004,** *47*, 18–21. (c) Yoshino, M. *Biochem. J.* **1987,** *248*, 815–820.
102. Johnson, J. A.; Gangemi, J. D. *Antimicrob. Agents Chemother.* **1999,** *43*, 1198–1205.
103. Memeo, M. G.; Lapolla, F.; Maga, G.; Quadrelli, P. *Tetrahedron Lett.* **2015,** *56*, 1986–1990.
104. (a) D'Abramo, C. M.; Archambault, J. *Open Virol. J.* **2011,** *5*, 80–95. (b) Kim, S. S.; Tam, J. K.; Wang, A. F.; Hegde, R. S. *J. Biol. Chem.* **2000,** *275*, 31245–31254.

(c) Wang, Y.; Coulombe, R.; Cameron, D. R.; Thauvette, L.; Massariol, M.-J.; Amon, L. M.; Fink, D.; Titolo, S.; Welchner, E.; Yoakim, C.; Archambault, J.; White, P. W. *J. Biol. Chem.* **2004,** *279*, 6976–6985.

105. Savion, M.; Memeo, M. G.; Bovio, B.; Grazioso, G.; Legnani, L.; Quadrelli, P. *Tetrahedron.* **2012,** *68*, 1845–1852.
106. De Clercq, E. *Biochem. Pharmacol.* **2004,** *68*, 2301–2315.
107. Al-Saad, D.; Memeo, M. G.; Quadrelli, P. *Sci. World J.* **2014,** 10 p Article ID 472373.
108. Al-Saad, D.; Memeo, M. G.; Quadrelli, P. *Lett. Org. Chem.* **2016,** *13*, 757–763.
109. Al-Saad, D.; Memeo, M. G.; Quadrelli, P. *Chem. Select* **2017,** *2*, 10340–10346.
110. Hrebabecky, H.; Dracinsky, M.; De Palma, A. M.; Neyts, J.; Holy, A. *Collect. Czech. Chem. Commun.* **2009,** *74*, 469–485.
111. Hrebabecky, H.; Dracinsky, M.; De Palma, A. M.; Neyts, J.; Holy, A. *Collect. Czech. Chem. Commun.* **2009,** *74*, 487–502.
112. Dejmek, M.; Hrebabecky, H.; Dracinsky, M.; De Palma, A. M.; Neyts, J.; Leyssen, P.; Mertlikova-Kaiserova, H.; Nenka, R. *Collect. Czech. Chem. Commun.* **2011,** *76*, 1549–1566.
113. Shing, T. K. M.; Wong, A. W. H.; Li, H.; Liu, Z. F.; Chan, P. K. S. *Org. Biomol. Chem.* **2014,** *12*, 9439–9445.
114. Pokorny, J.; Borkova, L.; Urban, M. *Curr. Med. Chem.* **2018,** *25*, 636–658.
115. Alvarez, R.; Velasquez, S.; San-félix, A.; Aquaro, S.; De Clercq, E.; Perno, C. F.; Karlsson, A.; Balzarini, J.; Camarasa, M. J. *J. Med. Chem.* **1994,** *37*, 4185–4194 and references cited therein.
116. Moukha-Chafiq, O.; Taha, M. L.; Lazrek, H. B.; Vasseur, J. J.; Pannecouque, C.; Witvrouw, M.; De Clercq, E. *Nucleosides Nucleotides Nucleic Acids* **2001,** *20*, 1811–1821.
117. Abdel-Rahmana, A. A. H.; Wada, T. *Z. Naturforsch.* **2009,** *64*, 163–166.
118. Montagu, A.; Roy, V.; Balzarini, J.; Snoeck, R.; Andrei, G.; Agrofoglio, L. A. *Eur. J. Med. Chem.* **2011,** *46*, 778–786.
119. Pinter, G.; Bereczki, I.; Batta, G.; Otvos, R.; Sztaricskai, F.; Roth, E.; Ostorhazi, E.; Rozgonyi, F.; Naesens, L.; Szarvas, M.; Boda, Z.; Herczegh, P. *Bioorg. Med. Chem. Lett.* **2010,** *20*, 2713–2717.
120. Lakshman, M. K.; Singh, M. K.; Parrish, D.; Balachandran, R.; Day, B. W. *J. Org. Chem.* **2010,** *75*, 2461–2473.
121. Lakshman, M. K.; Kumar, A.; Balachandran, R.; Day, B. W.; Andrei, G.; Snoeck, R.; Balzarini, J. *J. Org. Chem.* **2012,** *77*, 5870–5883.
122. Glowacka, I. E.; Balzarini, J.; Andrei, G.; Snoeck, R.; Schols, D.; Piotrowska, D. G. *Bioorg. Med. Chem.* **2014,** *22*, 3629–3641.
123. Glowacka, I. E.; Balzarini, J.; Piotrowska, D. G. *Arch. Pharm. Chem. Life Sci.* **2014,** *347*, 496–505.
124. Glowacka, I. E.; Balzarini, J.; Wroblewski, A. E. *Arch. Pharm. Chem. Life Sci.* **2014,** *347*, 506–514.
125. Senanayake, T. H.; Warren, G.; Wei, X.; Vinogradov, S. V. *J. Controlled Release* **2013,** *167*, 200–209.
126. Schulz, T.; Balzarini, J.; Meier, C. *ChemMedChem* **2014,** *9*, 762–775.
127. (a) Vasilyeva, S. V.; Shtil, A. A.; Petrova, A. S.; Balakhnin, S. M.; Achigecheva, P. Y.; Stetsenko, D. A.; Silnikov, V. N. *Bioorg. Med. Chem.* **2017,** *25*, 1696–1702. (b) Liu, Y.; Peng, Y.; Lu, J.; Wang, J.; Ma, H.; Song, C.; Liu, B.; Qiao, Y.; Yu, W.; Wu, J.; Chang, J. *Eur. J. Med. Chem.* **2018,** *143*, 137–149.

128. Vernekar, S. K. V.; Qiu, L.; Zacharias, J.; Geraghty, R. T. J.; Wang, Z. *Med. Chem. Commun.* **2014,** *5*, 603–608.
129. Munoz, A.; Sigwalt, D.; Iliescas, B. M.; Luczkowiak, J.; Rodriguez-Perez, L.; Nierengarten, I.; Holler, M.; Remy, J.-S.; Buffet, K.; Vincent, S. P.; Rojo, J.; Delgado, R.; Nierengarten, J.-F.; Martin, N. *Nat. Chem.* **2016,** *8*, 50–57.
130. Efthymiou, T.; Gong, W.; Desaulniers, J.-P. *Molecules* **2012,** *17*, 12665–12703.
131. Saini, M. S.; Dwivedi, J. *Int. J. Pharm. Sci. Res.* **2013,** *4*, 2866–2879.
132. Fan, F.; Cai, C.; Wang, W.; Gao, L.; Li, J.; Li, J.; Gu, F.; Sun, T.; Li, J.; Li, C.; Yu, G. *ACS Macro Lett.* **2018,** *7*, 330–335.
133. Giofré, S. V.; Romeo, R.; Chiacchio, U.; Romeo, G.; Chiacchio, M. A. *Mini-Rev. Org. Chem.* **2015,** *12*, 249–257.
134. Giofré, S. V.; Romeo, R.; Garozzo, A.; Cicero, N.; Campisi, A.; Lanza, G.; Chiacchio, M. A. *ARKIVOC* **2015,** *vii*, 253–269.
135. Kokosza, K.; Andrei, G.; Schols, D.; Snoeck, R.; Piotrowska, D. G. *Bioorg. Med. Chem* **2015,** *23*, 3135–3146.
136. Zhang, P.; Wei, C.; Wang, E.; Wanga, W.; Liu, M.; Yin, Q.; Chen, H.; Wang, K.; Li, X.; Zhang, J. *Carbohydr. Res.* **2012,** *351*, 7–16.
137. (a) Kim, H. O.; Schinazi, R. F.; Shanmuganathan, K.; Jeong, L. S.; Beach, J. W.; Nampalli, S.; Cannon, D. L.; Chu, C. K. *J. Med. Chem.* **1993,** *36*, 519–634. (b) Schinazi, R. F.; Gosselin, G.; Faraj, A.; Korba, B. E.; Liotta, D. C.; Chu, C. K.; Mathé, C.; Imbach, J.-L.; Sommadossi, J.-P. *Antimicrob. Agents Chemother.* **1994,** *38*, 2172–2174. (c) Jin, H.; Siddigui, M. A.; Evans, C. A.; Tse, H. L. A.; Mansour, T. S. *J. Org. Chem.* **1995,** *60*, 2621–2623. (d) Liang, C.; Lee, D. W.; Newton, M. G.; Chu, C. K. *J. Org. Chem.* **1995,** *60*, 1546–1553.
138. Gi, H.-J.; Xiang, Y.; Schinazi, R. F.; Zhao, K. *J. Org. Chem.* **1997,** *62*, 88–92.
139. Ortuño, R. M.; Moglioni, A. G.; Moltrasio, G. Y. *Curr. Org. Chem.* **2005,** *9*, 237–259 and references therein.

Chapter 2

Cycloaddition reactions for anticancer compounds

Paolo Quadrelli
Department of Chemistry, University of Pavia, Pavia, Italy

Cancer

The word "cancer" finds its origin in the Greek word *karkinos* (crab), and identifies several illnesses affecting both humans and animals, having in common the uncontrolled growth of abnormal cells in the organism. Carcinogenesis is a complex and multiphasic process and represents the main cause of mortality in developed countries. The modern technological instruments are able to conduct complex analyses on cellular molecular biology, increasing knowledge of the biological and chemical aspects of cancer cells. In vitro and in vivo studies demonstrated that cancer is a multiphasic process in which the alterations are determined by the human genome. The research target is to find the precise rules that dictate the transformation of normal cells into cancer cells. Cancer treatment is extremely complex: from diagnosis to surgery, from radiotherapy to chemotherapy, to immunotherapy. Recent discoveries and acquisitions on the cellular transmission signals design a complex map where a pivotal role is determined by the development of a molecular therapy to inhibit cancer growth signals.

A prevailing role in the development of a molecular therapy against cancer is determined by the growth signal inhibition. The growth signals are transmitted to the cell through transmembrane receptors that bind different classes of emitting signals molecules. The growth factor (GF) receptors, often acting through a tyrosine-kinase at the cytoplasm level, are overexpressed in several tumors. A first example of GF-depending proliferation is the interaction between hormones and estrogen or androgen receptors. Several hormone agents are available for the treatment of oncologic patients. These are mainly used in hormone-responsive tumor therapies such as in

Modern Applications of Cycloaddition Chemistry. DOI: https://doi.org/10.1016/B978-0-12-815273-7.00002-2

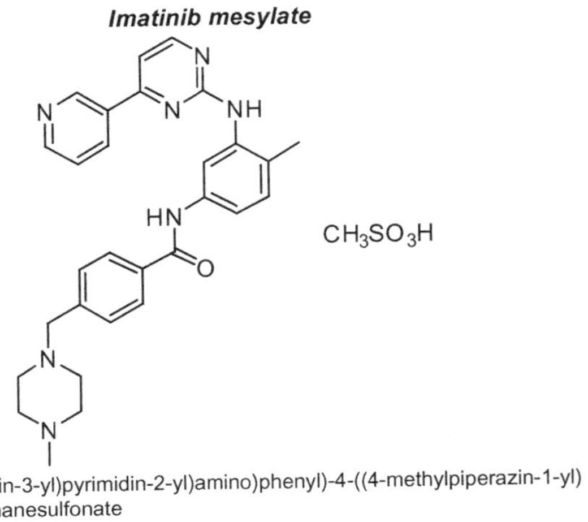

FIGURE 2.1 Structure of imatinib mesylate.

mammalian and prostatic cancers. New therapeutic agents have been developed for cancer treatment that identifies as target the key elements in signal transduction processes, such as kinase receptors. These receptors are composed by an extracellular portion, a transmembrane lipophilic part and an intracellular protein. Several therapeutic approaches target these parts of the kinases. Monoclonal antibodies and small-molecule inhibitors of the tyrosine-kinase activity have been proposed. The latter are generally reversible antagonists that bind the intracellular portion of the tyrosine-kinase. The best molecule is considered the imatinib mesylate (Fig. 2.1), which is a selective inhibitor of the intracellular kinase and is extremely efficient in patients affected by chronic myeloid leukemia, gastrointestinal tumors, and related others.

This research area is developing greatly, and several research groups around the world are involved in proposing new molecules for cancer treatment. This chapter is devoted to these studies and the use of cycloaddition reaction for the synthesis of new active molecules.[1]

Diels–Alder and hetero Diels–Alder reactions for anticancer compounds

In the introduction of Chapter 1, Cycloaddition reactions for antiviral compounds, we spoke about the Diels–Alder (DA) reactions as a valuable tool for the synthesis of antiviral compounds. The versatility of the method and the opportunities from the synthetic point of view given by the concerted or catalyzed processes called DA or hetero Diels–Alder (HDA) cycloaddition were also applied to the target that is the subject of this chapter: the anticancer compounds, their synthesis, and the useful application of cycloaddition processes to prepare valuable and effective compounds. Some interesting examples can be highlighted here to better show the importance of cycloadditions in this research area.

Indole arynes or indolynes were discovered and reported as a completely new class of reactive aryne intermediate.[2] A library of 6,7-annulated-4-substituted indoles was constructed, using a strategic combination of 6,7-indolyne cycloaddition and cross-coupling reactions under both Suzuki–Miyaura and Buchwald–Hartwig conditions.[3] Scheme 2.1 shows the synthetic strategy for the preparation of the target compounds.

The main 4,6,7-tribromoindole scaffold **2** was synthesized by the Bartoli reaction starting from the 1,2,5-tribromo-3-nitrobenzene **1**. The *N*-methyl-4,6,7-tribromoindole **2** scaffold was treated with n-butyllithium to generate the indole aryne **3**, which reacted with both cyclopentadiene ($X = CH_2$) and furan ($X = O$) to give the respective DA cycloadducts **4** in high yields. The C=C double bond in the cycloadducts **4** was reduced quantitatively with diimide prior to cross-coupling reactions. The 4-Br derivative **4** was reacted either with various boronic acids under a carefully prescribed set of conditions to give the Suzuki–Miyaura cross-coupled products **5** or with various commercially available secondary amines under a carefully prescribed set of conditions to give the Buchwald–Hartwig cross-coupled products **6**. These new annulated indoles had never been tested for their biological properties. The only exceptions are a few biologically active natural products, such as trikentrins, which are antibacterial agents; herbindoles, which possess

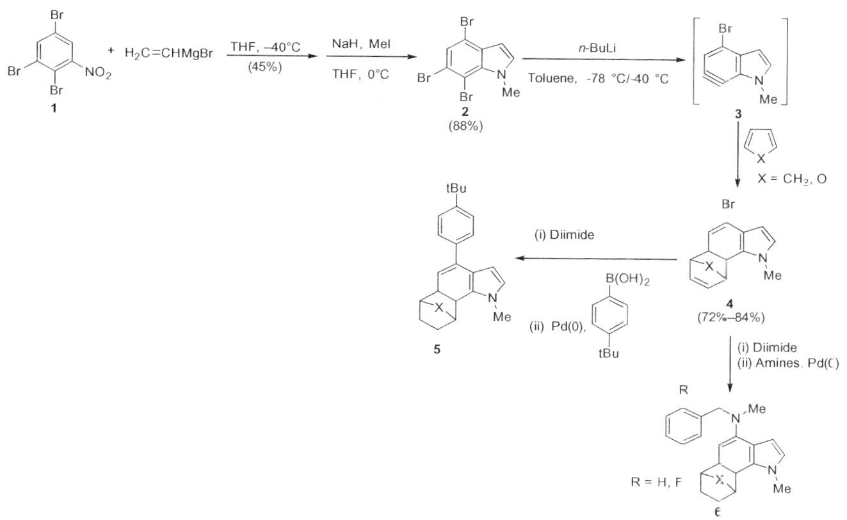

SCHEME 2.1 Synthetic strategy to indole derivatives **5** and **6**.

cytotoxic and antifeedant properties; teleocidins, which are tumor promoters used to study multistage carcinogenesis; and nodulisporic acids, which have antiparasitic activity.[4] Annulated indoles were therefore predicted to have unique chemical property space characteristics and a good probability of exhibiting interesting biological activities. Hence, the newly synthesized 6,7-annulated-4-substituted indoles **5** and **6** were evaluated for their antiproliferative activity in rapidly growing suspension cultures of L1210 leukemia cells, using tests of metabolic activity, DNA degradation, mitotic disruption, cytokinesis, and apoptosis in vitro.[5]

Most compounds inhibited the metabolic activity of leukemic cells in a time- and concentration-dependent manner but only a few of them were sufficiently potent to inhibit L1210 tumor cell proliferation by 50% in the low-μM range after 2 (IC_{50}: 4.5–20.4 μM) and 4 days (0.5–4.0 μM) in culture. However, the antiproliferative compounds that were the most effective at day 4 were not necessarily the most potent at day 2, suggesting different speeds of action. A 3 h treatment with antiproliferative annulated indole was sufficient to inhibit, in a concentration-dependent manner, the rate of DNA synthesis measured in L1210 cells over a 0.5 h period of pulse-labeling with 3*H*-thymidine. Some antiproliferative compounds induced DNA cleavage at 24 h in L1210 cells, containing 3*H*-thymidineprelabeled DNA, suggesting that these antitumor annulated indoles might trigger an apoptotic pathway of DNA fragmentation. Indeed the antiproliferative annulated indoles caused a time-dependent increase of caspase-3 activity with a peak at 6 h. Interestingly, the compounds with the most potent antiproliferative IC_{50} values at day 2 were consistently the most effective at inhibiting DNA synthesis at 3 h and inducing DNA fragmentation at 24 h. After 24–48 h, antiproliferative concentrations of annulated indoles increased the mitotic index of L1210 cells and stimulated the formation of many binucleated cells, multinucleated cells, apoptotic cells, and micronuclei, suggesting that these antitumor compounds might enhance mitotic abnormality, induce chromosomal damage or missegregation, and block cytokinesis to induce apoptosis. The authors conclude that although annulated indoles may have interesting bioactivity, novel derivatives with different substitutions must be synthesized to elucidate structure–activity relationships (SARs), identify more potent antitumor lead compounds, and investigate their molecular targets and mechanisms of action. A novel synthetic approach to bis(indolyl)methanes has been established, demonstrating the force of indole ring in anticancer compound synthesis. A one-pot synthetic strategy based on two consecutive HDA cycloaddition reactions of electrophilic conjugated nitrosoalkenes with indoles was extended to a range of new 1-hydroxyiminomethyl-bis(indolyl) methanes. The biological evaluation of the new bis(indolyl)methanes as anticancer agents was investigated, leading to promising activities.[5b]

The hasubanan and acutumine alkaloids are large families of structurally related natural products.[6] The hasubanan family contains over 70 alkaloids spanning four subclasses defined by the oxidation pattern of the A-ring. Fourteen additional acutumine alkaloids are also known. While the hasubanan alkaloids contain an aromatic ring, the acutumine alkaloids possess a spirocyclic cyclopentenone fragment and a secondary alkyl chloride function. Members of these families exhibit diverse biological effects, including antimicrobial and antiviral (HIV and hepatitis B) effects, opioid receptor binding activities, memory-enhancing properties, and selective inhibition of human T-cell proliferation. A general and versatile strategy to access the hasubanonine and stephamiersine classes of hasubanan alkaloids was recently proposed for the preparation of (−)-runanine (**15a**), (−)-hasubanonine (**15b**), (−)-delavayine (**19**), and (+)-periglaucine B (**20**) (Schemes 2.2 and 2.3). The pivotal step in the synthetic route was the DA reaction between 5-(trimethylsilyl)cyclopentadiene (**8**) and the quinone **7**, efficiently promoted in its enantioselective version by the Corey − Bakshi − Shibata oxazaborolidine (**9**, 22 mol%). The desired *endo*-adduct **10** was obtained in 78% yield, as a single diastereomer in 93% ee (Scheme 2.2). This was indicated by the authors as the first application of 5-(trimethylsilyl)cyclopentadiene (**8**) in an enantioselective DA reaction. Staudinger reduction of the *endo*-adduct **10** was conveniently effected using trimethylphosphine, to afford the protected iminoquinone **11** (99%). Due to the electron-donating ability of the trialkylsilyl substituent, the DA cycloadduct **10** and the protected iminoquinone **11**

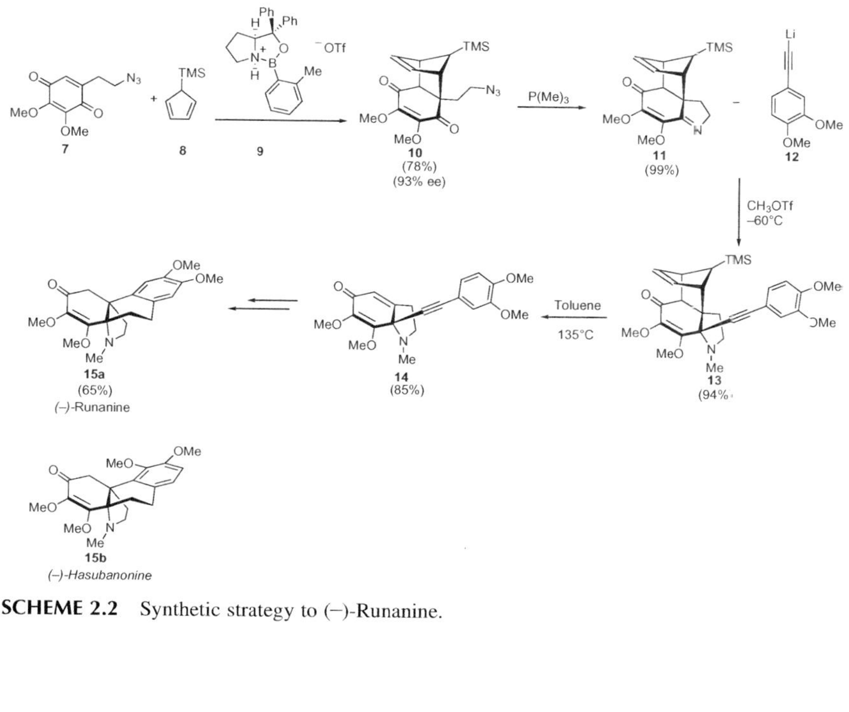

SCHEME 2.2 Synthetic strategy to (−)-Runanine.

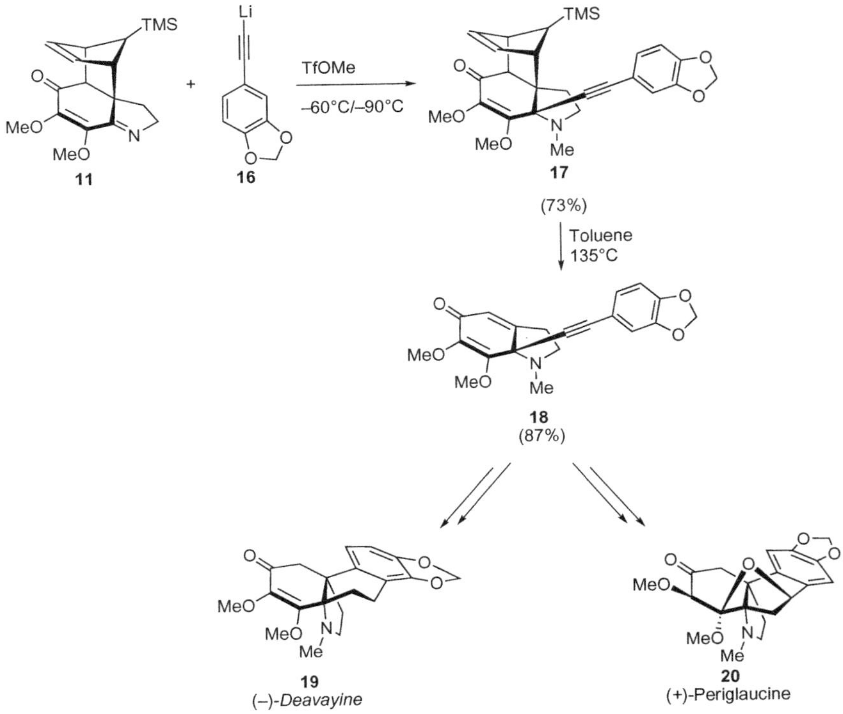

SCHEME 2.3 Synthetic strategy to (−)-*Deavayine* (+)-*Periglaucine*.

undergo retro-cycloaddition rapidly in the presence of acid and slowly under neutral conditions at ambient temperature.

To access the desired target, the acetylide **12** was employed. Addition of **12** to the *N*-methyliminium ion derived from **11** afforded the addition product **13** as a single detectable diastereomer (94%). The addition product **13** was transformed to (−)-runanine (**15a**) by a four-step sequence. First, the retro-cycloaddition was initiated by heating in toluene (135°C) to provide the dienone **14** (85%). Next, selective hydrogenation of the alkyne (Crabtree's catalyst) produced the *cis*-alkene. Acid-mediated cyclization of the *cis*-alkene provided the tetracycle derivative as a single isomer. Finally, hydrogenation of the styrenyl olefin of the tetracycle with Wilkinson' s catalyst afforded (−)-runanine (**15a**, 65%). (−)-Runanine (**15a**) and (−)-hasubanonine (**15b**) only differ in the positions of the aryl methoxy groups.

To establish methods to access stephamiersine alkaloids, the targeted hasubanonine alkaloid (−)-delavayine (**19**) has the same arene substitution as the stephamiersine alkaloid (+)-perigluacine B (**20**, Scheme 2.3). (−)-Delavayine (**19**) could be converted to (+)-perigluacine B (**20**) by a benzylic oxidation – 1,4-addition sequence. The first two steps to **18** are the same as previously described. Accordingly, a *cis*-alkene was prepared. Acid-mediated cyclization of the *cis*-alkene furnished the tetracycle isomer. Hydrogenation with Wilkinson's catalyst afforded (−)-delavayine (**19**, 73%). At this juncture, we recognized that the final intermediate might be converted directly to (+)-periglaucine B (**20**) by a redox-neutral hydration − 1,4-addition sequence. Heating a mixture of the tetracycle derivative and cobalt bis(acetylacetonate) in 2-propanol produced the hydration product. Addition of excess formic acid directly to the reaction mixture promoted cyclization. providing (+)-periglaucine B (**20**) in 55% yield.[7]

The synthesized alkaloids were tested against three human cancer cell lines: N87 (gastric), MDA-MB-361-DYT2 (breast), and HT-29 (colon). The most potent alkaloid, (−)-delavayine (**19**), possessed an IC_{50} value of 0.892 μM. (−)-Runanine (**15a**), (−)-hasubanonine (**15b**), and (+)-periglaucine B (**20**) exhibited IC_{50} values of 0.959, 0.970, and 0.992 μM, respectively. The similar potencies of the compounds suggest the α-methoxy ketone and amine functionality may be important for antiproliferative effects, while the arene substituents are not. For the MDA-MB-361-DYT2 and HT-29 cell lines, all alkaloids tested possessed IC_{50} values >1 μM.

A new family of merosesquiterpenes was synthesized by DA cycloaddition reaction between the labdane diene trans-communic acid, highly abundant in *Cupressus sempervirens*, or its methyl ester **21**, with the appropriate dienophile of type **22** (Scheme 2.4). Aromatization of the newly constructed six-membered ring **23** afforded the desired compounds of type **24**.

These compounds **24** demonstrated potent cytotoxic activity in vitro against human breast, colon, and lung tumor cells. An elevated activity (IC_{50}: 0.35 ± 0.10 mM) and specificity (TI: 9) was found for compounds containing a naphtoquinone moiety against the MCF-7 line, which corresponds to the most prevalent breast cancer cell subtype, luminal A. This compound exerts its antitumor action by inducing oxidative stress, arresting the cell cycle in stages G0eG1, and activating apoptosis, which are all associated with low cyclin D1 regulation, pRb hypophosphorylation, increased expression of p27 and p53, and poly(Adenosine diphosphate (ADP)-ribose) polymerase (PARP) fractioning. In addition, it markedly reduced tumor

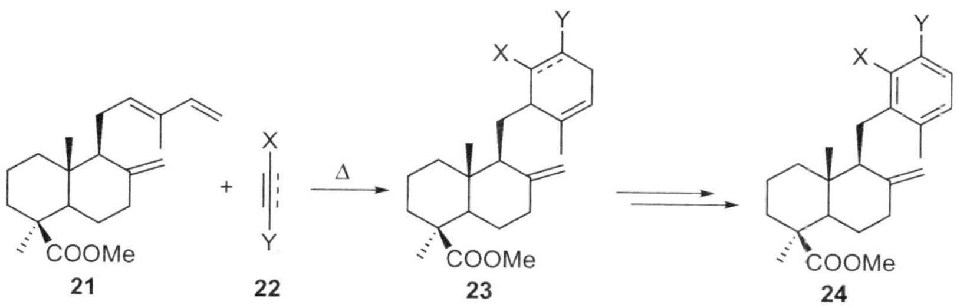

SCHEME 2.4 Synthetic strategy to merosesquiterpenes **24**.

development in immunocompetent C57BL/6 mice with allografts of E0771 mouse breast tumor cells (luminal A subtype). According to these findings, this new family of compounds may be highly useful in the treatment of human breast cancer.[8]

The synthetic transformation and modification of natural products intending to improve the biological properties is an area of current interest. The triterpenoids betulin and betulinic acid are very abundant in nature and now are commercially available. Starting from betulin and betulinic acid, allobetulone and betulonic acid were prepared in a few synthetic steps. The ketone function at the A-ring was used as the starting point for the synthesis of a series of 1,2,4-triazine-fused triterpenoids. The alkylation and Liebeskind-Srogl coupling were used for further substitution of 1,2,4-triazines, and the intramolecular HDA cycloaddition reaction leads to interesting fused thienopyridine derivatives (Scheme 2.5). 2-Oxobetulonic acid (**26**) was obtained by oxidation of betulonic acid (**25**) with oxygen in the presence of *t*-BuOK in *t*-BuOH. The condensation reaction of **26** with thiosemicarbazide in ethanol with potassium carbonate afforded the 1,2,4-triazine derivative of betulonic acid (**27**).

From triazine (**27**), further substitution with different alkyl groups was conducted to get two *S*-alkylated triazine derivatives (**28**, **29**). By heating the *S*-butynyl triazine derivative (**29**) in xylene under reflux, thienopyridine derivative (**30**) was obtained by intermolecular HDA reaction. The cytostatic

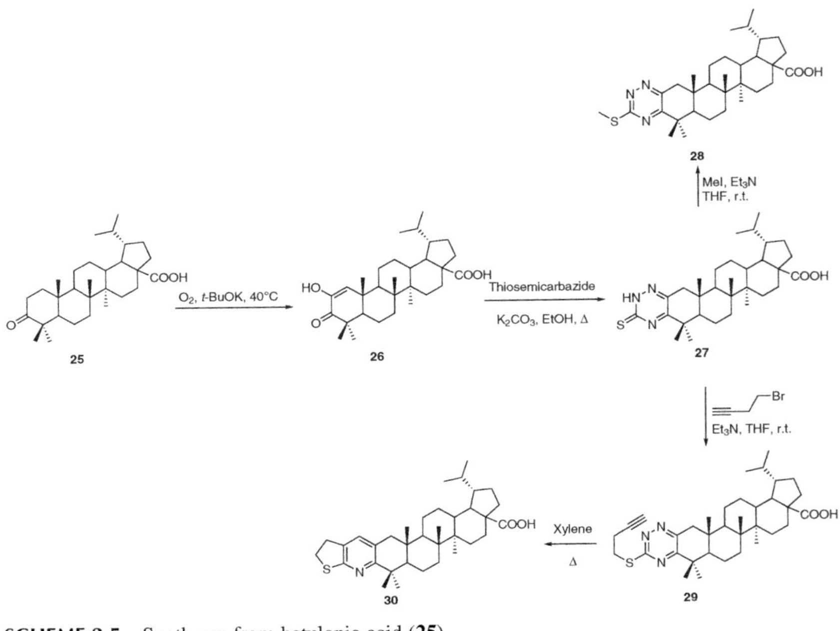

SCHEME 2.5 Syntheses from betulonic acid (**25**).

activity of five derivatives (**26–30**) was evaluated in murine leukemia L1210 and human cervix carcinoma HeLa and lymphoblast Contagious Equine Metritis (CEM) tumor cell cultures. Whereas **27** is devoid of significant cytostatic activity (IC_{50}: 160–204 μM), **26**, and in particular **28–30**, showed pronounced cytostatic activity (lower μM-range) with some preference for the human CEM lymphoblast tumor cells (IC_{50} range: 1.2–5 μM). The molecular mechanism of cytostatic action of these new derivatives is currently unclear and should be the subject of further study.[9]

A convenient synthesis of ferrocenylthymine-3,6-dihydro-2*H*-thiopyrane (**33a**) and ferrocenyl-4-thiothymine-3,6-dihydro-2*H*-thiopyrane (**33b**) by an HDA cycloaddition reaction was proposed as the access to a new class of ferrocenyl nucleobase derivatives. The 2*H*-thiopyrane ring mimics the sugar subunit present in natural nucleosides and nucleotides. Target complexes **33a,b** were synthesized in two steps according to Scheme 2.6.

The synthesis of ferrocenyl-4-thiothymine-3,6-dihydro-2*H*-thiopyran **33b** involved first the preparation of the thioketone **32b** by a thionation reaction between the ketone **31** and Lawesson's reagent (path b). This reaction afforded thioketone **32b** as a blue solid in 77% yield. The subsequent thermally induced HDA cycloaddition reaction with 2,3-dimethyl-1,3-butadiene gave the target complex **33b** as a yellow solid in 77% yield. It is worth noting that the thionation reaction yielded a trace amount of thioketone **32a**, along with **32b** as major product. All attempts to obtain **32a** as the major product with Lawesson's reagent as the thionation agent failed. This observation can be rationalized by the (stepwise) mechanism of the thionation reaction with the first step predominantly occurring at the exocyclic carbonyl group of ketone **31**. Accordingly, the initially formed thioketone **32a** undergoes rapid conversion into product **32b**. To obtain thioketone **32a** as a sole isolated product, diphosphorus pentasulfide (path a) instead of Lawesson's reagent was successfully employed as thionation reagent. Thus finally, complex **32a** was obtained as a blue solid in 68% yield. Unexpectedly, HDA cycloaddition reaction of thioketone **32a** with 2,3-dimethyl-1,3-butadiene under conditions previously applied afforded compound **33a** in a trace amount. To overcome this problem, the cycloaddition reaction conditions were systematically optimized. Finally it was found that addition of lithium

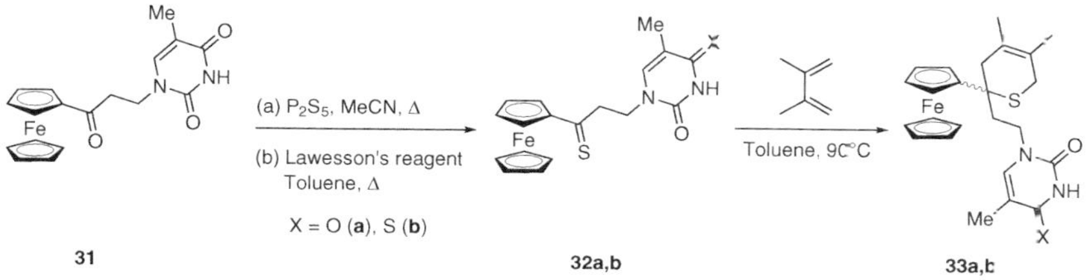

SCHEME 2.6 Synthesis of ferrocenyl derivatives **33a,b**.

chloride resulted in reaction acceleration and allowed us to obtain ferrocenylthymine-3,6-dihydro-2*H*-thiopyran **33a** as a yellow solid in 65% yield. Stereochemical outcome of the HDA cycloaddition reaction of thioketones **32a,b** with 2,3-dimethyl-1,3-butadiene was not investigated in this study.[10] Compounds **33a,b** were tested in vitro for their antiproliferative activity against human colon carcinoma HT-29, estrogen receptor-responsive human breast adenocarcinoma MCF-7, estrogen-negative human breast adenocarcinoma MDA-MB-231, human promyelocytic leukemia HL-60, and human monocytic MonoMac6 cancer cells. Ferrocenyl-4-thiothymine-3,6-dihydro-2*H*-thiopyrane **33b** showed a cytostatic effect on MonoMac6, MCF-7, and HL-60 cell lines, but was devoid of activity on MDA-MB-231 and HT-29 cells. In contrast, all cell lines tested were sensitive toward thymine derivative **33a**. Pro-apoptotic activity of **33a,b** was confirmed by means of acridine orange/ethidium bromide double-staining assay.

Anticancer compounds through 1,3-dipolar cycloaddition reactions

Azides

1,3-Dipolar cycloaddition reactions also concur to the synthesis of anticancer compounds and in this section the use of some 1,3-dipoles will be illustrated, remarking the application in the synthesis of active compounds; hence just some dipoles will be cited in dependence of the biological activity of the target compounds. As already mentioned in the previous chapter, azides are also protagonists in anticancer research and quite a large number of examples are reported here, displaying the potentialities of this type of 1,3-dipole.

Mammalian cell-surface oligosaccharides have been identified to function in cell–cell communication, cell adhesion, immune response, and molecular recognition. In particular, sialic acids containing glycoconjugates are involved in different disease states, such as inflammation, metastasis of tumor cells, and invasion of pathogens. The influenza virus utilizes structures containing sialic acid adhering to the cell prior to infection and releasing newly formed virus from the infected cells sequentially.[11] The presence of sialic acid can influence not only protein–protein interactions through its negative charge but also ligand–receptor interactions mediated by sialylation. For instance, the change of specific carbohydrate residues on key T-cell proteins can alter function and intermolecular interactions, which affect the development, survival, or reactivity of T cells.[12] Sialyltransferases, a family of glycosyltransferase enzymes, play a critical role in molecular recognition by the sialic acid binding proteins of pathogen. Their activities have been demonstrated as potential markers of tumorogenesis.[13] Therefore, the design of potent and specific inhibitors of sialyltransferase has become a promising strategy for cancer treatment. Scheme 2.7 reports the use of azide in the

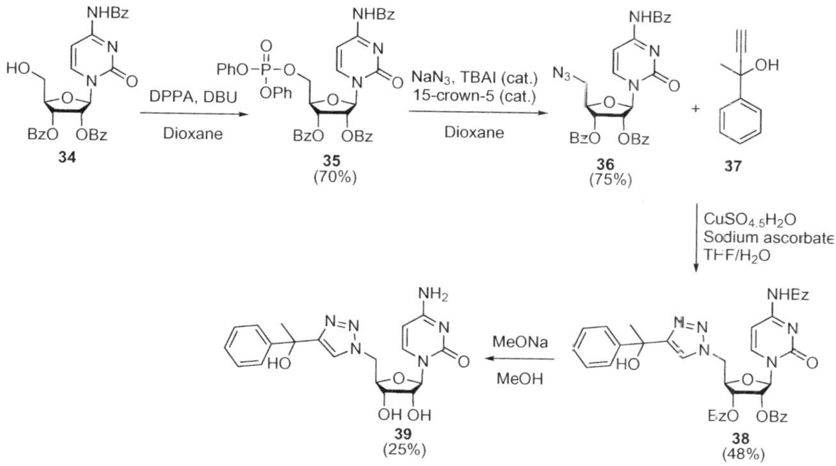

SCHEME 2.7 Azido strategy to compound **39**.

synthetic approach to an active molecule. The requisite 5′-hydroxyl of benzoyl protected cytidine **34** was prepared in three steps by selective protection of the 5′-alcohol of cytidine as the *tert*-butyldiphenylsilyl ether and subsequent benzoylation of the 2′- and 3′-alcohols and the C4-amine, followed by tetrabutylammonium fluoride-mediated liberation of the 5′-alcohol. Treatment of **34** with diphenyl phosphoryl azide and 1,8-diazabicyclo[5.4.0]undec-7-ene resulted in approximately 70% recovery of the 5′-phosphate ester intermediate **35**.

The azide compound **36** could be prepared efficiently in 75% isolated yield by reacting **35** with excess sodium azide (NaN_3), a catalytic amount of tetrabutylammonium iodide, and 15-crown-5. Employing the strategy of click reaction, Cu(I)-catalyzed 1,3-cycloaddition of organic azides with terminal alkynes, the azide **36** was treated with alkyne **37**, $CuSO_4 \cdot 5H_2O$, and sodium ascorbate to afford the corresponding 5′-triazole **38** in 48% yield. Alkaline deprotection of **38** provided the 5′-triazole cytidines **39** in 25% yield. The final compound **39** displays an IC_{50} of 37.5 μM against α-2,3-sialyltransferase with 98% inhibition at 300 μM.[14]

An efficient synthesis of 4-substituted triazolyl-nucleosides is shown in Scheme 2.8; it is based on a straightforward 1,3-dipolar cycloaddition between 1-azido-ribose **40** and a terminal alkyne, the 1-decyne, under a cooperative effect of microwave activation and copper (I) catalysis.

The cycloadduct **41** was obtained in nearly quantitative yield after a short reaction time (1–2 min). After removal of acetyl protecting groups, the free nucleoside **42** was evaluated against L1210, Molt4/C8, and CEM tumor cell lines. The target compound showed IC_{50} values against L1210, Molt4/C8, and CEM, respectively, of 56, 50, and 44 μM. SAR study shows that the

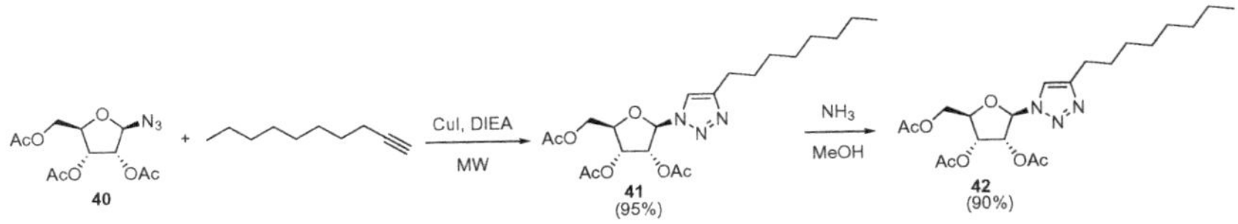

SCHEME 2.8 Synthesis of 4-substituted triazolyl-nucleosides.

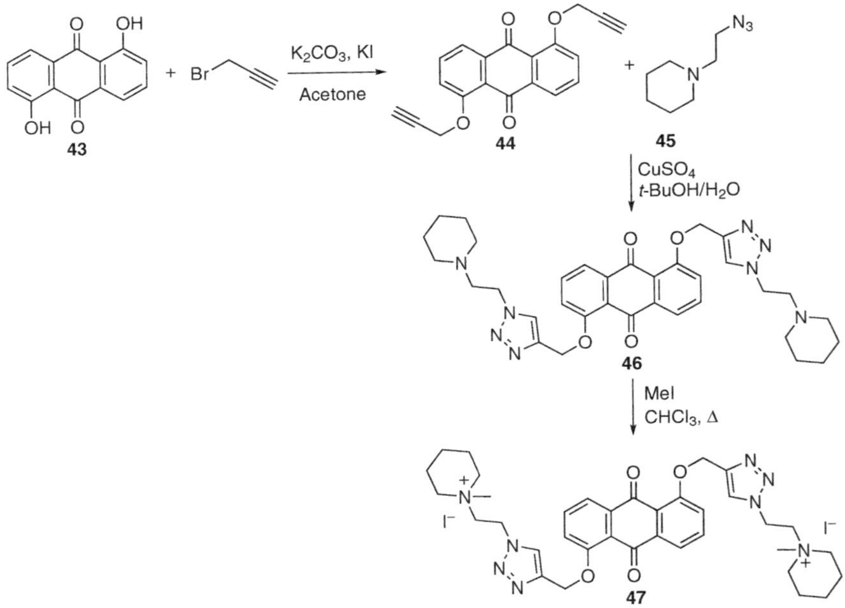

SCHEME 2.9 Synthetic strategy with anthracyclines.

substituent on the triazole ring has a major effect since the nucleoside **42** contains a long alkyl chain. These types of substituents were found in the most active compounds in the series prepared by the authors.[15]

Anthracycline antibiotics are an important class of drugs for cancer chemotherapy, and several have been successfully administered in the clinic, including daunomycin and adriamycin.[16] However, long-term exposure to these drugs could lead to multidrug resistance. As such, modified anthracycline derivatives are being studied as alternate, second-line therapeutics in order to improve the overall effectiveness of this drug class. Three water-soluble anthraquinone derivatives were designed and synthesized employing click chemistry to prepare novel and potent antitumor drugs. The synthesis is reported in Scheme 2.9 starting from 1,5-dihydroxyanthracene-9,10-dione (**43**), suitably derivatized with triple bonds with propargyl bromide to give the dipolarophile **44**. 1,3-Dipolar cycloaddition reaction with the

1-(2-azidoethyl)piperidine (**45**) afforded the cycloadduct **46** that was methylated to give **47**.

An 3-(4,5-dimethylthiazol-2-yl)-2,5-diphenyltetrazolium bromide (MTT) assay indicated that compound **47** had significant inhibitory activity against Bovine Granulosa Cell (BGC) gastric cancer cells in vitro. Apoptosis induced by this compound was observed by flow cytometry and laser confocal microscopy. Mechanistic analysis showed that **47** induced the generation of several reactive oxygen species, the loss of mitochondrial membrane potential, the transition of mitochondrial permeability, and the release of cytochrome C from the mitochondrion to cytoplasm. These results suggest that the anthraquinones might be potential lead compounds for cancer chemotherapy.[17]

Colchicine, the main alkaloid isolated from *Colchicum autumnale* (meadow saffron), has long been known for its remarkable antimitotic activity. Its biological activity mainly results from its ability to selectively bind to tubulin, leading to inhibition of microtubule formation (tubulin polymerization) and mitotic arrest in the metaphase. Allocolchicine and its analogs (with a six-membered aromatic ring C) have also emerged as promising antitumor agents. Scheme 2.10 shows the synthesis of the azide **49** that served as a late intermediate for the preparation of interesting new compounds. The azide **49** is a particularly attractive precursor for the synthesis of various C7-modified colchicinoids of type **50** through Huisgen 1,3-dipolar cycloaddition. The copper(I)-catalyzed version of this reaction, also often referred to as click chemistry, has recently evolved as a particularly powerful tool in drug discovery.

Thus the diversification of colchicine through this approach would add a new dimension to colchicine research, greatly facilitating the discovery of

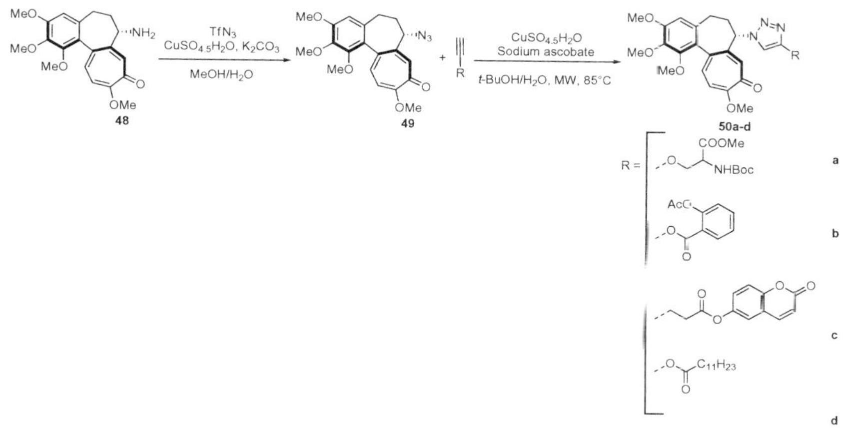

SCHEME 2.10 Synthesis of compounds **50a–d**.

new active compounds. All triazoles of type **50** synthesized were evaluated for their biological activities and compounds **50a–d** were found to be the best active products. Nearly all compounds displayed significant inhibition of tubulin polymerization in an in vitro polymerization assay. To specify whether the new compounds indeed target the colchicine-binding site of tubulin, a binding assay was performed in which the competitive displacement of fluorescein labeled colchicine by the compounds (tenfold excess) was measured. Under these conditions, the triazoles **50a,c** showed very strong binding (~80% displacement). The results from these two initial assays clearly indicate that the 1,2,3-triazole moiety represents a suitable linker to attach various substituents at the C-7 position of colchicine without interfering with the binding of the compound to the colchicine-binding site of tubulin. Next, the novel compounds were studied on the morphology of microtubules using HeLa cervix carcinoma cells. Particularly pronounced morphological effects at concentrations below 1 μm were observed for some of the compounds tested. Significant examples are **50c** (500 nm) and **50d** (100 nm). The antitumor properties of the new compounds were investigated using BJAB tumor cells (Burkitt-like lymphoma cells). The induction of apoptosis (AC_{50}) was determined using a DNA fragmentation assay, while proliferation inhibition (IC_{50}) was determined by cell counting. All compounds exhibit antitumor activity, some of them in a remarkable manner. Those compounds that exhibited good efficacy (lower micromolar or even nanomolar range) also showed only low unspecific cytotoxicity as determined by LDH-release measurements after 1 h. The fatty acid ester **50d** was identified as a particularly potent new colchicinoid. This compound strongly inhibits cell growth even at concentrations below 10 nm. In agreement with the results of the protein assays, the allo-analogs showed the weakest (however, still potentially useful) activities. Noteworthy is the fact that **50d** only showed 60% displacement in the protein assay (competition binding), while its biological activity in the whole-cell assay exceeded that of colchicine by a factor of three. Here, the importance of secondary effects, such as membrane transport or metabolism, becomes apparent.[18]

Nucleoside analogs comprise a significant class of anticancer agents; many of them are halogenated nucleosides or nucleobases. Ever since the demonstrated effectiveness of 5-fluorouracil (5-FU), many fluorinated nucleic acid analogs have been developed, including loxuridine, capecitamine, gemcitabine, and fludarabine, to improve the bioavailability, metabolic stability, and interactions with enzymes. The fluorine atom has unique stereoelectronic properties that make it an effective bioisostere for both hydrogen and oxygen atoms. The electronegativity and van der Waals radius of fluorine (4.0, 1.47 Å) are similar to those of oxygen (3.5, 1.57 Å), making it an effective mimic of the hydroxyl groups, even though C–F bonds participate in hydrogen bonds only very weakly. Nevertheless, the presence of fluorine atoms contributes to the electrostatic interactions and improved binding affinity of a substrate in the active site of an enzyme. A Korean research group has proposed a simple approach to analogs and the strategy is sketched in Scheme 2.11.

Starting from the iodo-uracil derivative **51**, the synthesis proceeds quite simply through OH groups protection, Sonogashira coupling to insert the triple bond and fluorinated-azide 1,3-dipolar cycloaddition reaction to afford the cycloadduct **56**, deprotected to give the final compound **57**. All the steps are performed in very good yields.

Compound **57**, which features a perfluorodecyl-substituted triazole unit, inhibited the cancer cell (PC-3, MDA-MB-231, ACHN) growth rate by more than 80% at a concentration of 10 μM. The compounds presenting longer (e.g., 1*H*,1*H*,2*H*,2*H*-perfluorododecyl) or shorter (e.g., perfluorooctyl) perfluoroalkyl chains exhibited highly reduced or no anticancer activity. Therefore, it appears that the perfluorodecyltriazole unit plays an important role in inhibiting cancer cell growth. It seems that the mode of action of **57**, among the several metabolic effects of fluoropyrimidines, might be related to its inhibition of the key enzyme thymidylate synthase.[19]

Starting from the same intermediate **54**, the synthesis and antiviral evaluation of a series of C5-(1,4- and 1,5-disubstituted-1,2,3-triazolo)-nucleoside derivatives of type **60** was described by Agrofolio. The key step is again the regioselective Huisgen's 1,3-dipolar cycloaddition, using either copper-catalyzed azide-alkyne cycloaddition (CuAAC) or ruthenium-catalyzed azide-alkyne cycloaddition under microwave activation (Scheme 2.12).

Some of the synthesized compounds possess activity against herpes simplex viruses 1 and 2, varicella-zoster virus, human cytomegalovirus, and vaccinia virus. Their cytostatic activities were determined against murine leukemia cells, human T-lymphocyte cells, and cervix carcinoma cells. Compounds were also evaluated on a wide panel of RNA viruses, including vesicular stomatitis virus, influenza viruses of types A (H1N1 and H3N2) and B in Madin–Darby canine kidney (MDCK) cell cultures, parainfluenza-3 virus, reovirus-1, Sindbis virus, and Punta Toro virus in Vero cell cultures and vesicular stomatitis, Coxsackie B4, and respiratory syncytial virus, with no specific antiviral effect. However, the in vitro inhibitory effects of compounds of type **60** on the proliferation of murine leukemia cells (L1210), human T-lymphocyte cells (CEM), and cervix carcinoma cells (HeLa) demonstrated that several of them showed cytostatic activity. The 4-phenoxyphenyl triazolyl derivative **60** exhibits a pronounced cytostatic activity with all IC_{50} values <1 μM ($IC_{50} = 0.5$ μM for L1210, $IC_{50} = 0.8$ μM for CEM, $IC_{50} = 0.28$ μM for HeLa). Compound **60** has a comparable antiproliferative activity as 5-FU against L1210 and HeLa cells. It is superior to 5-FU against CEM cells.[20]

Novel nucleoside-Cinchona alkaloid conjugates were synthesized using a click chemistry approach based on the copper(I) catalyzed Huisgen azide alkyne cycloaddition as depicted in Scheme 2.13. The reaction is easy and relies upon the cycloaddition between the azido-derivative **61** and the

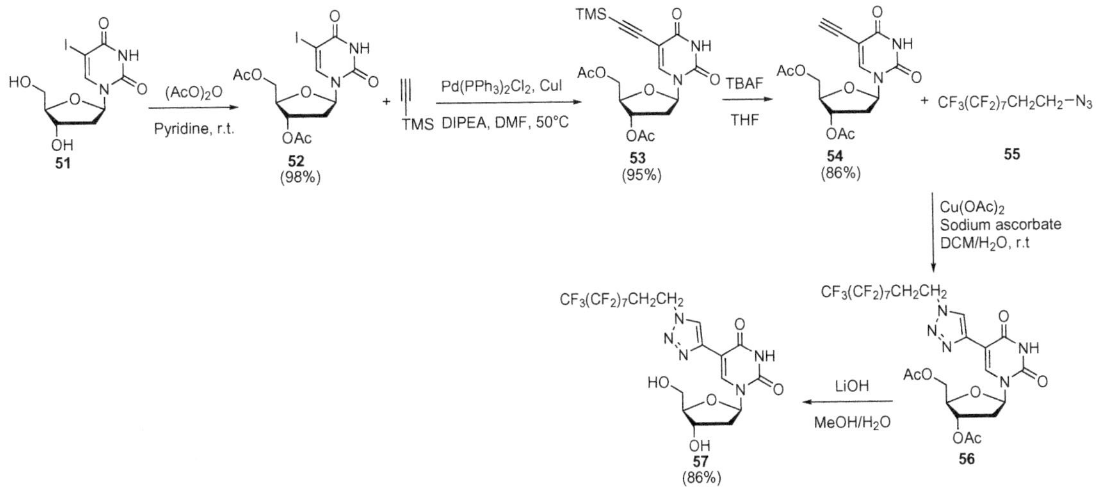

SCHEME 2.11 Synthesis of analog **57**.

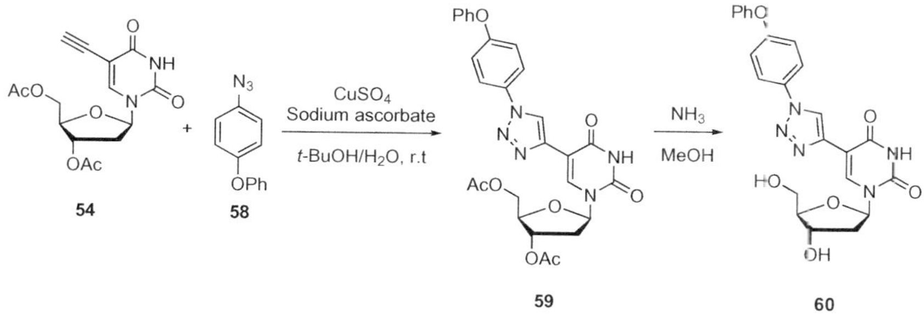

SCHEME 2.12 Azido synthesis of compound **60**.

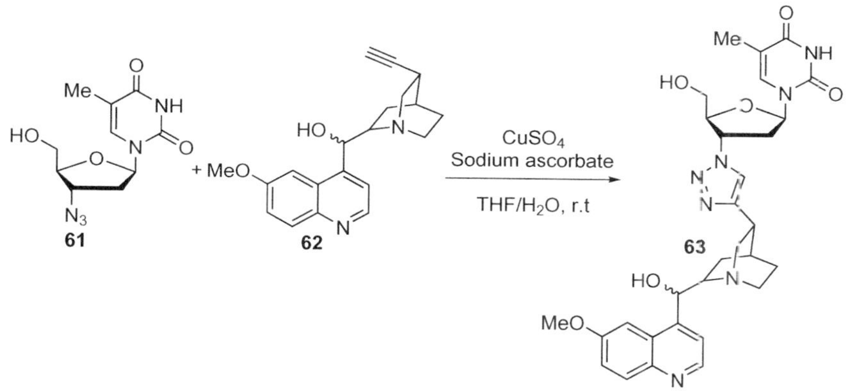

SCHEME 2.13 Novel nucleoside-Cinchona alkaloid conjugates synthesis.

dipolarophile **62** to afford the most active compound **63** of the series prepared by the Polish authors.

All obtained conjugates showed strong fluorescence emission and some of them exhibited marked cytotoxic activity in vitro. The in vitro cytotoxic activity of conjugates was studied in human KB (nasopharynx carcinoma) tumor tissue culture using cytosine arabinoside ($ED_{50} = 0.99$ μg/mL) as a standard. The most active was compound **63** ($ED_{50} = 1.3$ μg/mL); the other compounds showed only a moderate activity ($ED_{50} = 11.6-21$ mg/mL).[21]

Some research groups were involved in the synthesis of nucleoside analogs bearing a bitriazolyl moiety; this is because their more expanded aromatic surface and increased rigidity of the nucleobase may favor interactions with biological targets to induce potent anticancer and antiviral activities.[22] Here, the synthesis of a novel bitriazolyl acyclonucleoside is reported, endowed with dual anticancer and immunomodulatory activity. This compound belongs to a new family of bitriazolyl acyclonucleosides and the synthetic strategy is shown in Scheme 2.14: the key step is again the Cu(I)-catalyzed click process starting

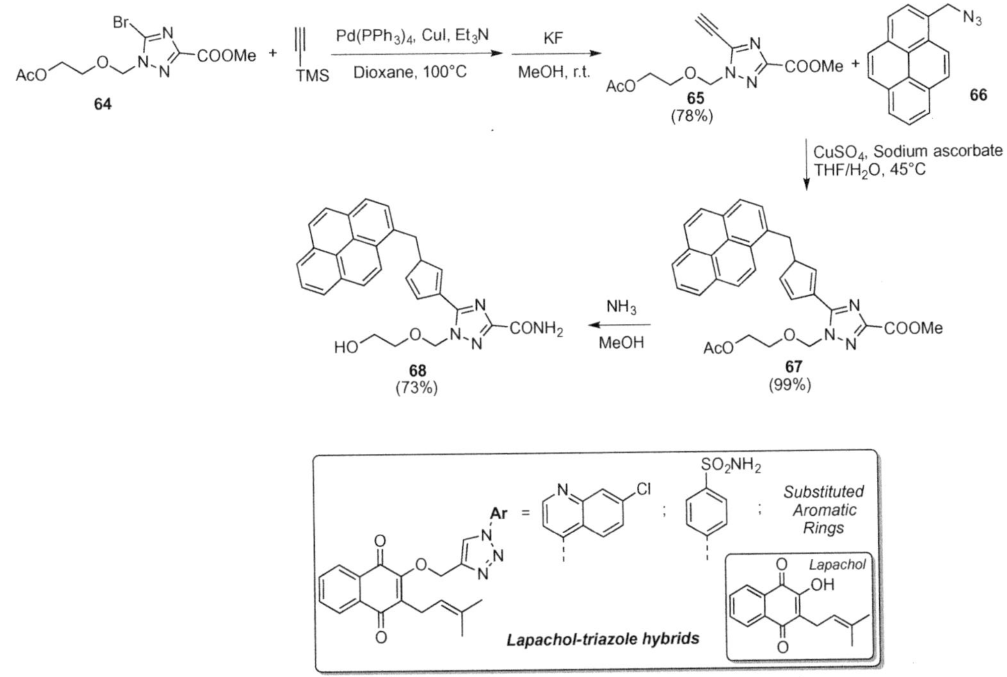

SCHEME 2.14 Pd-catalyzed synthesis of compound **68**.

from terminal alkyne-containing triazole acyclonucleoside **65**, easily prepared from the bromine derivative **64** through classical Sonogashira coupling and deprotection of the triple bond. Compound **67** was obtained in excellent yield from the pyrenyl-azide **66**; treatment with ammonia afforded the derivative **68**. Cycloadduct **67** displayed potent antiproliferative activity against different cancer cell lines and was able to activate caspase-dependent apoptosis in the drug-resistant pancreatic cancer MiaPaCa-2 cell line. Most importantly, this compound was also able to stimulate the immune response in mouse dendritic cells through the activation of TLR7. This dual-acting anticancer and immunomodulating compound may constitute a potentially interesting drug candidate for cancer therapy.

The initial biological screening of the bitriazolyl acyclonucleosides **67** and **68** was performed in human drug-resistant pancreatic cancer MiaPaCa-2 cells by measuring their antiproliferation activity using MTT assay. Cycloadduct **67** showed exceptional cell growth suppressing activity that was superior to the currently recommended anticancer drug gemcitabine. Cycloadduct **67** was able to inhibit the proliferation of the human pancreatic cancer Capan-2 and Panc-1, prostate cancer PC-3, cervical cancer HeLa, and breast cancer MCF-7 cell lines. Moreover, a dose-dependent assay with cycloadduct **67** in the previously mentioned cell lines conducted at 50 nM to 200 μM revealed a 50% growth inhibition (IC_{50}) of most cell lines at a dose of 10 μM or below.[23a] Considering the potent anticancer activity shown by cycloadduct **67**, we performed a SAR analysis using pancreatic cancer MiaPaCa-2 cells as the test model. Compound **68**, the deprotected form of cycloadduct **67**, showed significantly decreased activity. This implies that alteration of the sugar part of **68** was detrimental to anticancer activity. We then turned our attention to the contribution of the pyrene group to the anticancer activity of **68** and prepared compounds bearing naphthalenyl and anthracenyl groups, both of which are less bulky comparatively. Altogether the obtained results highlighted **67** as possessing the highest antiproliferative activity among all the newly synthesized compounds described in the work at hand. Closely related from the synthetic strategy point of view are the examples inspired by the structure of Lapachol, an abundant prenyl naphthoquinone occurring in Brazilian Bignoniaceae, clinically used, in former times, as an antimalarial drug, despite its moderate effect (see the inset in Scheme 2.14). Aiming to search for potentially better antimalarials, a series of 1,2,3-triazole hydrids was synthesized. Alkylation of the hydroxyl group gave its propargyl ether which, via CuAAC click chemistry with different organic aromatic and heteroaromatic azides, afforded the desired derivatives. All the synthetic compounds were evaluated for their in vitro activity against chloroquine resistant *Plasmodium falciparum* and for cytotoxicity to Hep-G2 cells. Compounds containing the naphthoquinolyl triazole moieties showed higher antimalarial activity than Lapachol (IC_{50} 123.5 μM) and selectivity index (SI) values in the range of 4.5–197.7.[23b]

A simple and efficient synthesis of modified 1,2,3-triazole nucleosides was developed, taking advantage of a sequential one-pot acetylation-azidation-cycloaddition procedure. It was found to be highly effective under a cooperative effect of ultrasound activation and iron/copper catalysis. The reactions were carried out under both conventional and ultrasonic irradiation conditions. In general, improvement in rates and yields were observed when reactions were carried out under sonication compared with conventional conditions. This one-pot procedure provides several advantages such as operational simplicity, high yield, safety, and environment friendly protocol. Scheme 2.15 shows the preparation of two highly active compounds **70a,b** from the sugar **69**.

The resulting substituted nucleosides **70a,b** were evaluated for their anticancer activity against the K562 chronic myelogenous leukemia (CML) cell line. Derivatives **70a** (methoxynaphthyl) and **70b** (pyridyl) were found to be the most active in K562 CML cells. They showed appropriate decrease of cell viability and promising therapeutic potential in CML compared to 5-Aminoimidazole-4-carboxamide 1-β-D-ribofuranoside (AICAR). Indeed, these compounds kill 77%–93% of CML cancer cells at 0.5 μM concentration compare to AICAR (35%). The authors also observed that compounds featuring ribose sugar, close analogs of AICAR, are more active than those with glucose or 2-deoxyribose. It is worth noting that the substituent on triazole ring also has a dramatic effect on CML activity since compounds featuring methoxynaphthyl and pyridyl exhibit high activity and are two- to threefold more active than AICAR, whereas triazolyl-nucleosides with aromatic (thienyl, phenyl, and methoxyphenyl) or alkyl substituents have only low activity.[24]

Development of fluorescent chemical probes that can recognize biomacromolecules with high specificity and interfere with cellular processes is an emerging trend in chemical biology. Within this context, G-quadruplex DNA has received considerable attention as a prospective target for designing anticancer drugs. G-quadruplex structures are widespread in the genome, found at the end of telomeres, in the promoter regions protooncogenes, and in the untranslated regions of mRNAs. These structures are

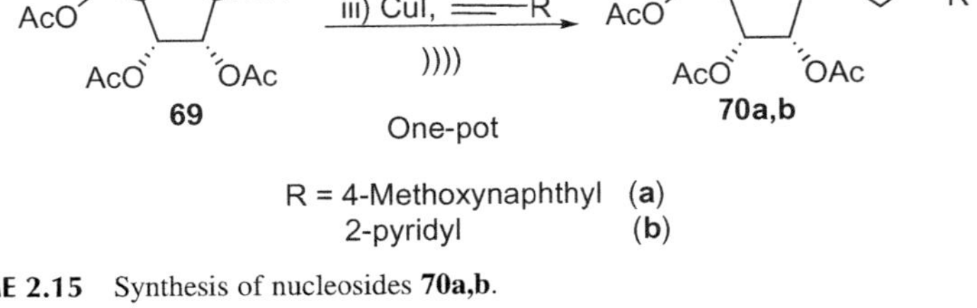

SCHEME 2.15 Synthesis of nucleosides **70a,b**.

believed to play key roles in the human genome, such as telomere maintenance and gene regulation.[25] Biomolecular logic gate systems based on the interaction of fluorescent molecular probes with the G-quadruplex DNA were proposed, and some authors reported the design and synthesis of two molecules containing guanine bases and their possible self-assembly to form artificial G-quartets as a nature-inspired strategy to interact with the G-quadruplex.[26] The synthesis of G-quadruplex binding ligand using click chemistry has been proposed for the fabrication of complex nanoarchitectures using supramolecular self-assembly of guanosine derivatives.[27] Among the five natural nucleobases, guanine and its derivatives have received considerable interest in supramolecular chemistry and nanotechnology. Taking inspiration from the natural self-organization of nucleosides, the design of a flexible ligand by linking a biocompatible fluorescent tag between two guanosine units using CuI-catalyzed 1,3-dipolar azide-alkyne cycloaddition is shown in Scheme 2.16.

It was hypothesized that the guanosine units can interact with the G-quartet of the quadruplex by means of hydrogen bonding and the flexible linker can tether in the groove region of quadruplex sequences. Since quadruplex sequences vary in the groove and loop regions, the ligand may show selectivity for a particular quadruplex sequence, for which the groove region would be maximally occupied. Further the ligand based on a guanosine nucleoside can be incorporated into DNA and the attached fluorophore

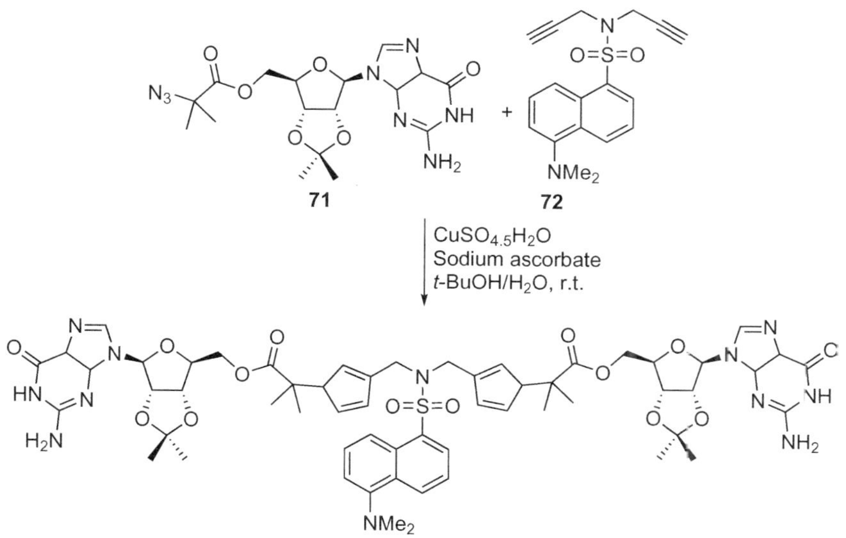

71 **72**

73
(61%)

SCHEME 2.16 Synthesis of a biocompatible fluorescent tag.

would enable visualization of the nucleus in living cells. Guanosine azide **71** was prepared in high yields and submitted to a double click reaction with the dansyl dialkyne **72** in the presence of sodium ascorbate and Cu(II) to afford the desired bis-cycloadduct **73** in 61% yield. The fluorescence enhancement of the dansyl guanosine derivative in the presence of *c-myc* DNA occurs in real time, which could potentially be developed as a biomarker for the quadruplex in the nanomolar concentration range. Dinucleoside DDG **73** induces cell cycle arrest at G2/M phase, exhibits specific cytotoxicity to the human melanoma A375 cells over normal human keratinocyte cells, and promotes cell death by apoptosis.[28]

A series of novel 1,2,3-triazoleedithiocarbamateeurea hybrids of type **78** were designed and synthesized, and their antiproliferative activities against four selected human cancer cell lines were evaluated. Scheme 2.17 shows the synthetic strategy based upon functionalization with a triple bond of the piperazine **74** followed by 1,3-dipolar cycloaddition of the azides **76a,b** to afford the derivatives **77a,b**, subsequently treated with diisopropyl amine to give the final adducts **78a,b**.

The results showed that a number of the hybrids exhibited potent activity in selected human cancer cell lines. Among them, compounds **78a** and **78b** showed broad spectrum anticancer activity with IC_{50} values ranging from 1.62 to 20.84 μM and 0.76 to 13.55 μM, respectively. Interestingly, compounds **78a,b**, being very potent against MGC-803 cells, exhibited no significant cytotoxicity against normal human embryonic kidney cells at up to 55 and 70 μM, respectively. Evidence of cell cycle arrest and apoptosis induction was obtained for the most effective compounds **78a,b** by means of flow cytometry and microscopic techniques.[29] The same authors also prepared other carbamate derivatives of 4b-(1,2,3-triazol-1-yl)podophyllotoxin by means of click chemistry, and their cytotoxicities against human cancer cell lines HL-60, A-549, HeLa, and HCT-8 were evaluated. Some compounds were more potent than the anticancer drug etoposide, inducing cell cycle arrest in the G2/M phase accompanied by apoptosis in A-549 cells. Furthermore, the formation of microtubules in A-549 cells was inhibited and the inhibition of DNA topoisomerase-II was confirmed.[30a] Analogously, a simple and efficient Cu-catalyzed decarboxylative/click reaction for the preparation of 1,4-disubstituted 5-arylselanyl-1,2,3-triazoles from propiolic acids, diselenides, and azides found application in anticancer therapy. The mechanistic study revealed that the intermolecular Azide-Alkyne Cycloaddition (AAC) reaction of an alkynyl selenium intermediate occurred. The resulting multisubstituted 5-seleno-1,2,3-triazoles (see examples in the inset of Scheme 2.17) were tested for in vitro anticancer activity by MTT assay, and the tested compounds showed potent cancer cell-growth inhibition activities.[30b]

A facile synthesis of glycoprotein-based glycoligands and their binding with influenza hemagglutinin and human galectin-3 was reported by a group

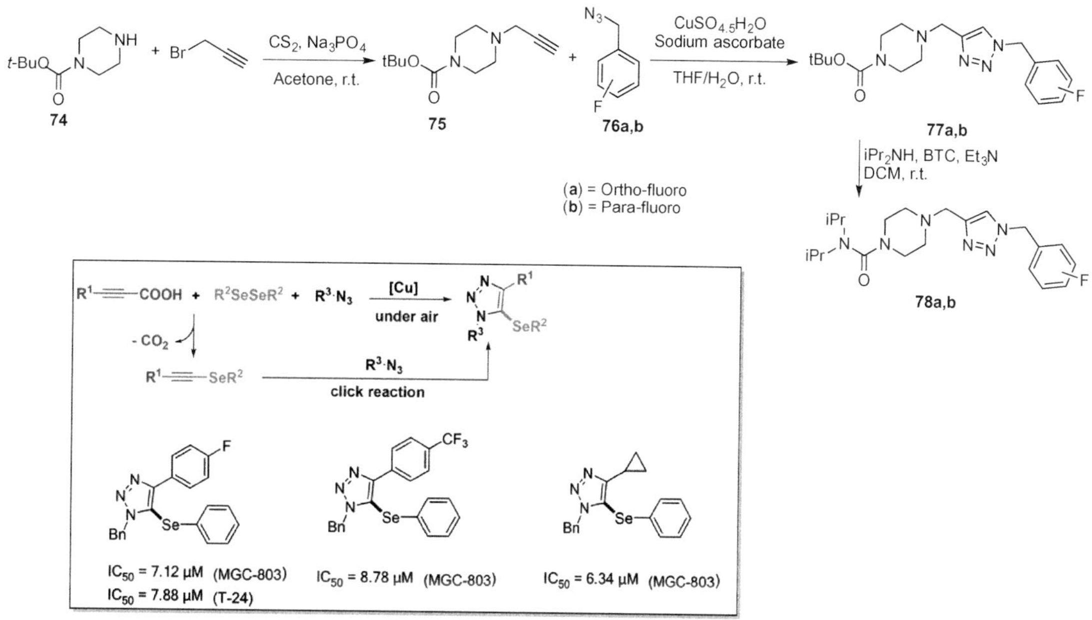

SCHEME 2.17 Synthesis of 1,2,3-triazoledithiocarbamateurea hybrids.

of American researchers. Human serum albumin (HSA) was used as the scaffold and an Asn-linked complex type *N*-glycan prepared from chicken eggs was used as the glycan building block. It was found that CuAAC reaction between the alkyne-labeled glycan and the azide-tagged HSA led to an efficient formation of the glycoconjugates (Scheme 2.18).

The density of glycan ligands on the protein scaffold was readily varied by changing the molar ratios of the two reactants. Binding studies indicated that the sialylated and desialylated multivalent glycoligands could selectively bind to influenza hemagglutinin and human galectin-3, respectively, with high affinity. In the two glycan–lectin interactions, a clear multivalent effect was observed. Moreover, a cell-based assay showed that the synthetic multivalent glycoligands could efficiently inhibit the attachment of galectin-3 to human prostate cancer and lung cancer cell lines. This study suggests that the synthetic glycoprotein-based glycoligands can be useful for different applications, including blocking the function of galectin-3 in cancer metastasis.[31]

MLN8237 (Fig. 2.2) is a highly potent and presumably selective inhibitor of aurora kinase A (AKA) and has shown promising antitumor activities. Like other kinase inhibitors that target the ATP-binding site of kinases, MLN8237 might be expected to have potential cellular off-targets. It was reported as the first photoaffinity-based, small-molecule AKA probe capable of both live-cell imaging of AKA activities and in situ proteome profiling of potential off-targets of MLN8237 (including AKA-associating proteins). By using two mutually compatible, bioorthogonal reactions (CuAAC chemistry and trans-Cyclooctene (TCO)-tetrazine ligation), small molecule-based multiplex bioimaging for simultaneous in situ monitoring of two important cell-cycle regulating kinases (AKA and CDK1) was demonstrated.

A broad range of proteins, as potential off-targets of MLN8237 and AKA-interacting partners, is subsequently identified by affinity-based proteome profiling coupled with large-scale LC–MS/MS analysis. From these studies, novel AKA interactions were discovered, which were further validated by cell-based immune precipitation experiments. The structures of MLN8054 and MLN8237 are highly homologous, with the only differences being at the two peripheral aromatic rings (i.e., arrowed in Fig. 2.2B). MLN8237 was chosen over MLN8054 in the probe design, for the obvious reasons that (1) MLN8237 has thus far shown more favorable biological and clinical profiles, and (2) MLN8237 is a more selective AKA inhibitor (thus a better candidate) for conversion into an AKA-specific imaging probe. Since the X-ray structure of the MLN8237/AKA complex was not available, MLN8054/AKA complex was selected instead for close examination (Fig. 2.2C): the CO_2H group of MLN8054 was shown to project toward the solvent-accessible surface, indicating modifications at this position with a suitable linker would minimize interference of its binding to the ATP-binding pocket of AKA. As shown in Fig. 2.2B, MLN-2 was obtained from

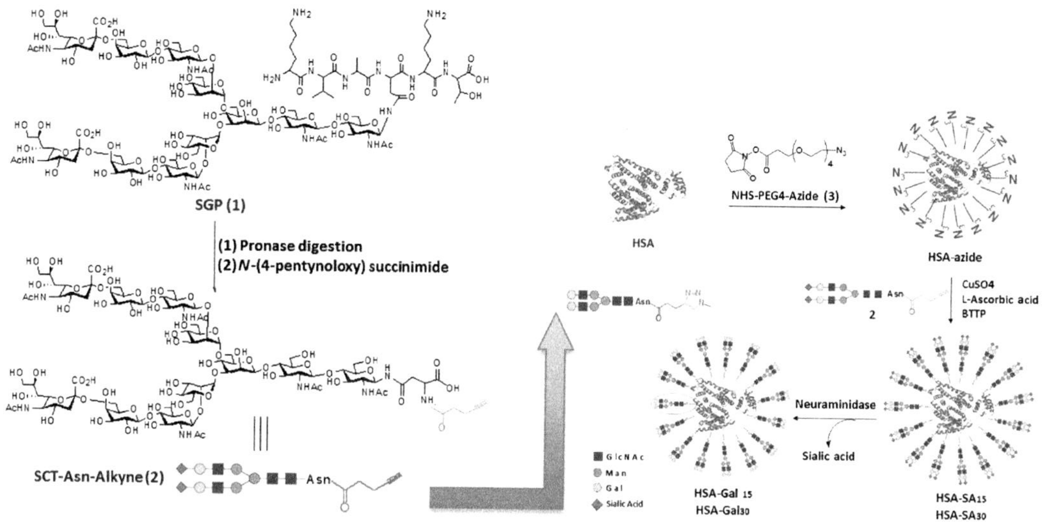

SCHEME 2.18 Synthetic strategy to glycoconjugates. *Adapted from: Wang, H.; Huang, W.; Orwenyo, J.; Banerjee, A.; Vasta, G. R.; Wang, L.-X. Bioorg. Med. Chem.* ***2013****, 21, 2037–2044.*

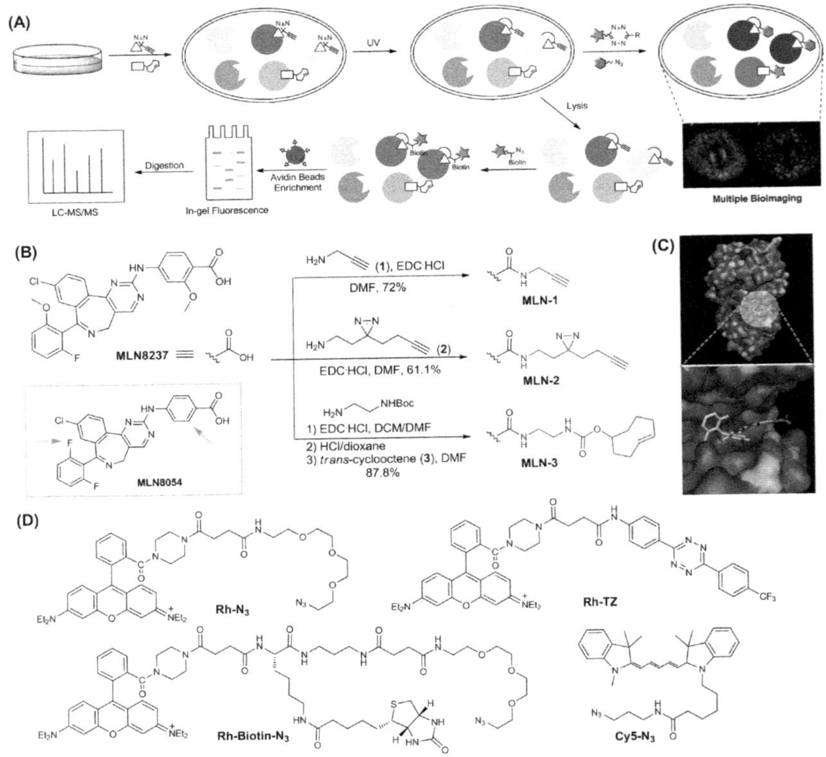

FIGURE 2.2 Overall workflow and structures of MLN8237 probes and reporters.
(A) Overall workflow of in situ cell-based proteome profiling and multiplex bioimaging.
(B) Synthetic scheme and structures of the three AKA probes (i.e., MLN-1, MLN-2, and MLN-3). For comparison, the structure of MLN8054 was shown (*boxed*), in which the differences were highlighted (*arrow*). (C) X-ray complex of AKA/MLN8054 (top; pdb code 2X8133), in which the ATP-binding pocket was highlighted (bottom). MLN8054 was shown in sticks with the CO_2H group highlighted. (D) Structures of different functional reporters used in this study. *From: Su, Y.; Pan, S.; Li, Z.; Li, L.; Wu, X.; Hao, P.; Kwan Sze, S.; Yao, S. Q. Sci. Rep. 2015, 5, 7724.*

MLN8237 by coupling its nonessential carboxylic acid to recently developed minimalist linker (2), which possesses a photoreactive diazirine moiety flanked by a terminal alkyne and a primary amine. The aliphatic diazirine has advantages over other photoreactive groups due to its small size and short UV irradiation time. A terminal alkyne as the chemically tractable tag was used, as it is small, thus minimizing interference of the probe upon binding to its intended protein targets. The tag is also readily amendable for subsequent cellular imaging, in situ proteome profiling, and large-scale pull-down/target identification via the well-established bioorthogonal click

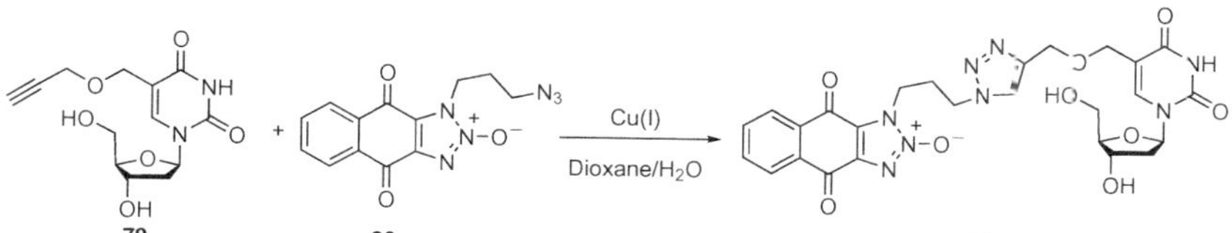

SCHEME 2.19 Synthesis of fluorescent nucleosides.

chemistry with suitable functional reporters (Rh-N_3, Rh-Biotin-N_3, or Cy5-N_3; Fig. 2.2D).[32]

A series of novel fluorescent pyrimidine nucleosides containing naphtho [1,2,3-*cd*]indole-6-(2*H*)-one fragments was designed and synthesized according to the sequence illustrated in Scheme 2.19. Introduction of fluorescent fragments into the position 5 of the uridine or cytidine heterocycles was carried out in two ways: by Sonogashira coupling reaction and CuI-catalyzed cycloaddition.

The obtained nucleoside derivatives became fluorescent due to the inserted fragments. The excitation wavelength (440–450 nm) was outside the absorption band of many biomolecules and significantly differed from the emission wavelength (560–600 nm). In addition, the intended nucleoside analogs were shown to kill cultured human tumor cells at submicromolar concentrations. The antitumor activity of new compounds was studied in a standard cytotoxicity assay after a continuous exposure of human K562 (leukemia) or HCT116 (colon carcinoma) cell lines followed by MTT-test. Compound **81**, obtained through the azide **80** cycloaddition to the dipolarophile **79**, showed a high cytotoxic potency: IC_{50} was $0.5 \pm 0.1\ \mu M$. This value is within the same order of magnitude with conventional chemotherapeutic drugs such as doxorubicin.[33]

The regioselective postsynthetic modifications of nucleic acids are essential to studies of these molecules for science and applications. A facile universal approach is reported by harnessing versatile phosphoramidation reactions to regioselectively incorporate alkynyl/azido groups into postsynthetic nucleic acids primed with phosphate at the 5′ termini. With and without the presence of copper, the modified nucleic acids were subjected to azide-alkyne cycloaddition to afford various nucleic acid conjugates including a peptide–oligonucleotide conjugate (POC) with high yield. The POC was inoculated with human A-549 cells and demonstrated excellent cell-penetrating ability despite cell deformation caused by a small amount of residual copper chelated to the POC. The combination of phosphoramidation and azide-alkyne cycloaddition reactions thus provides a universal regioselective strategy to postsynthetically modify nucleic acids.

The synthesis of additional hetero-bifunctional cross-linkers, Cu(I) ligands and other azido-/alkynyl-containing molecules starts by performing nucleic acid modifications by the CuAAC reactions. For example, in order to have azide-modified nucleic acids in one of the CuAAC reactions (Scheme 2.20A), several azido carboxylic acid succinimidyl esters were prepared and provide the best yield for the CuAAC reaction. In addition, tris(3-hydroxy-propyltriazolylmethyl)amine (THPTA) and tris(benzimidazol-2-ylmethyl) amine (NTB) were synthesized as Cu(I)-stabilizing ligands to accelerate the CuAAC reactions and prevent biomolecule degradation. However, studies of THPTA and NTB indicated that only THPTA was capable of stabilizing DNA/RNA structures in the CuAAC reactions and THPTA was thus the sole ligand used in subsequent CuAAC reactions for nucleic acid modifications. NTB was not a good ligand in CuAAC for nucleic acid modifications because nucleic acids were completely degraded by the end of the reaction. This study also explicated the toxicity of residual copper in synthesized bio-conjugates destined for biological systems.[34]

Synthetic oligonucleotides enable therapeutic intervention by using, among others, RNA interference and antisense methods.[35] The first systemically active antisense oligonucleotide was admitted to the market,[36] which

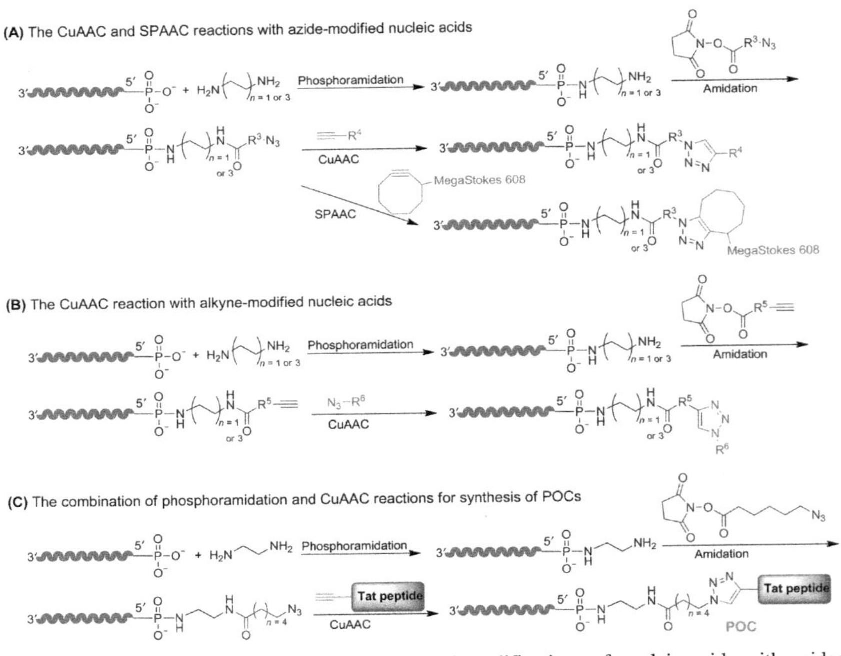

SCHEME 2.20 Phosphoramidation reaction-based modifications of nucleic acids with azides and alkynes for subsequent synthesis of nucleic acid conjugates including POCs by the CuAAC and SPAAC reactions. R3 and R6, substituents in azides; R4 and R5, substituents in alkynes. *From: Su, Y.-C.; Lo, Y.-L.; Hwang, C.-C.; Wang, L.-F.; Wu, M.-H.; Wang, E.-C.; Wang, Y.-M.; Wang, T.-P. Org. Biomol. Chem.* ***2014****, 12, 6624–6633.*

contains several phosphorothioate linkages that confer the required stability against degradation by nucleases. Despite the recent reports of success, there are still concerns about target specificity, tissue distribution, and cellular uptake of the phosphorothioate antisense oligonucleotides (PSAOs). These problems prompted the exploration of peptide-PSAO conjugates. However, the nucleophilicity of phosphorothioates poses a challenge. The reported conjugation methods have mostly been restricted to disulfide exchange, peptide coupling reactions, and DA reactions. Low conjugation yields and the need for laborious multistep procedures are recurring problems, which probably explain the scarcity of reports describing the conjugation of peptides with PSAOs. A new method involving a combination of oxime ligation and strain-promoted click reactions was recently proposed and allows rapid and efficient conjugation reactions between unprotected phosphorothioate oligonucleotides and peptides after solid-phase synthesis. Scheme 2.21 shows the combination of copper-promoted and strain-promoted [2 + 3] click reactions for the synthesis of the POCs.

A dibenzocyclooctyne (Dibo) unit was tethered to the peptide-based XIAP inhibitor AEG 40730 (Scheme 2.21; Pep1, Pep2). It is difficult to access the required reaction partners, the azido-modified oligonucleotides, directly by solid-phase oligonucleotide synthesis because of the incompatibility of the chemistry of azide and phosphoramidite groups. Instead, the supported alkynyloligonucleotide ON-CPG was used to install the azido group through Cu-catalyzed 1,3-dipolar cycloaddition reaction with di(4-azidophenyl)-ether. After cleavage with ammonia and purification, the azide-containing oligonucleotides N_3–ON were applied in strain-promoted click reactions with the dibopeptide conjugates (Pep1a, Pep1b, Pep2a). The solution phase coupling provided facile access to the peptide–oligonucleotide conjugates Pep-ON in 68%–84% yield after purification (Scheme 2.21b). A competition assay was used to assess the affinity of the peptide–oligonucleotide complexes for the BIR2-BIR3 domain of XIAP. The bivalent complexes (Pep2a-ON5n) showed higher binding affinities than the monovalent conjugates (IC_{50} = 100–500 nm vs 500–2200 nm). The comparison of three different peptide conjugates exposed a trend: the highest affinity was observed for bivalent complexes with the smallest distance between the two peptide appendages (e.g., Pep2a-ON50, IC_{50} = 100 nm).[37]

A facile synthetic approach to construct the *O*-propargyl derivatives of podophyllotoxin and 40- epipodophyllotoxin was proposed by means of click chemistry. Scheme 2.22 shows the simple synthetic approach where the propargyl derivative **83** is coupled with the 2-cyanophenylazide to afford the cycloadduct **84** whose cytotoxicity was measured against diverse human cancer cell lines viz. PC-3, PANC-1, COLO-205, and A-549 by MTT assay.

The most potent molecule discovered **84** exhibited the highest cytotoxicity on all the four cancer cell lines with IC_{50} values of 3.8–22 nM. The

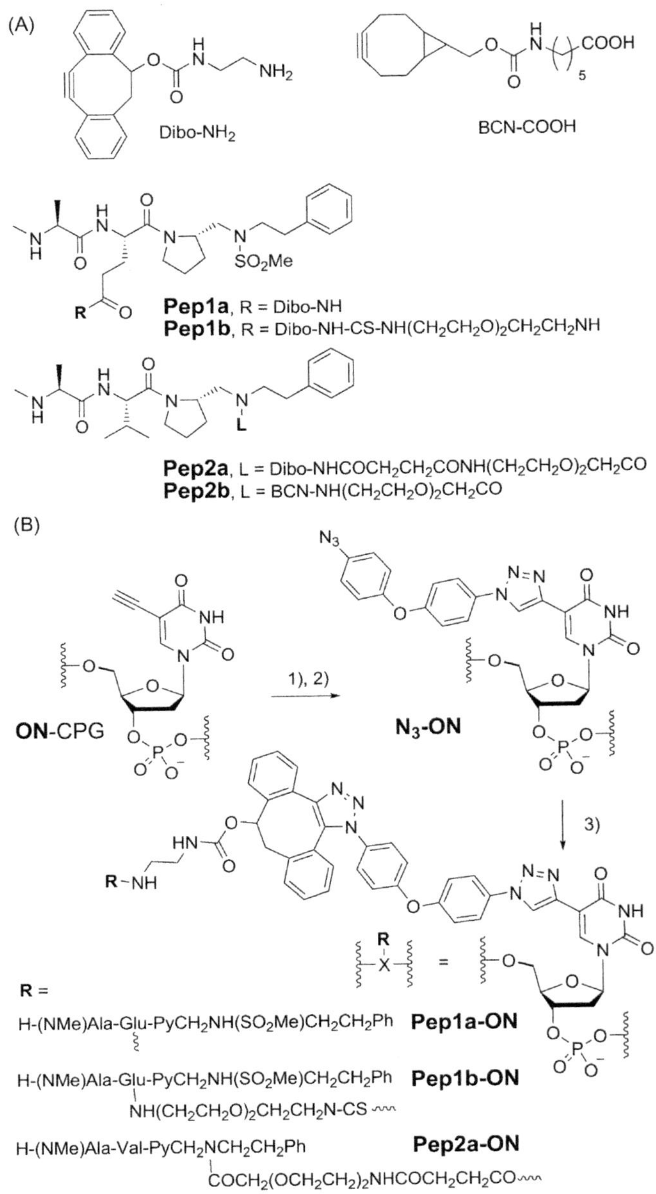

SCHEME 2.21 (A) Structures of the cyclooctyne-modified Smac-mimetic peptide conjugates (Pep) used in this study. (B) Synthesis of peptide–oligonucleotide conjugates: (1) 20 equiv $(N_3C_6H_4)_2O$, 0.2 equiv CuBr, 0.4 equiv 3-[tris(3-hydroxypropyltriazolylmethyl)amine], THF/DMF (1:1), 20°C, 18 h; (2) ammonia cleavage; (3) 5–10 equiv Pep1a, Pep1b, or Pep2a, pH 7.5 triethylammonium acetate buffer: MeCN (10:1), 20°C, 18 h. *Modified from: Abendroth, F.; Seitz, O. Angew. Chem. Int. Ed.* ***2014****, 53, 10504–10509.*

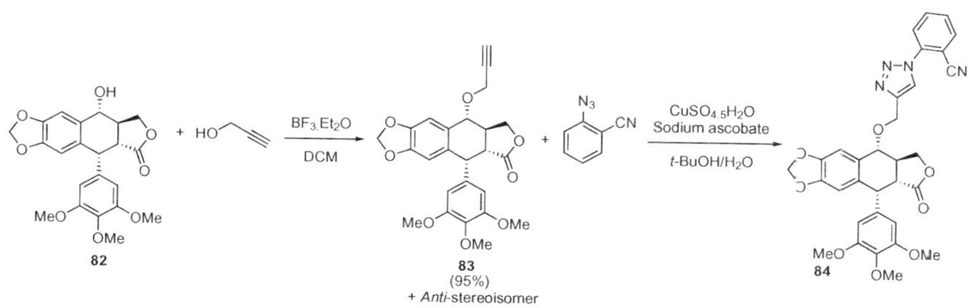

SCHEME 2.22 Synthesis of compound **84**.

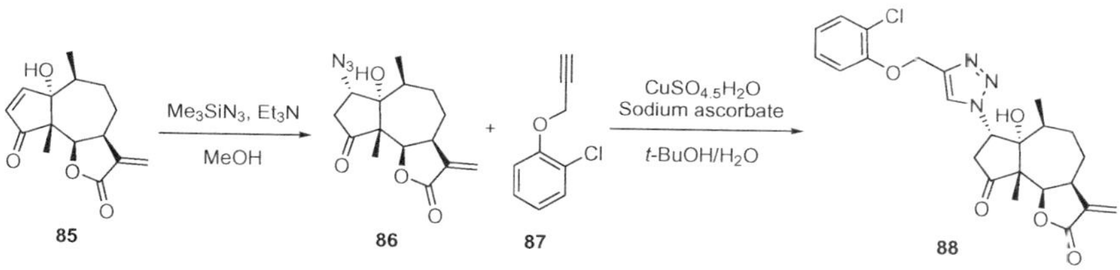

SCHEME 2.23 Synthesis of 1,2,3-triazole coronopilin congeners.

compound was further found to induce apoptosis and strongly hindered the motility of aggressive prostate cancer PC-3 cells.[38]

A series of 1,2,3-triazole coronopilin congeners have been designed and synthesized by employing a click chemistry approach starting from parthenin **85** through azide functionalization and coupling with a variety of dipolarophiles. The dipolarophile **87** afforded the cycloaddutct **88** that displayed the highest potency in the biological evaluation (Scheme 2.23).

Compound **88** was evaluated for its cytotoxicity against a panel of six human cancer cell lines (PC-3, THP-1, HCT-15, HeLa, A-549, and MCF-7). While many compounds exhibited significant anticancer activity, compound **88** was found to be the most promising analog in this series with IC_{50} values of 3.1 μM on PC-3 cell line. Flow-cytometric studies showed that 1,2,3-triazole derivative **88** induced dose-dependent apoptosis in the sub-G1 phase. This lead molecule **88** was further studied for NF-kB (p65) transcription factor inhibitory activity using ELISA and Western blotting analysis, which confirmed concentration-dependent inhibitory activity against NF-kB, p65 with 80% inhibition in 24 h at 100 μM.[39]

A novel family of betulinic acid analogs, carrying a triazole unit at C-3 attached through a linker, was synthesized by the application of azide-alkyne click reaction. Betulinic acid **89** used in this study was isolated in bulk quantity (∼50 g) from methanolic extracts of the *Dillenia indica* fruits. The derivatization studies choose the C3 hydroxyl group of **89** as a functionalization site since it was found to be more suited for showing anticancer activities compared to other functional groups (at C28 and C20).[40] Thus compound **90** was prepared (Scheme 2.24) by treating **89** with chloro acetyl chloride in the presence of DIPEA and DMAP (catalytic amount). Azidation of **90** using sodium azide led to the isolation of the intermediate **91**, which served as a useful substrate for click chemistry. Several dipolarophile were prepared and reacted with the azido-derivative **91**.

These were screened for their anticancer activity against different cancer cells and normal human peripheral blood mononuclear cell (PBMC) by MTT assay. Compound **92** [(3*S*)-3-{2-(4-(hydroxymethyl-1*H*-1,2,3-triazol-1-yl) acetyloxy)}-lup-20(29)-en-28-oic acid] was found as the most potent inhibitor of cell line HT-29 with IC_{50} value 14.9 μM. Its role as an inducer of apoptosis was investigated in this cell line by Annexin V/PI binding assay and by following its capability for ROS generation, depolarization of mitochondrial transmembrane potential, activation of caspases, PARP cleavage, nuclear degradation, and expression of pro- and antiapoptotic proteins. It exhibited much higher cytotoxicity than the standard drug 5-FU but showed negligible cytotoxicity toward normal PBMC. Elevated level of ROS generation, activation of caspase-3 and caspase-9, DNA fragmentation, higher expression of Bax and Bad, lower expression of Bcl-2 and Bcl-xl, and increased level of Bax/Bcl-xl ratio identified **92** as a promising inducer of apoptosis that follows a mitochondria-dependent pathway. Biophysical studies indicate that compound **92** acts as a minor groove binder to the DNA.[41] Similarly, a Vietnamese research group, in collaboration with Belgian researchers, prepared a variety of triterpenoid propargyl amides (derived from **89**) from the corresponding acids that were used as substrates for the construction of novel triazole-tethered triterpenoid-AZT conjugates via a click chemistry–mediated coupling with azidothymidine (AZT). Thus 19

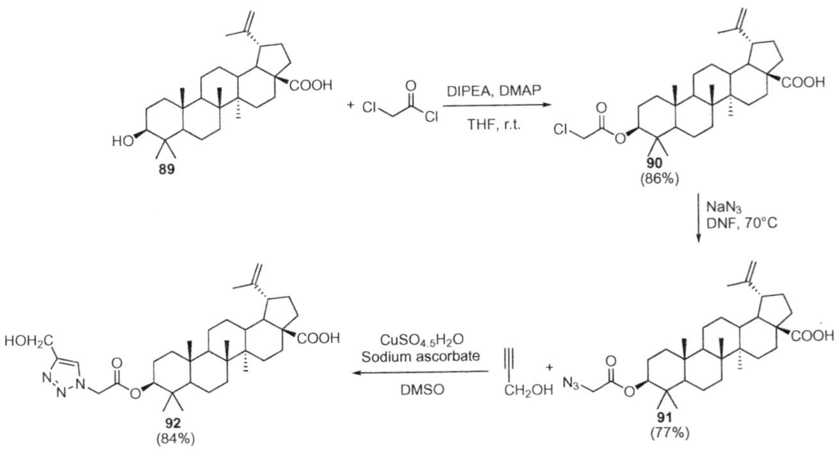

SCHEME 2.24 Synthesis from betulinic acid of azido derivatives.

new hybrids were successfully prepared and evaluated as cytotoxic agents, revealing an interesting anticancer activity of 4 triterpenoid-AZT hybrids on KB and Hep-G2 tumor cell lines.[42]

Closely related to these studies, a series of novel oleanolic acid coupled 1,2,3-triazole derivatives have been designed and synthesized by employing a Cu(I) catalyzed Huisgen 1,3-dipolar cycloaddition reaction. The antiproliferative evaluation indicated that some compounds exhibited excellent anticancer activity against the examined cancer cell lines. Some derivatives were found to possess the best inhibitory activity against HT1080 cells. A series of pharmacology experiments also showed significantly induced HT1080 cell apoptosis.[43]

The antitumor effect of most of antimitotic agents is to disrupt tubulin polymerization and depolymerization, thus interfering with the mitotic spindle formation that occurs prior to the cell division.[44] Colchicine is a cytostatic agent isolated from *C. autumnale*. It is a powerful inhibitor of polymerization of tubulin, a protein that is a basic structural unit of microtubules comprising the mitotic spindle. Although colchicine has been medically used since ancient times for the treatment of gout and Mediterranean fever, its application in anticancer therapy is limited by high nonspecific toxicity (neurotoxicity) of this agent. Nevertheless, colchicine and several structural analogs such as allocolchicine, combretastatin A-4, and 4-arylcoumarins are considered promising structures for the design of new antitumor agents that interact with the colchicine-binding site of tubulin.[45]

It was shown that the lipophilic biodegradable triazole-containing pyrroloallocolchicine derivative **96** bearing the oleic acid moiety efficiently inhibits Nalm 6 cell division and causes high levels of apoptosis at nanomolar concentrations. Compound **96** was synthesized starting from the suitably prepared alcohol **93**, functionalized with the azido group to give **94**. This latter was allowed to react with the dipolarophile **95**, containing the oleic substituent, according to a 1,3-dipolar cycloaddition reaction to afford the desired compound. This is just an example of the numerous products synthesized by this strategy (Scheme 2.25). Derivative **96** was the most interesting among the others and can be included in the phospholipid bilayer of nanosized liposomes.

The application of a therapeutic agent incorporated into selective liposomal drug delivery systems leads to a substantial decrease in systemic toxicity and an increase in the targeted action of this antitumor agent ($IC_{50} = 0.25\ \mu M$).[46]

A series of pyrazolo-triazole hybrids were designed and synthesized by combining the 1,3-diphenyl pyrazole and triazole scaffolds to obtain (1-benzyl-1*H*-1,2,3-triazol-4-yl)(1,3-diphenyl-1*H*-pyrazol-4-yl)-methanones (Fig. 2.3).

All the synthesized compounds, bearing different substituents on the two out of three phenyl rings, were screened for their anticancer activity against

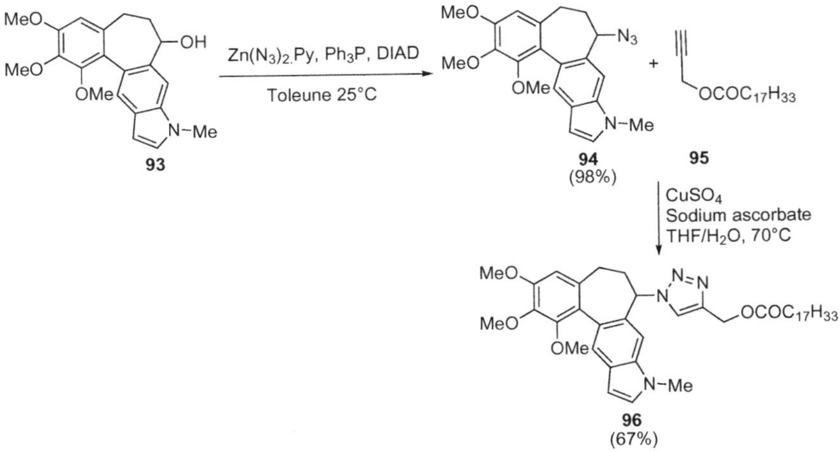

SCHEME 2.25 Azido strategy to compound **96**.

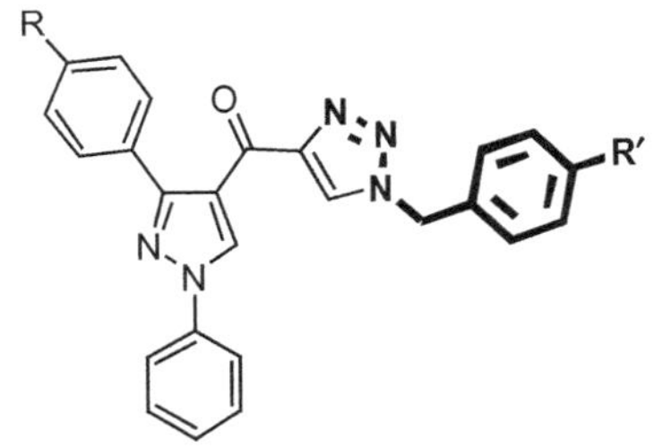

FIGURE 2.3 Structure of (1-benzyl-1*H*-1,2,3-triazol-4-yl)(1,3-diphenyl-1*H*-pyrazol-4-yl) methanone.

four tumor cell lines, viz. HT-29 (colon), PC-3 (prostate), A-549 (lung), and U87MG (glioblastoma) cells. Most of the tested compounds showed moderate to potent cell growth inhibition on different cancer cells; in particular, some compounds exhibited promising cytotoxicity against these cell lines with the IC_{50} values in the range of 0.86–3.72 μM. In addition, the potential mechanism of cell growth inhibition and apoptotic induction by these compounds was investigated in U87MG cancer cells using cell-based assays, including wound healing assay, flow cytometry, Hoechst staining, acridine orange/ethidium bromide staining, Annexin V-FITC/propidium Iodide dual staining, Rhodamine 123 staining, and carboxy-DCFDA staining. The results indicate that the compounds induce apoptosis in U87MG cells via mitochondrial pathway through upregulation of pro-apoptotic (Bax) and downregulation of antiapoptotic (Bcl-2) genes.[47]

A new series of 4,5,6,7-tetrabromobenzotriazole derivatives (Fig. 2.4) was synthesized and characterized as CK2 inhibitors.

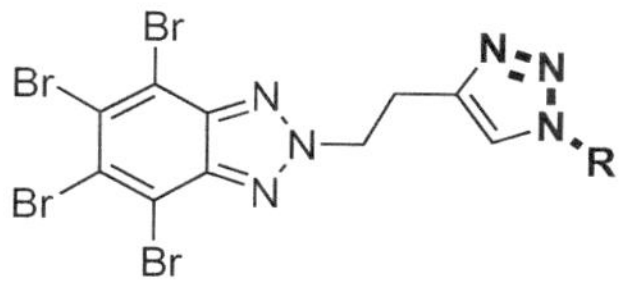

FIGURE 2.4 Structure of 2-(2-(1*H*-1,2,3-triazol-4-yl)ethyl)-4,5,6,7-tetrabromo-2*H*-benzo[*d*][1,2,3]triazole.

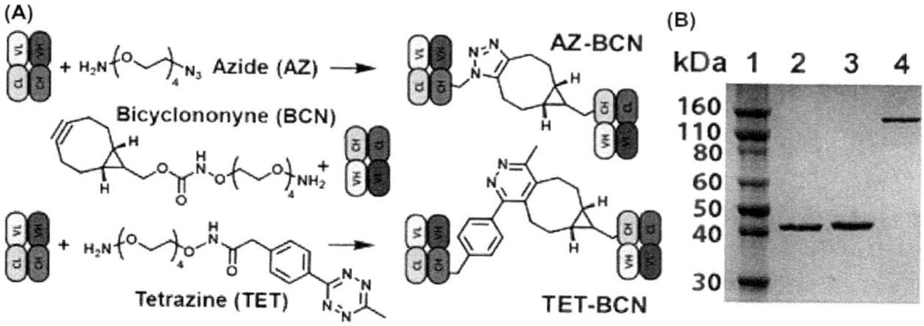

FIGURE 2.5 Synthesis and characterization of BiFabs.
(A) Structures of linkers ligated to Fabs for BiFab conjugation: (top) previous strategy using AZ- and BCN-based linkers; (bottom) strategy in this work using TET- and BCN-based linkers. (B) Characterization of monomers and conjugates by nonreducing gel electrophoresis: (1) protein ladder; (2) αBCMA Fab; (3) αCD3 Fab; (4) BiFab-(TETBCN)-BCMA. *From: Ramadoss, N.; Schulman, A. D.; Choi, S.; Rodgers, D. T.; Kazane, S. A.; Kim, C. H.; Lawson, B. R.; Young, T. S. J. Am. Chem. Soc.* ***2015****, 137, 5288–5291.*

They were readily synthesized using a click-chemistry approach based on a Cu(I)-catalyzed azide-alkyne 1,3-dipolar cycloaddition (CuAAC). Some of the synthesized compounds, bearing free or Boc-protected aliphatic amine moieties, present interesting inhibitory activities using an in vitro assay, with K_i values in the low micromolar range and a high degree of selectivity against a panel of 24 kinases. Selected compounds were tested for their antiproliferative effect on several cancer cell lines, and for their pro-apoptotic activity toward human Jurkat T-leukemia and MCF-7 breast adenocarcinoma cells, showing that they can be proposed as promising anticancer agents. Docking studies as well as crystallographic analysis allowed us to identify ligand-CK2 interactions that account for the molecular recognition process, and can help to further optimize this family of compounds as CK2 inhibitors.[48]

The development of immunotherapies for multiple myeloma is critical to provide new treatment strategies to combat drug resistance. A bispecific antibody against B cell maturation antigen (BiFab-BCMA) is reported to display potently and specifically redirect T cells to lyse malignant multiple myeloma cells (Fig. 2.5).

The conjugation of BiFabs through the site-specific modification of Fabs with paired eterobifunctional linkers containing an azide (AZ) or bicyclononyne (BCN) was already reported.[49] These linkers undergo [3 + 2] Huisgen cycloaddition to provide conjugation without the requirement of a catalyst (Fig. 2.5A). To improve the rate of the conjugation, a new strategy was developed in which tetrazine (TET) replaces AZ of one of the linkers. Tetrazines undergo an inverse-electron-demand DA reaction with BCN that is up to 104-fold faster than strain-promoted alkyne-azide cycloadditions (AZ-BCN). Toward this end, the AZ, BCN, and TET linkers were synthesized to compare the conjugation rates in the context of BiFab generation. Each linker contained an aminooxy functionality to enable oxime ligation to the ketone of the unnatural amino acid p-acetylphenylalanine (p-AcF) (Fig. 2.5A).

BiFab-BCMA lysed target BCMA-positive cell lines up to 20-fold more potently than a CS1-targeting bispecific antibody (BiFab-CS1) developed in an analogous fashion. Furthermore, BiFab-BCMA robustly activated T cells in vitro and mediated rapid tumor regression in an orthotopic xenograft model of multiple myeloma. The in vitro and in vivo activities of BiFab-BCMA are comparable to those of anti-BCMA chimeric antigen receptor T-cell therapy (CAR-T-BCMA), for which two clinical trials have been initiated. A BCMA-targeted bispecific antibody presents a promising treatment option for multiple myeloma.[50]

A simple and efficient synthesis of novel, D-ring–substituted estrone derivatives (Fig. 2.6) containing a 16α-triazolyl moiety relies upon the preparation of two epimeric azido alcohols (16α-azido-17α-hydroxy and 16α-azido-17β-hydroxy) of estra-1,3,5(10)-triene-3-methyl ether, followed by CuAAC with various terminal alkynes.

The steroidal triazoles were obtained in high yields and their activities against three human cancer cell lines (HeLa, MCF-7, and A431) were screened. The most effective analogs were submitted to additional experiments in order to characterize their antiproliferative properties. As evidenced by flow cytometry, the selected steroids induced a disturbance in the HeLa cell cycle in a concentration- and exposure-dependent manner, through an increase of the hypodiploid population (sub-G1) and a cell cycle arrest in the G2/M phase. A noncancerous human fibroblast cell line (MRC5) was used to determine the selectivities of these compounds. Fluorescent microscopy after Hoechst33258-propidium iodide double-staining revealed nuclear

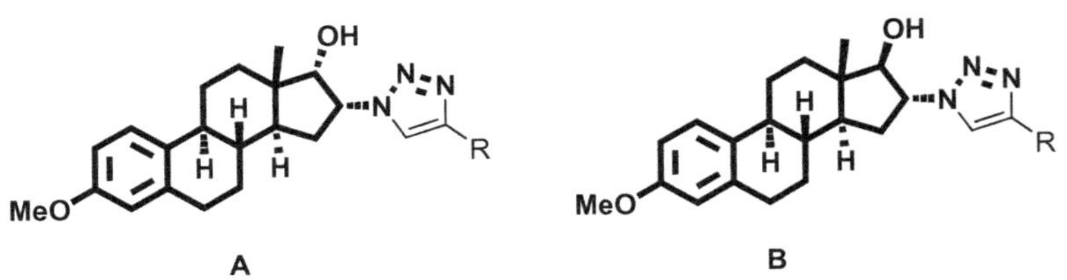

FIGURE 2.6 Azido cycloadducts to steroids.

condensation and disturbed cell membrane integrity. The enhanced activities of caspase-3 and caspase-9 without activation of caspase-8 in the treated cells indicated the activation of the intrinsic pathway of apoptosis. The levels of cell cycle regulators (CDK1, cyclin B1/B2 and cdc25B) were decreased and the ratio Bax/Bcl-2 was increased 24 h after the treatment of HeLa cells (determined at an mRNA level by means of an RT-PCR technique). Under the same conditions, two agents elicited substantially increased degrees of phosphorylation of stathmin, as evidenced by Western blotting. The presented results demonstrate that these steroids, suitably derivatized, can be regarded as appropriate structural scaffolds for the design and synthesis of further steroid analogs as innovative drug candidates with good efficacy.[51]

The synthesis of purine/triazole 6′-isonucleosides and of glucuronic acid/glucuronamide-derived *N*-glycosylsulfonohydrazides through efficient and stereo- or regioselective methodologies is another example of azide chemistry application (Fig. 2.7). Their structures were envisaged to mimic nucleosides, sugar phosphates, or nucleotides, and were expected to provide potential inhibitors of therapeutically relevant enzymes, the active sites of which could potentially bind their structural fragments or functional groups. Such enzymes include cholinesterases, carbonic anhydrase II (CA-II), and cyclin-dependent kinase 2 (CDK-2). A (triazolyl)methyl amide-linked disaccharide nucleoside, based on a new prospective structural framework for analogs of nucleoside diphosphate sugars, was synthesized. The synthetic strategies employed unprotected or partially protected carbohydrate derivatives as precursors, including ribose, glucuronic acid, glucuronolactone, and glycopyranosides and relied on stereoselective *N*-glycosylation, regioselective Mitsunobu coupling, and click chemistry approaches.

Some 6′-isonucleosides and triazole-containing glycoderivatives displayed moderate selective acetylcholinesterase inhibitory activities. The best inhibitor was an aminomethyltriazole 6′-isonucleoside with a K_i value of 11.9 μM. *N*-Glucuronylsulfonohydrazide showed good inhibition of CA-II (K_i = 9.5 μM). Molecular docking of the most active compounds into the

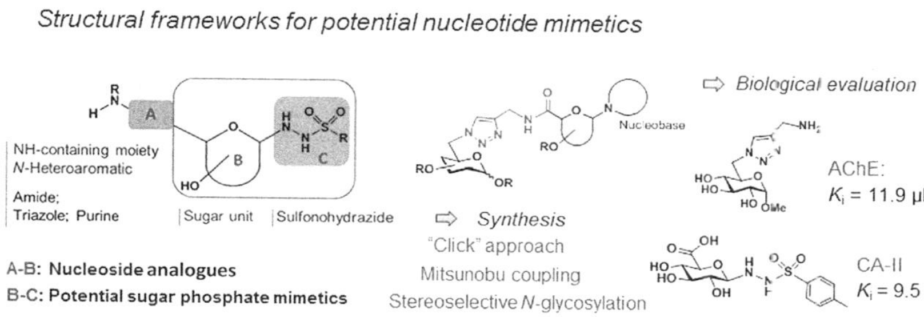

FIGURE 2.7 Synthetic strategy for potential nucleotide mimetics. *From: Xavier, N. M.; Lucas, S. D.; Jorda, R.; Schwarz, S.; Loesche, A.; Csuk, R.; Conceição Oliveira, M. Synlett* ***2015****, 26, 2663–2672.*

effected enzymes showed interactions with key amino acid residues for substrate recognition. In addition, the tested compounds did not show toxicity to normal cells.[52]

A series of steroidal hybrids with different terminal bioactive scaffolds were synthesized using the molecular hybridization approach and further evaluated for their antiproliferative activity against several cancer cell lines of different origins using the MTT assay. Scheme 2.26 shows the strategy and the structure of the most potent compound **99** that bears an isatin residue on the dipolarophile moiety derived from **97**.

The preliminary results indicated that insertion of a terminal isatin motif determine a remarkably sensitive activity of the compounds to SH-SY5Y cells, thereby exerting potent growth inhibition in vitro. This selectivity is possibly attributed to LSD1 inactivation ($IC_{50} = 3.18\ \mu M$). Besides, the chloro atom at the 7-position on the isatin core was beneficial for the activity through the SARs studies. Among this series, compound **99** showed the best inhibitory activity ($IC_{50} = 4.06\ \mu M$) against SH-SY5Y cells, which was comparable to that of 5-FU. Compound **99** arrested cell cycle at G2/M phase induced apoptosis accompanied with decrease of mitochondrial membrane potential, and inhibited LSD1 potently ($IC_{50} = 3.18\ \mu M$). Docking studies showed that compound **99** formed interactions with surrounding amino acid residues and the steroid nucleus occupied the tubular hydrophobic cavity of the active site.[53]

From four molecules, inspired by the structural features of fascaplysin, with an interesting potential to inhibit CDKs, a new series of triheterocyclic derivatives based on 1*H*-pyrrolo[2,3-b]pyridine (7-azaindole) and triazole heterocycles were designed and prepared (Fig. 2.8).

Using a Huisgen type [3 + 2] cycloaddition reaction as the convergent key step, 24 derivatives were synthesized and their biological activities were evaluated. Comparative molecular field analysis, based on three-dimensional quantitative SAR (3D-QSAR) studies, was conducted on a series of 30 compounds from the literature with high to low known inhibitory activity toward

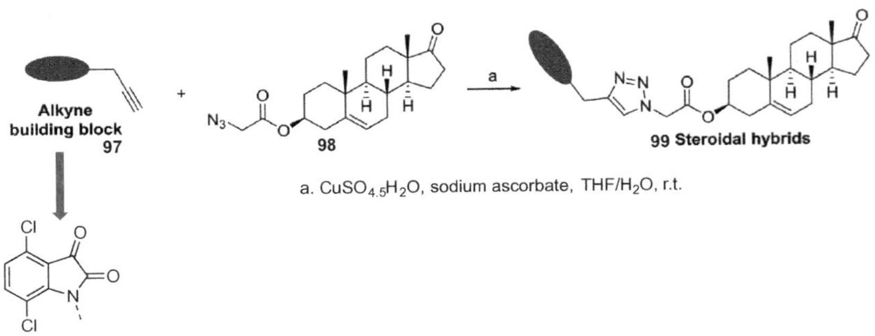

SCHEME 2.26 Synthesis of steroidal hybrids through azido strategy.

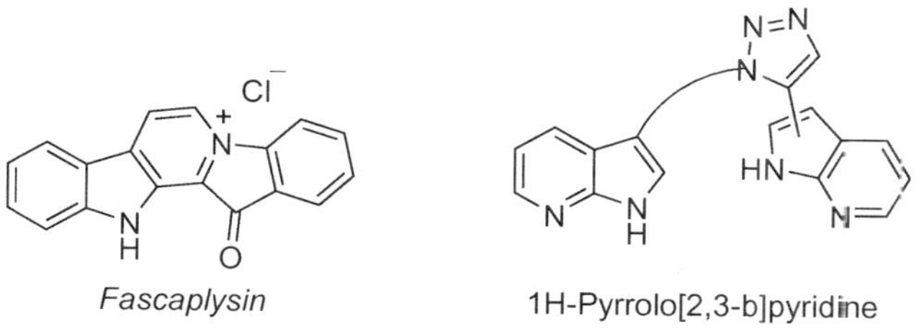

FIGURE 2.8 Structures of *fascaplysin* and 1*H*-pyrrolo[2,3-*b*]pyridine.

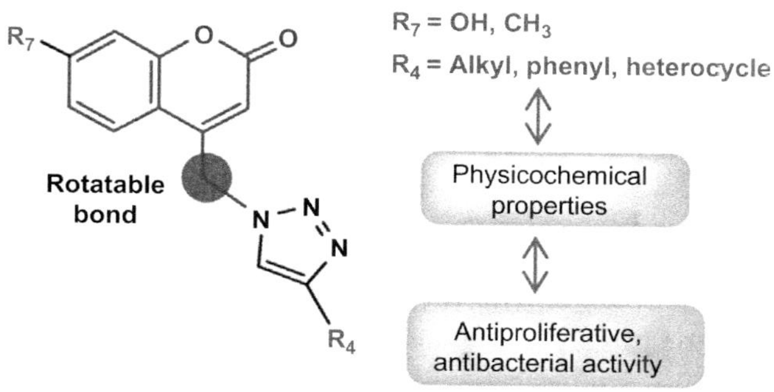

FIGURE 2.9 Conformationally unrestricted 4-substituted 1,2,3-triazoleecoumarin hybrids.

CDK-2/cyclin E and was validated by a test set of 5 compounds giving satisfactory predictive values. Remarkably, it also gave a good prediction of pIC_{50} for our triheterocyclic series, which reinforce the validation of this model for the pIC_{50} prediction of external set compounds. The most promising compound, belonging to the series whose structure is shown in Fig. 2.8, showed a micromolar range inhibitory activity against CDK-2/cyclin E and also an antiproliferative and pro-apoptotic activity against a panel of cancer cell lines.[54]

Following the same synthetic strategy, 4-substituted 1,2,3-triazole core in designed coumarin hybrids (Fig. 2.9) with diverse physicochemical properties were prepared by eco-friendly copper(I)-catalyzed Huisgen 1,3-dipolar cycloaddition reaction under MW irradiation. Coumarine1,2,3-triazoleebenzofused heterocycle hybrids emerged as the class of compounds exhibiting the highest antiproliferative activity. The strong relationship between lipophilicity and antiproliferative activities was observed indicating that lipophilic 1,2,3-triazoleecoumarin hybrids containing phenylethyl, 3,5-difluorophenyl, 5-iodoindole, and benzimidazole subunits showed the most potent cytostatic effects.

The 7-methylcoumarine1,2,3-triazole-2-methylbenzimidazole hybrid can be highlighted as a lead that exerted the highest cytotoxicity against

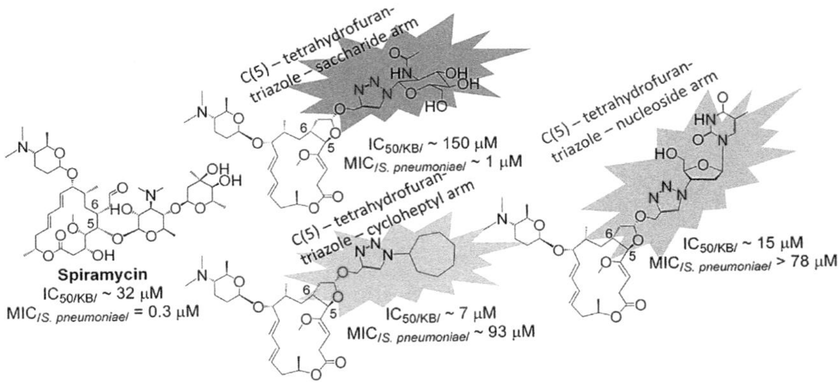

FIGURE 2.10 Spiramycin and related compounds structures with IC_{50} and MIC values. *From: Klich, K.; Pyta, K.; Kubicka, M. M.; Ruszkowski, P.; Celewicz, L.; Gajecka, M.; Przybylski, P. J. Med. Chem.* ***2016****, 59, 7963–7973.*

hepatocellular carcinoma (HCC) Hep-G2 cells with IC_{50} value of 0.9 μM and high selectivity (SI = 50). This compound induced cell death, mainly due to early apoptosis. A strong antiproliferative effect could be associated with inhibition of 5-lipoxygenase (5-LO) activity and perturbation of sphingolipid signaling by interfering with intracellular acid ceramidase activity. These findings indicate that coumarine 1,2,3-triazole could be used as the scaffold for structural optimization to develop more potent and selective anticancer agents.[55]

Huisgen cycloaddition of azides allowed obtaining of novel triazole-bridged antibiotics with the reconstructed C(5) arm of spiramycin (Fig. 2.10).

Combined analysis of biological data together with experimentally determined lipophilicity (clogP) and solubility show the importance of the chemical nature of the newly introduced triazole C(5) arm in the presence of attractive antibacterial and anticancer potency. The most cytotoxic active triazole conjugates having a hydrophobic and bulky C(5) arm showed higher selectivity toward cancer cell lines (HeLa, KB, MCF-7, Hep-G2, and U87) relative to HDF normal cells than that of the parent spiramycin. These studies demonstrated that the aldehyde group of spiramycin is not crucial for the presence of interesting antibacterial [$MIC_{(S.\ pneumoniae)} \sim 1.2$ μM] and anticancer [$IC_{50(Hep\text{-}G2)} \sim 6$ μM] properties of 16-membered lactone macrolides based on spiramycin's aglycone.[56]

The development of protein conjugate therapeutics requires control over the site of modification to allow for reproducible generation of a product with the desired potency, pharmacokinetic, and safety profile. Placement of a single nonnatural amino acid at the desired modification site of a recombinant protein, followed by a bioorthogonal reaction, can provide complete control. To this end, Allozyne Pharmaceuticals, Inc., researchers described

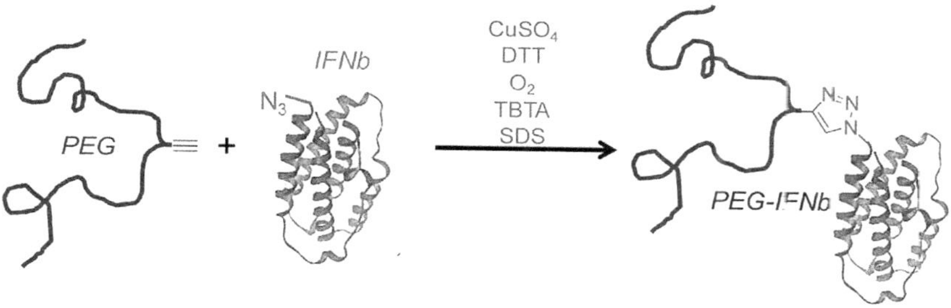

FIGURE 2.11 Optimized reaction conditions for the cycloaddition between propargyl-activated PEGs, TBTA, copper sulfate, and DTT in the presence of SDS. *From: Nairn, N. W.; Shanebeck, K. D.; Wang, A.; Graddis, T. J.; VanBrunt, M. P.; Thornton, K. C.; Grabstein K. Bioconj. Chem.* ***2012****, 23, 2087–2097.*

the development of CuAAC (a click chemistry reaction) for site-specific PEGylation of interferon β-1b (IFNb) containing azidohomoalanine (Aha) at the *N*-terminus. Reaction conditions were optimized using various propargyl-activated PEGs, tris(benzyltriazolylmethyl)amine, copper sulfate, and dithiothreitol in the presence of Sodium Dodecyl Sulphate (SDS) (Fig. 2.11).

The requirement for air in order to advance the redox potential of the reaction was investigated. The addition of unreactive PEG diol reduced the required molar ratio to 2:1 PEG–alkyne to IFNb. The resultant method produced high conversion of Aha-containing IFNb to the single desired product. PEG-IFNbs with 10, 20, 30, and 40 kDa linear or 40 kDa branched PEGs were produced with these methods and compared. Increasing PEG size yielded decreasing in vitro antiviral activities along with concomitant increases in elimination half-life, AUC, and bioavailability when administered in rats or monkeys. A Daudi tumor xenograft model provided comparative evaluation of these combined effects, wherein a 40 kDa branched PEG-IFNb was much more effective than conjugates with smaller PEGs or unPEGylated IFNb at preventing tumor growth in spite of dosing with fewer units and lesser frequency. The results demonstrate the capability of site-specific nonnatural amino acid incorporation to generate novel biomolecule conjugates with increased in vivo efficacy.[57]

A new class of tricyclic heterocycles 4*H*-benzo[*g*][1,2,3]triazolo[5,1-*c*][1,4]oxazocines of type **103** was synthesized through a Knoevenagel condensation/azide-alkyne cycloaddition reaction cascade in one-pot operation (Scheme 2.27). The easy functionalization of the aldehydes **100** with the azides **101** leads to the not-isolated intermediates **102** that undergo intramolecular 1,3-dipolar cycloaddition reaction to afford the desired compounds.

These eight-membered rings containing heterocycles exhibited moderately high anticancer activity against our cancer cell lines; human cervix cancer cell line (HeLa), human prostate cancer cell line (DU145), human breast cancer cell line (MCF-7), and human breast adenocarcinoma epithelial cell

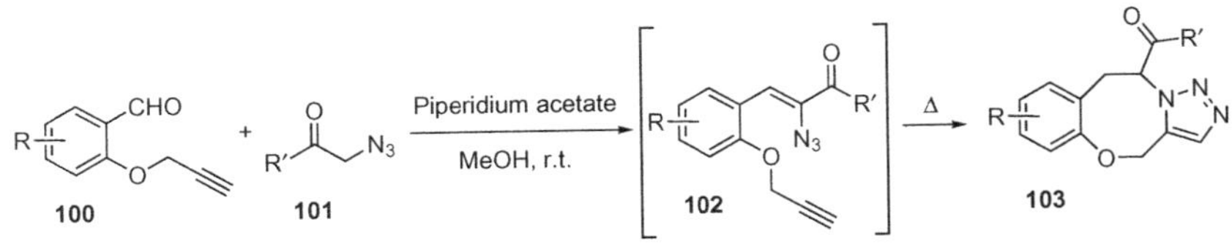

SCHEME 2.27 Synthesis of 4*H*-benzo[*g*][1,2,3]triazolo[5,1-*c*][1,4]oxazocines.

line (MDA-MB-231). These results indicate that these compounds have a weak cytotoxic effect on normal human mammary epithelial cell line (MCF-10A). Cell cycle and apoptosis assay indicate that they inhibit the cell cycle at the G2/M phase and induce apoptosis. Through the RED100 assay, it is evident that they have potential to inhibit pBR 322 plasmid DNA cleavage by BamH1. In spite of the modest results, it represents a good example of intramolecular cycloaddition reaction for the construction of potentially active compounds.[58]

A series of nucleoside analogs bearing a 1,4,5-trisubstituted-1,2,3-triazole aglycone was synthesized using a straightforward click/electrophilic addition or click/oxidative coupling tandem procedures. SAR analysis, using cell culture assays, led to the discovery of a series of compounds belonging to the 5-alkynyl-1,2,3-triazole family that exhibits potent antileukemic effects on several hematologic malignancies including CML and myelodysplastic syndrome (MDS) either sensitive or resistant to their respective therapy. Ethyl 5-(3-ethoxy-3-oxoprop-1-yn-1-yl)-1-((3a*R*,4*R*,6*R*,6a*R*)-6-(hydroxymethyl)-2,2-dimethyltetrahydrofuro[3,4-*d*][1,3]dioxol-4-yl)-1*H*-1,2,3-triazole-4-carboxylate, whose structure is reported in Fig. 2.12, also proved efficient in vivo on mice xenografted with SKM1-R MDS cell line.

Additionally, some insights in its mode of action revealed that this compound induced cell death by caspase and autophagy induction.[59]

These examples of the use of azides in the synthesis of anticancer molecules represent a summary of the myriads of cases reported in modern literature and cover the main synthetic approaches toward active compounds. Similar or different approaches that led, however, to inactive or moderately active compounds were not reported here. The reader can gain further insights on this specific research area on the chemistry of azides through some dedicated reviews papers. The synthesis of 7-deazapurine (pyrrolo[2,3-*d*]pyrimidine) 2′-deoxyribonucleosides, including β-D- and β-L-enantiomers, fluoro derivatives, and 2′,3′-dideoxyribonucleosides was reviewed in 2012.[60] It covers the various aspects of convergent nucleoside synthesis. Stereochemically defined α-D and α-L 2′-deoxyribonucleosides as well as sugar derivatives were prepared by nucleobase anion glycosylation. This glycosylation reaction is regioselective for the pyrrole nitrogen and stereoselective for β-nucleoside formation. A broad range of reporter groups were introduced by the Sonogashira cross coupling or the copper(I)-catalyzed

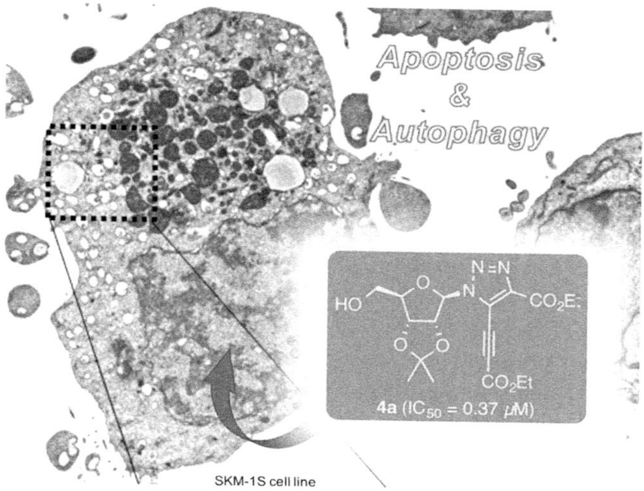

FIGURE 2.12 Ethyl 5-(3-ethoxy-3-oxoprop-1-yn-1-yl)-1-((3a*R*,4*R*,6*R*,6a*R*)-6-(hydroxymethyl)-2,2-dimethyltetrahydrofuro[3,4-*d*][1,3]dioxol-4-yl)-1*H*-1,2,3-triazole-4-carboxylate induces the formation of autolysosomal vacuoles and disorganization of intramitochondrial structures. *Modified from: Amdouni, H.; Robert, G.; Driowya, M.; Furstoss, N.; Métier, C.; Dubois, A.; Dufies, M.; Zerhouni, M.; Orange, F.; Lacas-Gervais, S.; Bougrin, K.; Martin, A. R.; Auberger, P.; Benhida, R. J. Med. Chem.* ***2017****, 60, 1523–1533.*

Huisgen–Meldal–Sharpless click reaction. The application of 7-deazapurine nucleosides as antiviral or anticancer agents and the use of 7-deazapurine nucleoside triphosphates in the Sanger dideoxy DNA-sequencing are also reported.

Another review is dedicated to the synthesis of galectin-3 inhibitors that are of utmost importance for development of new antitumor therapeutic strategies. In this review an updated selection of synthetic glycoconjugates inhibitors of tumor-related galectin-3 is presented, properly addressed as monosaccharide- and disaccharide-based inhibitors, and multivalent-based inhibitors, disclosing relevant methods for their synthesis along with their inhibitory activities toward galectin-3. In general, Cu(I)-assisted 1,3-dipolar azide-alkyne cycloaddition (CuAAC) reactions were predominantly applied for the synthesis of the described inhibitors, which had their inhibitory activities against galectin-3 evaluated by fluorescence polarization, surface plasmon resonance, hemagglutination, ELISA and cell imaging assays. Overall, the presented synthetic glycoconjugates represent frontline galectin-3 inhibitors, finding important biomedical applications in cancer.[61]

Finally, the synthesis of azapeptides was accounted in 2017; these peptidomimetics have the α-carbon(s) of one or more amino acid residues substituted by nitrogen. The resulting semicarbazide analogs have been shown to reinforce β-turn conformation through the combination of lone-pair–lone-pair repulsion of the adjacent hydrazine nitrogen and urea planarity. The applications of azapeptides include receptor ligands, enzyme inhibitors, prodrugs, probes, and imaging agents. In brief, aza residues are introduced into the peptide chain using the submonomer strategy by semicarbazone incorporation, deprotonation, *N*-alkylation, and orthogonal deprotection. Amino acylation of the resulting semicarbazide and elongation gives the desired azapeptide. A number of chemical transformations have taken advantage of the orthogonal chemistry of semicarbazone residues (e.g., Michael additions and *N*-arylations). In addition, libraries have been synthesized from libraries by diversification of aza-propargylglycine (e.g., coupling reactions, [1,3]-dipolar cycloadditions, and 5-exo-dig cyclizations) and aza-chloroalkylglycine residues. Oxidation of aza-glycine residues has afforded azopeptides that react in pericyclic reactions (e.g., DA and ene reactions). Azapeptides have been employed to make a series of conformationally constrained second mitochondria-derived activator of caspase (Smac) mimetics that exhibit promising apoptosis-inducing activity in cancer cells. The synthesis of cyclic azapeptide derivatives was used to make an aza scan to study the conformation-activity relationships of the anticancer agent cilengitide, cyclo(RGDf-N(Me)V), and its parent counterpart cyclo(RGDfV), which exhibit potency against human tumor metastasis and tumor-induced angiogenesis.[62]

Nitrones and nitrile oxides

As in the case of antiviral compounds, nitrones have received a great deal of attention also for the synthesis of anticancer compounds.

A series of new compounds, endowed with a relevant biological activity, can be originated from alteration of the carbohydrate fragment: in this context, the design of novel ribose rings has resulted in the discovery of effective biological agents with lower toxicity and higher biological effects. Promising results in fact have been obtained from a new generation of nucleoside analogs in which the furanose ring has been replaced by alternative carbo- or heterocyclic rings.[63] Uracil, thymine, cytosine, adenine, and guanine nucleosides possessing an isoxazolidine moiety have been synthesized, as racemic mixtures, to investigate their pharmacological activities; in particular (±)-AdT has been reported to inhibit HIV replication in C 8166, with an activity inversely related to the multiplicity of infection used.[64] However, no data about the toxicity of this compound appeared in the literature while these compounds were synthesized and tested. The consideration that both enantiomeric purity and absolute configuration are key factors in determining the physiological activity of these molecules suggested the synthesis of enantiomerically pure *N,O*-nucleosides in order to investigate their biological features. Scheme 2.28 shows the synthetic strategy that relies upon the reaction of the ribofuranose oxime **104** with

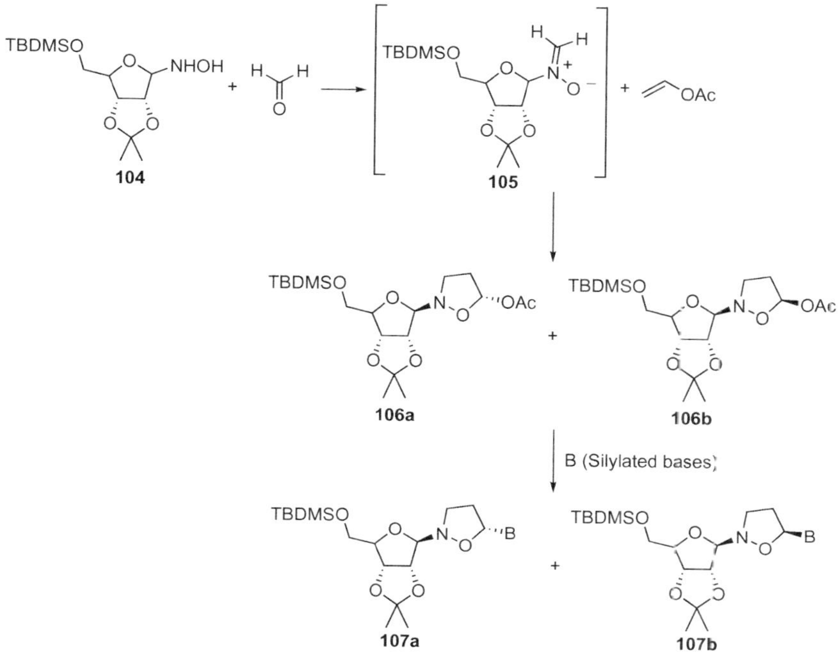

SCHEME 2.28 Nitrone strategy to compounds **107a,b**.

formaldehyde to give the corresponding nitrone **105**, not isolated, and in situ allowed to react with vinyl acetate to afford the diastereoisomeric cycloadducts **106a,b**.

Upon adapting the Vorbrüggen glycosidation protocol with the use of silylated heterobases of purine and pyrimidine type, the target *N,O*-nucleosides **107a,b** were achieved in a simple and straightforward way. Cytotoxicity and apoptotic activities were then investigated. (5′*S*)-5-Fluoro-1-isoxazolidin-5-yl-1*H*-pyrimidine-2,4-dione [(−)-AdFU], one of the compounds synthesized, while showing low level of cytotoxicity, was found to be a good inductor of apoptosis on lymphoid and monocytoic cells, acting as a strong potentiator of Fas-induced cell death.[65]

The same authors applied the same strategy based on nitrone cycloaddition to prepare 5-iodo *N,O*-nucleosides of type **110**, obtained from 1,3-dipolar cycloaddition of methyl-phenyl-nitrone **108** with vinyl acetate followed by coupling with silylated iodo-uracil (Scheme 2.29).

A series of 5-alkynyl *N,O*-nucleosides were synthesized via a palladium-catalyzed (Sonogashira) coupling reaction. The cytotoxic activity of modified nucleosides against HEp-2 cell lines was determined in vitro. The 5-ethynyl *N,O*-nucleoside **111**, the only compound in the series containing a terminal acetylene unit, was found to be the most active compound.[66]

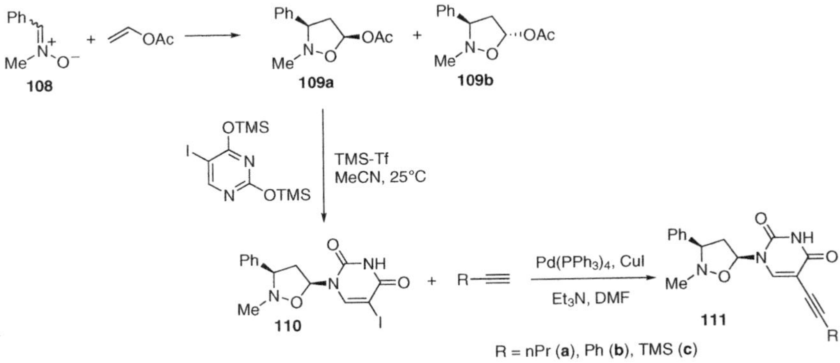

SCHEME 2.29 Synthesis of isoxazolidine through nitrone cycloaddition and functionalization to nucleoside analogs.

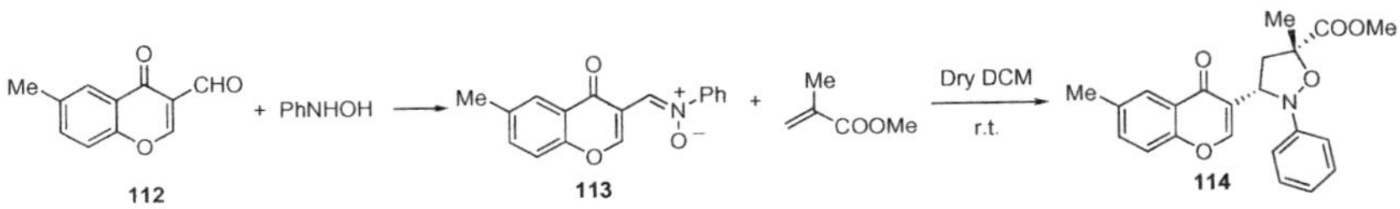

SCHEME 2.30 Synthesis of *N*-phenyl-3′-(chrom-4-one-3-yl)-isoxazolidines.

Regio- and stereoselective 1,3-dipolar cycloaddition reaction of the *C*-(chrom-4-one-3-yl)-*N*-phenylnitrones (**113**) with different monosubstituted, disubstituted, and cyclic dipolarophiles were carried out to obtain substituted *N*-phenyl-3′-(chrom-4-one-3-yl)-isoxazolidines. Some of the obtained isoxazolidines displayed significant cytotoxicity against several human cancer cell lines. Scheme 2.30 reports the synthetic approach to one of the most active compounds.

The aldehyde **112** was allowed to react with the phenylhydroxylamine to afford the desired nitrone **113**, whose reaction with the methyl metacrylate gives the isoxazolidine **114**. The preliminary bioassay results revealed that compound **114** showed higher cytotoxicity than paclitaxel against the A-549 cell line with an IC_{50} value of 0.7 μM. Compound **114** induces apoptosis through a mitochondrial-dependent pathway in human promyelocytic leukemia (HL-60) cells as revealed by induced apoptotic bodies' formation, DNA ladder, increased Sub-G0 DNA fraction, and loss of mitochondrial membrane potential in HL-60 cells.[67]

Chromone derivatives are naturally occurring compounds possessing a wide spectrum of biological activities such as antiinflammatory, antifungal, antimicrobial, antiviral, and anticancer. The *N* and *O* containing five-membered heterocycles, isoxazolidines, and isoxazoline derivatives have been shown to display useful anticancer and antiviral properties. Scheme 2.31 reports another example of chemical moieties possessing anticancer/

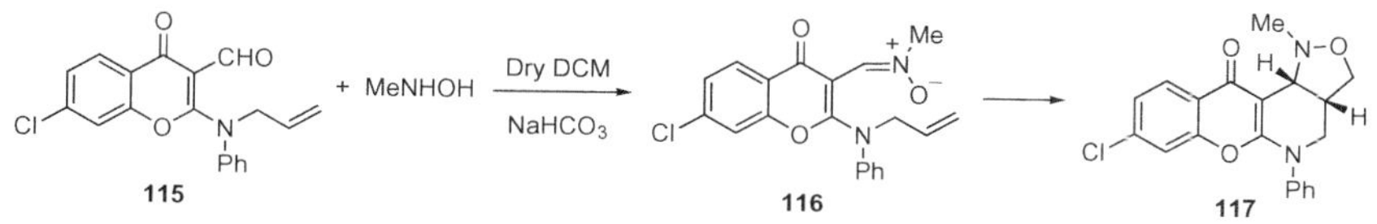

SCHEME 2.31 Intramolecular nitrone cycloaddition synthesis of compound **117**.

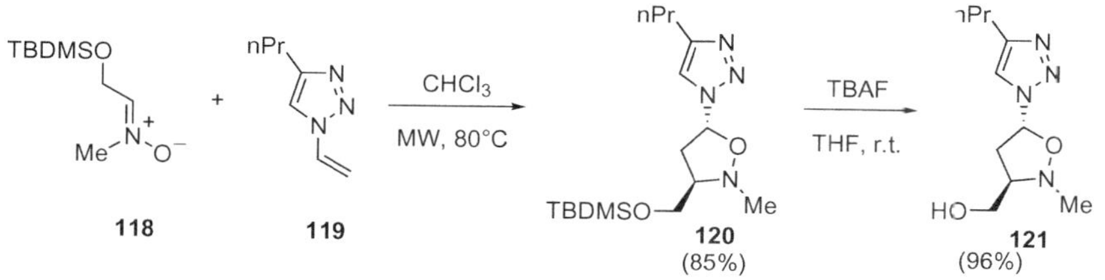

SCHEME 2.32 Nitrone strategy to compound **121**.

cytotoxic potential, through the intramolecular 1,3-dipolar cycloaddition reaction of a nitrone suitably substituted with a dipolarophilic moiety to screen the synthesized chromano-piperidine fused isoxazolidines for in vitro cytotoxic potential against different human cancer cell lines. The method relies upon the synthesis of the nitrone starting from the aldehyde **115** by reaction with *N*-methylhydroxylamine in dichloromethane to afford the desired 1,3-dipole **116**.

Intramolecular 1,3-dipolar cycloaddition of the in situ generated nitrone **116** yielded the chromano-piperidine fused isoxazolidines **117**. This and the other synthesized compounds were evaluated for their cytotoxic potential against various human cancer cell lines. Different responses were obtained from in vitro tests to human cancer cell lines. Compound **117** was found be most active against neuroblastoma revealed by IC_{50} of 10.07 μM. These investigations provided an easy access to novel chromone derivatives bearing fused isoxazolidine moiety possessing significant cytotoxic potential, further marking the importance of these heterocycles in this research area.[68]

A synthetic approach toward a series of 3-hydroxymethyl-5-(1*H*-1,2,3-triazol)isoxazolidines was reported by Romeo, Giofré, et al., by taking advantage of a procedure based on the cycloaddition reaction, under microwave irradiation, of a nitrone with 1-vinyl triazoles, prepared by a click reaction of azides with alkynes. The nitrone **118** was allowed to react with the dipolarophile **119** to afford the protected triazole-isoxazolidine **120** (Scheme 2.32). The final deprotection of the hydroxyl group led to the compound **121** in excellent yields.

Biological tests show that the synthesized compound is able to inhibit proliferation of follicular and anaplastic human thyroid cancer cell lines, with an IC_{50} value of 3.87 μM. The obtained compounds induce caspase-3 activation and DNA fragmentation prevalently in follicular human thyroid cancer cell lines.[69]

1,2,3-Triazole-appended *N,O*-nucleosides find in the approach that combines a 1,3-dipolar cycloaddition process and an alkyne-azide click chemistry reaction to be the best way for their preparation. The same authors as earlier performed the biological assays on six tumor cell lines, and demonstrated the antiproliferative activity of this type of compound. The effect was more evident in the U87MG human glioblastoma cell line. Thus these nucleoside analogs are indeed promising scaffolds for the construction of new anticancer agents.[70] Moreover, a novel series of 2′-oxa-3′-aza-4′a-carbanucleosides, featured with a triazole linker at the 5′-position, was also developed by exploiting a click chemistry reaction of 5′-azido-2′-oxa-3′-aza-4′a-carbanucleosides with substituted alkynes. Biological tests indicate antitumor activity for the synthesized compounds: most of them inhibit cell proliferation of Vero, BS-C-1, HEp-2, MDCK, and HFF cells with a CC_{50} in the range of 5.0–40 μM. The synthesized compounds do not show any antiviral activity.[71]

Moving from nitrone to nitrile oxides a few examples can be cited. Nitrile oxides and their chemistry have been continuously developed over the past three decades and from the very first and simplest nitrile oxide, the fulminic acid, known since the end of the 19th century, a huge number of new 1,3-dipoles have been prepared and used in organic synthesis. Several simple and convenient methods for the preparation of nitrile oxides are reported in literature. Many convenient syntheses were also reported for obtaining isoxazoles, isoxazolines, and isoxazolidines. A series of norcanthadirin analogs were synthesized by the 1,3-dipolar cycloaddition reactions of norcanthadirin derivatives of substituted amines with nitrile oxides. Scheme 2.33 reports the synthesis of one of these analogs that displayed the best results in terms of biological activity.

Derivative **122** was treated with the benzalchloroxime **123**, the precursor of benzonitrile oxide, in the presence of chloramine T to afford the

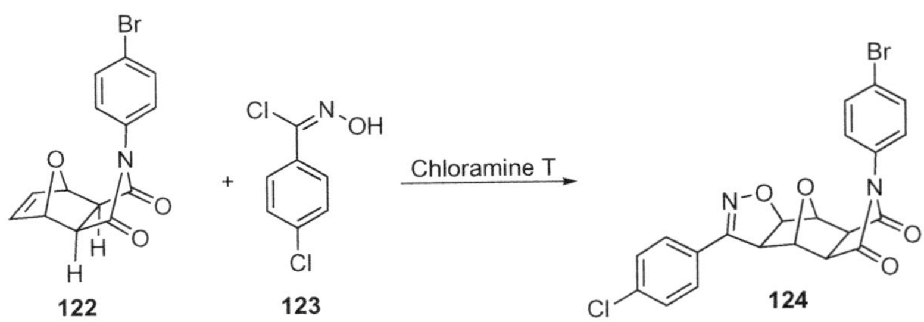

122 **123** **124**

SCHEME 2.33 Nitrile oxide strategy to compound **124**.

compound **124**. This latter and the other synthesized compounds were screened for their antiproliferative activity in vitro against a panel of tumor cell lines: KB, SGC-7901, HL-60, Bel-7402, HO-8910, and ECA109, producing IC_{50} values from 0.36 to >100 μM. In particular, compound **124** showed potency for the treatment of hepatoma, with IC_{50} value to Bel-7402 cell line comparable to that of norcantharidin.[72]

Three new types of aryl diketo acid (ADK) isosteres were designed by conversion of the biologically labile 1,3-diketo unit into heteroaromatic motif such as isoxazole, isothiazole, or 1*H*-pyrazole to improve the physicochemical property of ADK-based HIV-1 integrase (IN) inhibitors. The synthesis of the heteroaromatic carboxylic acids was established by employing phenyl β-diketoester or benzaldehyde as the starting material and 1,3-dipolar cycloaddition as the key reaction. Scheme 2.34 shows the strategy followed for the synthesis of one of the most representative and active compounds of the series.

The aldehyde **125** was converted into the corresponding chloroxime **126** through standard methodologies and the cycloaddition to a triple bond ester afforded the desired compound **127**. The inhibitory activity of this new class of ADK isosteres on IN catalytic function was evaluated. These heteroaromatic carboxylic acids, the phenyl-substituted isoxazole-carboxylic acids, exhibited best potency against the strand transfer reaction. In particular, the 3-(3-(benzyloxy)phenyl)isoxazole-5-carboxylic acid (**127**) showed a $IC_{50} = 81$ μM in inhibiting the strand transfer reaction.[73]

The ubiquitin-proteasome pathway (UPP) influences essential cellular functions including cell growth, differentiation, apoptosis, signal transduction, antigen processing, and inflammatory responses. The main proteolytic component of the UPP is the 26S proteasome, which is responsible for the turnover of many cellular proteins and represents an attractive target for the treatment of pathologies such as cancer, as well as inflammatory, immune, and neurodegenerative diseases. Natural and synthetic proteasome inhibitors having different chemical structures and potency have been

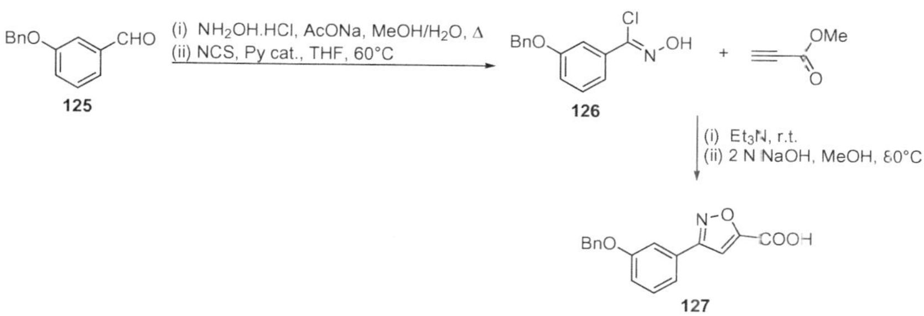

SCHEME 2.34 Synthesis of compound **127** via nitrile oxide cycloaddition chemistry.

proposed and the synthesis, proteasome inhibition, and modeling studies of novel C-terminal isoxazoline vinyl ester (**IsxVE**) pseudopeptides have been illustrated by some Italian researchers. Some new compounds that contain a C-terminal extended conjugation inhibit β1 and especially β5 proteasomal catalytic subunits with IC_{50} values ranging from 10 to 100 μM. These results will permit further optimization based on these structural moieties to develop more active and selective molecules. Scheme 2.35 shows the strategy followed for the synthesis of these compounds.

The aldehyde **128**, converted into the oxime **129**, reacted with the dipolarophile **130** in the presence of NCS and triethylamine for the in situ generation of the corresponding nitrile oxide. The 1,3-dipolar cycloaddition afforded the isoxazoline **131**. Trifluoroacetic acid deprotection of the acetal group gave the formyl derivative **132** that was converted into the unsaturated ester **134** by Wittig reaction with **133**. The **H-Lue-IsxVE** compound **134** was used for the synthesis of the **R-Ser-Leu-IsxVE** by addition of a Boc-protected serine and of the **R-Val-Ser-Leu-IsxVE** by subsequent addition of a Boc-protected valine. This new series of oligopetidic molecules with a C-terminal isoxazoline vinyl ester as pharmacophore aimed to extend conjugation, resulting in a favorable substrate for the nucleophilic attack by the catalytic *N*-terminal Thr1. The inhibitory activity determined for the separated isomers of selected compounds and molecular modeling studies showed that probably the *S* configuration at carbon 5 of the isoxazoline moiety produce active analogs.[74]

It was demonstrated that pyranone derivatives significantly inhibited human gynecological tumors, prostate tumors, digestive system tumors, and hematological tumors without affecting the normal hematopoietic and immune systems. Additionally, this type of compound can effectively inhibit tumor growth in H22 mice with tumor while having no effect on peripheral blood leukocytes, ratio of peripheral blood granulocyte cell and lymphocyte, thymus index, and spleen index. An interesting example of a synthetic approach to these compounds is that reported in Scheme 2.36.

The pivotal step on the synthesis is represented by the 1,3-dipolar cycloaddition reaction of the nitrile oxide obtained from the dehydration of an aliphatic nitro derivative **136**, prepared from the bromo-derivative **135** by nitro group substitution. In the presence of phenylisocyanate under the typical experimental conditions for the in situ generation of the corresponding nitrile oxide, the 1,3-dipolar cycloaddition occurred with the acetylene-anisole **137**, affording the cycloadduct **138**. Cleavage of the N–O bond with $Mo(CO)_6$ led to the enaminocketone **139**, whose treatment with formic acid gave the pyranone **140**. The hydroxyl functionality in **142** was introduced by oxidation with PIFA of the phenol derivative **141**. Compound **142** was found to inhibit the growth of drug-resistant breast cancer cells, inducing apoptosis, downregulation of the ATP level, and inhibition of the expression of P-gp (IC_{50} 0.75 μM in MCF-7 cell line).[75]

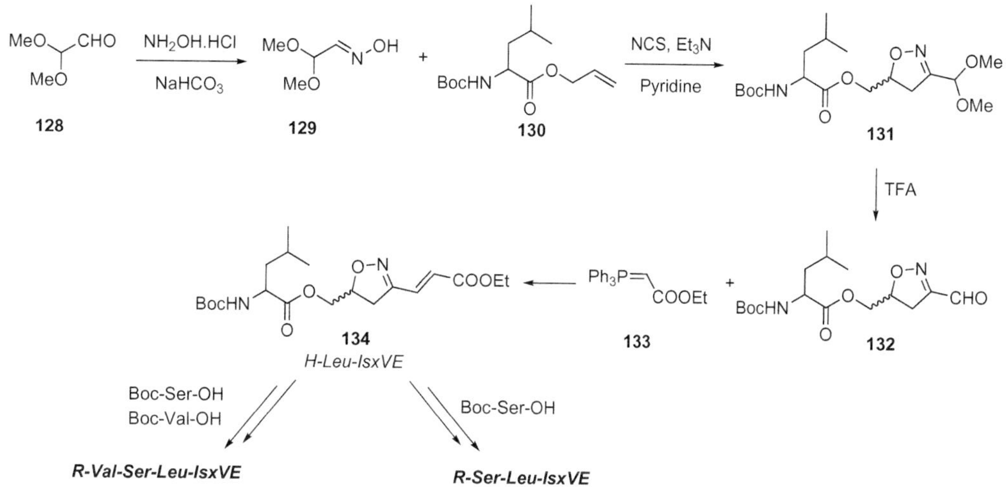

SCHEME 2.35 Nitrile oxide synthesis of oligopetidic molecules.

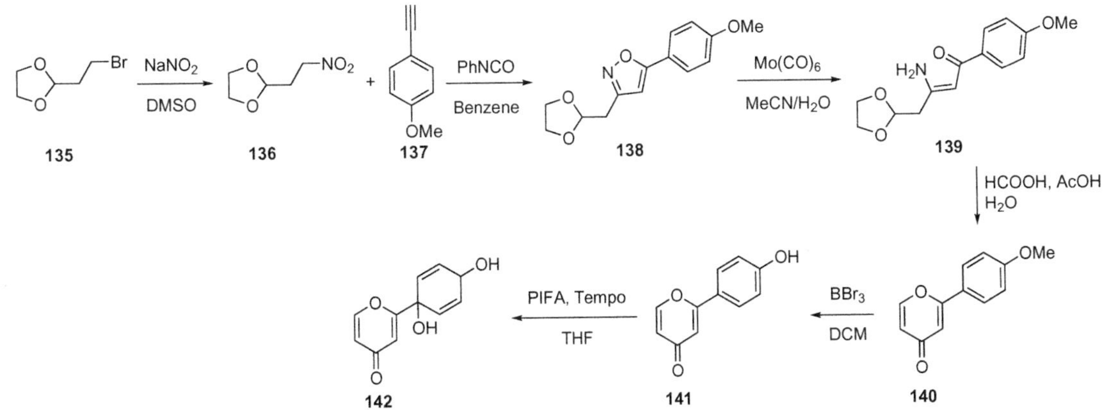

SCHEME 2.36 Synthesis of pyranone derivatives through nitrile oxide cycloaddition reactions.

Nitrile imines

Closely related to nitrile oxide from the reactivity point of view, the nitrile imines have been employed in some cases for the synthesis of biologically active compounds. The optimization, by molecular modeling, of small pyrazole derivatives, as kinase inhibitors obtained through a 1,3-dipolar cycloaddition between nitrile imines and functionalized acetylenes has been described. Scheme 2.37 shows the synthesis of two pyrazoles obtained from a common dipolarophile, the *tert*-butyl (4-propiolamidophenyl)carbamate (**143**). Chloro-aryl hydrazones **144** and **145** are the precursors of the nitril imines that, generated in situ, add to the triple bond of **143** to afford the pyrazoles **146** and **147**, respectively. The target compounds **148** and **149** were easily obtained deprotection and reduction of the nitro group, accordingly.

The two compounds, selected as potential agents active against HCC were then evaluated in vitro for their biological activity on HCC-derived cell lines. The compounds show a promising inhibitory growth efficacy (IC_{50} 50−100 μM) in SNU449 cell line, as well as block of cell cycle progression and induction of apoptosis, and were considered lead compounds for further SAR developments.[76]

9-Substituted (pyrazol-5-yl)methyl- or (2-pyrazolin-5-yl)methyl-9*H*-purines were synthesized from 9-allyl-6-chloro-9*H*-purine through the 1,3-dipolar cycloaddition reaction with nitrile imines, prepared in situ from the corresponding hydrazine **150** and NBS/Et_3N under MW or from hydrazinoylchloride and Et_3N under reflux (Scheme 2.38). The coupling of new 6-chloropurines with amines in H_2O under microwaves resulted quantitatively to modified pyrazol-5-yl- or 2-pyrazolin-5-yl adenine homo-*N*-nucleosides.

The new compounds **153** and **154** were tested in vitro for their ability to (1) interact with 1,1-diphenyl-2-picryl-hydrazyl, (2) inhibit lipid peroxidation, (3) inhibit the activity of soybean lipoxygenase, and (4) inhibit in vitro thrombin; and for their antiproliferative and cytotoxic activity. Pyrazolines were found to be more potent in vitro. Compound **153** exhibited satisfactory combined antioxidant and antilipid peroxidation activity, inhibition of lipoxygenase (89%), and thrombin inhibitory ability, whereas compound **154** exhibited high lipoxygenase inhibitory activity in combination with significant antithrombin activity. No compound exhibited a significant cytotoxic activity, while all showed moderate antiproliferative activity. These studies confirm that the presence of a pyrazolinyl substituted purine is an important structural scaffold for the antioxidant, anti-LO, and antithrombin activity. The presence of the substituted 9-chloropurine seems to be highly implicated in the antiproliferative response. The suggested structural variations could affect both efficacy and their tolerability partly due to differences in their physicochemical properties, which determine their distribution in the body and their ability to pass through and enter into the interior membranes. These compounds may prove useful for treating a variety of inflammatory and coronary artery diseases and may lead to the development of new drugs.[77]

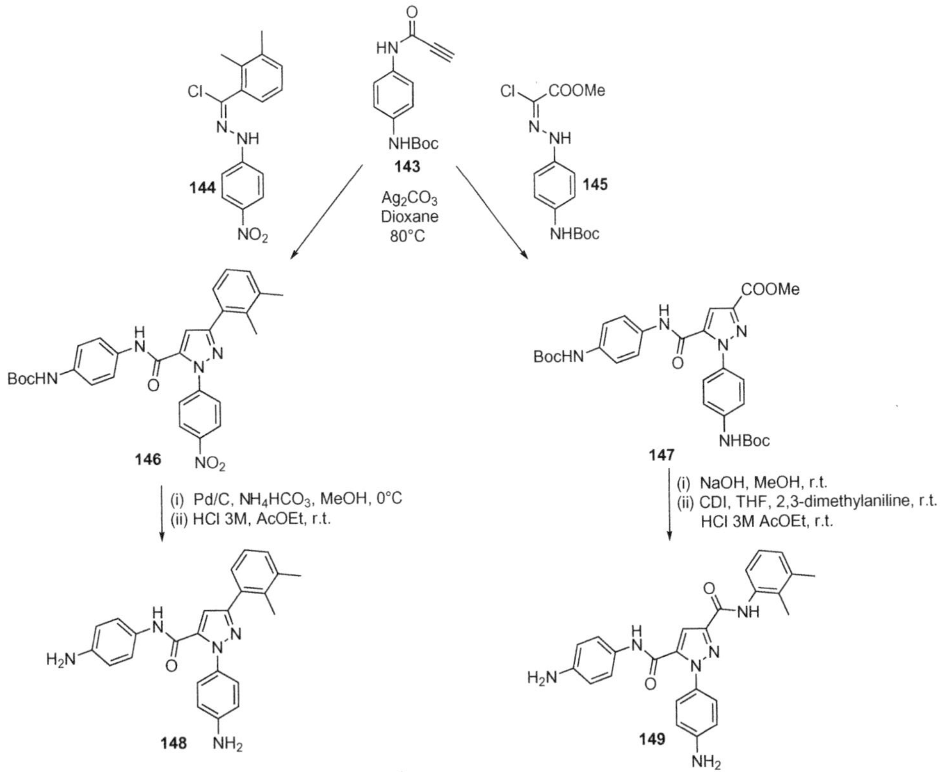

SCHEME 2.37 Nitrile imine synthetic strategy to compounds **148** and **149**.

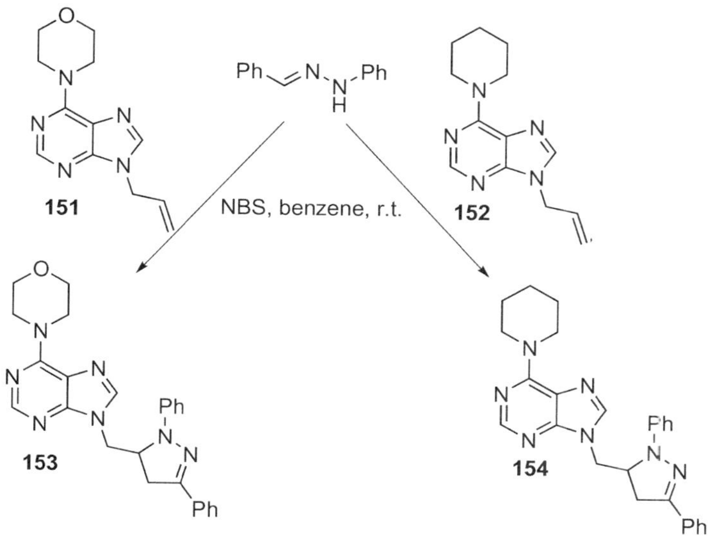

SCHEME 2.38 Nitrile imine synthesis of nucleoside analogs.

Azomethine ylides

Some other 1,3-dipoles can be found at work in the key synthetic steps toward anticancer compounds; here we also report some examples where azomethine ylides have been used. These 1,3-dipoles offer a remarkable solution to the synthetic problems arising when the complexity of molecular targets from natural and unnatural sources demand chemistry for the construction of nitrogen-containing five-membered heterocycles. Ylides represent the most conceptually simple and efficient method. Their generation is usually in situ, followed by dipolarophile attack and furnishes pyrrolidines and pyrrolines with operational simplicity.

The construction of a spiro-oxindole-based 2,5-dihydropyrrole scaffold with potential bioactivity has been established via an efficient three-component reaction of isatin, amino-ester, and alkyne. This protocol represents the first 1,3-dipolar cycloaddition of electron-deficient alkynes with isatin-derived azomethine ylides, providing an easy access to spiro-oxindole-based 2,5-dihydropyrroles with structural diversity in high yields (up to 99%). Scheme 2.39 shows the structures of the reagents and obtained compound **156**.

In the inset of Scheme 2.39 the transition state of the cycloaddition is also shown. The condensation of isatin and diethyl 2-aminomalonate **155** in the presence of trifluoroacetic acid generated the corresponding azomethine ylide, which then participated in 1,3-dipolar cycloadditions with ethyl acetylendicarboxylate to afford product **156** via a sequential Michael addition and

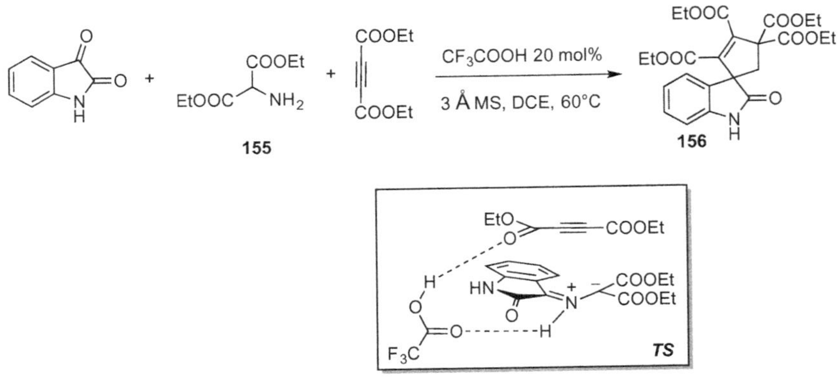

SCHEME 2.39 Spiro-oxindole synthesis and transition state structure.

Mannich-type cyclization rather than a concerted pathway. The trifluoroacetic acid served as a Brønsted acid/Lewis base bifunctional catalyst to simultaneously activate the isatin-derived azomethine ylides and but-2-ynedioats by H-bonding interactions, thereby accelerating the desired reaction to build up the structurally rigid spiro skeleton. The bioscreen of the new spiro-dihydropyrrole has led to the finding of compound **156** as the best promising for the level of cytotoxicity to MCF-7 cells. It displayed the highest inhibition rate of 20.8%, opening the way to further structural modifications to improve the performances.[78a] Compounds with spirofused 11*H*-indeno[1,2-*b*]quinoxaline and cyclopropa[*a*]pyrrolizine or azabicyclo [3.1.0]hexane moieties were recently prepared by a simple and efficient method. All the products were synthesized in good to high yields and excellent diastereoselectivity by one-pot three-component 1,3-dipolar cycloaddition reactions of various derivatives of cyclopropene with azomethine ylides. Azomethine ylides were in situ generated from 11*H*-indeno [1,2-b]quinoxalin-11-one derivatives and amines. The anticancer activity of some of the obtained compounds against the human leukemia K562 cell line displayed promising results.[78b]

Functionalized oligomeric organic compounds with well-defined β-proline scaffold have been synthesized by a cycloadditive oligomerization approach in racemic and enantiopure forms. The structure of the novel β-peptides was investigated by NMR spectroscopic and X-ray methods determining the conformational shapes of the β-proline oligomers in solution and solid states. The main structural elements subject to conformational switches are β-peptide bonds between 5-arylpyrrolidine-2-carboxylic acid units existing in Z/E configurations. The substituted proline **157** (Scheme 2.40) was derivatized at the nitrogen atom with acryloyl chloride to give **158** that undergoes 1,3-dipolar cycloaddition with the in situ generated azomethine ylide from the imine **159**. The cycloadduct **160** was further derivatized with acryloyl chloride to give **161**.

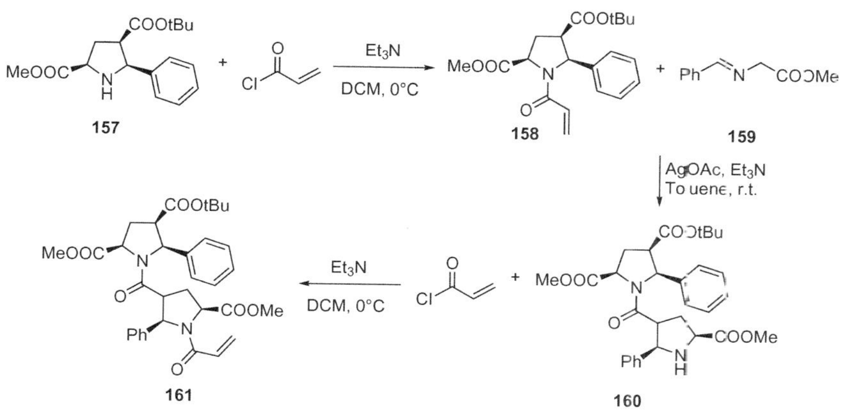

SCHEME 2.40 Azomethine ylide synthesis of compound **161**.

This compound belongs to a library of short β-peptides and intermediate acrylamides that were tested on antiproliferative activity toward the hormone-refractory prostate cancer cell line PC-3 revealing several oligomeric compounds with low micromolar and submicromolar activities.[79]

A series of novel steroidal spirooxindoles **165a–c** (Scheme 2.41) were synthesized from pregnenolone **162** in a high regioselective manner using the 1,3-dipolar cycloaddition as the key step. This protocol resulted in the formation of two C–C bonds, one C–N bond, and the creation of one pyrrolidine ring and three contiguous stereocenters in a single operation.

Pregnenolone **162** was condensed with three substituted benzaldehydes **163a–c** according to a typical aldol condensation KF/Al_2O_3 promoted to afford the calcone derivatives **164a–c**. Upon refluxing this intermediate in the presence of isatin and sarcosine the in situ formed azomethine ylide added to the calcone C=C double bond to give the final compounds **165a–c**. Biological evaluation showed that these synthesized steroidal spirooxindoles exhibited moderate to good antiproliferative activity against the tested cell lines, and some of them were more potent than 5-FU. Compounds **165b** and **165c** displayed the best antiproliferative activity against MCF-7 cells with the IC_{50} values of 4.0 and 3.9 μM, respectively. Flow cytometry analysis demonstrated that compound **165a** caused the cellular apoptosis and cell cycle arrest at G2/M phase in a concentration-dependent manner. Docking results indicated that compound **165a** fitted well into the MDM2 active site 1RV1 by interacting with Lys94 and Thr101 residues (Fig. 2.13).

The docking results would help us visualize the potential receptor–ligand interactions of steroidal spirooxindoles with MDM2 and guide further SAR studies. The crystal structure of MDM2 (PDB: 1RV1) was extracted from the RCSB Protein Data Bank; the target protein structure of 1RV1 was docked with the most active compound **165a**. The lowest energy pose of compound

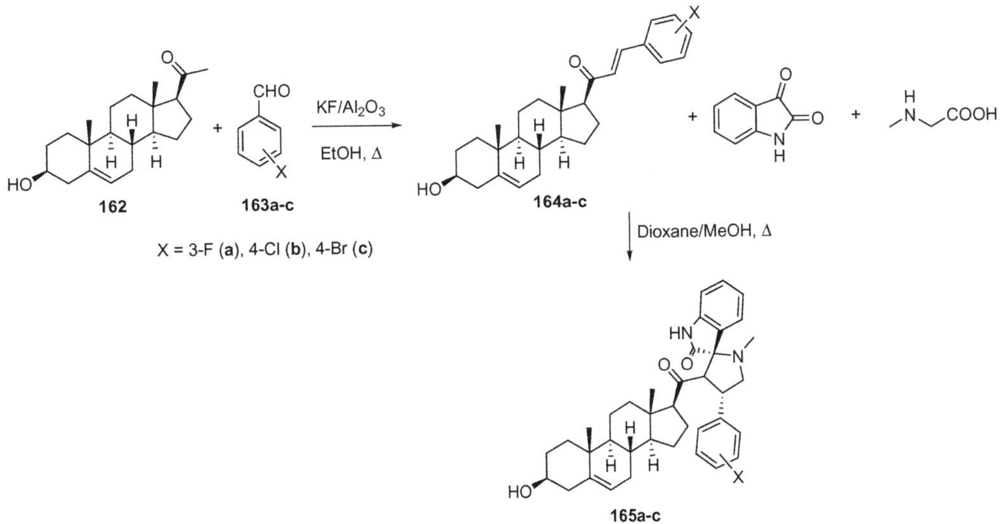

SCHEME 2.41 Synthesis of steroidal spiro-oxindole derivatives via azomethine ylide.

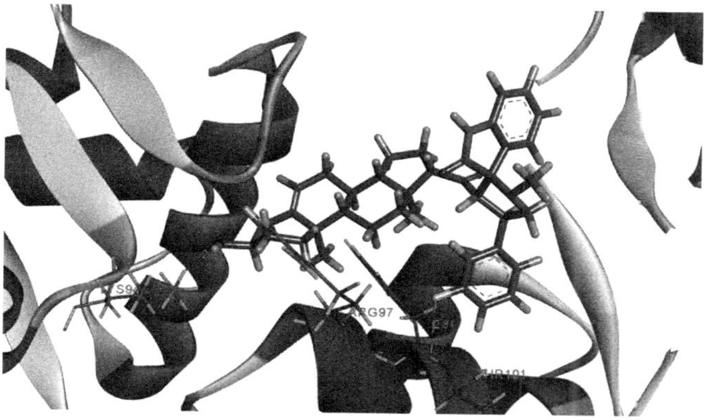

FIGURE 2.13 Binding model of compound **165a** with MDM2 active site 1RV1. *From: Yu, B.; Sun, X.-N.; Shi, X.-J.; Qi, P.-P.; Zheng, Y.-C.; Yu, D.-Q.; Liu, H.-M. Steroids* ***2015****, 102, 92–100.*

165a at MDM2's active site shows that the compound at hand fit well into this pocket showing two hydrogen binding interactions with Lys94 and Thr101 residues. The F–H and N–O distances are 2.81 and 2.90 Å, respectively.[80]

A facile method via 1,3-dipolar cycloaddition of substituted benzylidene-2-phenyloxazolone under mild conditions with azomethine ylides, which were generated in situ by a decarboxylative route from a common set of diverse isatins and amino acid derivatives, was developed for a 15-membered library of regio- and stereoselective oxazolones-grafted spiro-oxindole-pyrrolidine, pyrrolizidines, and pyrrolothiazoles. After screening their cytotoxic activities against a spectrum of cell lines, compound **167** (Scheme 2.42) was identified as a potent antitumor agent and inducing apoptosis.

The active compound was obtained in 91% yield starting from isatin, proline, and the oxazolidinone **166** by heating the reaction mixture at 50°C in ethanol as the solvent. Compound **167** was tested against many important human colon cancers or HCC cell lines. Gefitinib and Sorafenib frequently used in colon cancer or HCC therapeutic were taken as positive control. Compound **167** exhibited anticancer activities against human breast cancer cells with IC_{50} values ranging from 3.4 to 8.1 μM and against HCC cells with IC_{50} values ranging from 7.2 to 8.7 μM in very close similarity with Gefitinib and Sorafenib; thus it could be considered a leading compound to treat antihuman colon cancer or HCC. The SAR analysis exhibited that the scaffords of spirooxindoles with pyrrolizine showed better biological activity, especially compounds with halogen groups or strong electron withdrawing groups (–CF3). To rationalize the anticancer effects of the studied compound, docking studies on compound **167** showed that it interacted with the active sites of the enzymes through H-bonds with Ser864, Trp 323, Tyr324, Gly54, His92, Asp292 Ser64, and amino-acid residues, respectively (Fig. 2.14).

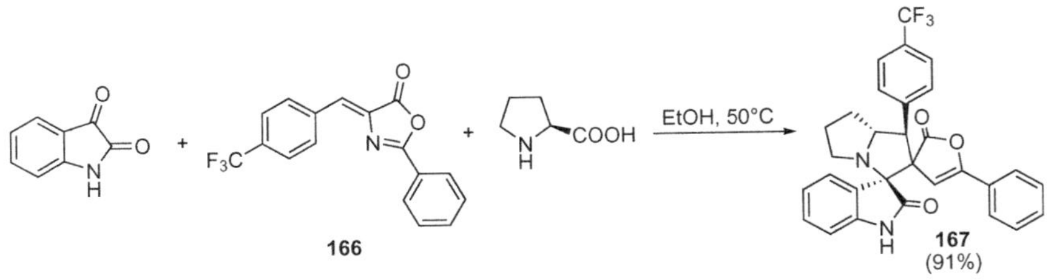

SCHEME 2.42 Synthesis of compound **167**.

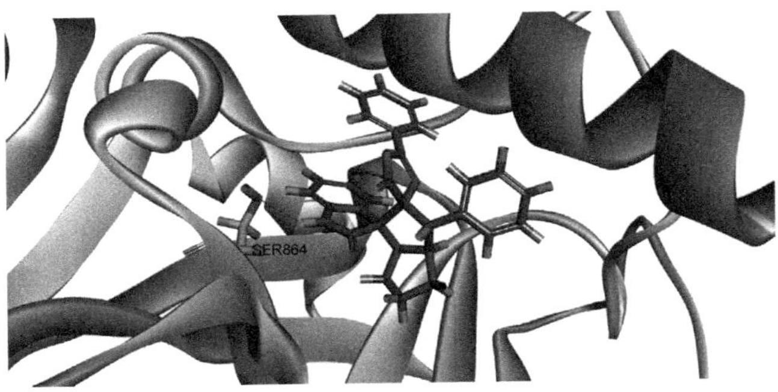

FIGURE 2.14 Compound **176** docked in the active site of PARP-1. The score of C-docker energy was −27.5578 while C-docker interaction energy was 46.1337. *From: Dong, H.; Song, S.; Li, J.; Xu, C.; Zhang, H.; Ouyang, L. Bioorg. Med. Chem. Lett.* ***2015****, 25, 3585–3591.*

These docking results supported the potential of these compounds to interfere with protein synthesis and DNA replication, and induce apoptosis in cancer cells, ultimately leading to death.[81]

These examples have demonstrated the relevance of azomethine ylides in the approach to complex molecules and specifically in targeting the construction of nitrogen-containing, five-membered heterocycles; such ylides represent indeed the most conceptually simple and efficient method. As the ylides are usually in situ generated and instantly trapped by suitable dipolarophiles, this operational simplicity offers a valuable entry to pyrrolidines and pyrrolines. Several other synthetic strategies are reported in literature[82] for the synthesis of biological active compounds such as alkaloids and other complex monocyclic and bicyclic heterocyclic compounds.[83] The reported examples represent somewhat the most intriguing within recent literature on this topic.

Anticancer compounds through [2 + 2] cycloaddition reactions

We conclude this chapter dedicated to anticancer compounds by citing the application of [2 + 2] cycloaddition reactions to this topic. An early example of the [2 + 2] cycloaddition strategy to anticancer compounds was offered by the synthesis of analogs of the oxetanocin A, a nucleoside antibiotic having an oxetanosyl-*N*-glycoside linkage, which exhibits antiviral and antitumor activity. Since it was isolated in 1986, continuous studies have been made concerning the chemistry and biological activity of this unique four-membered ring containing a nucleoside and its synthetic congeners. The carbocyclic analogs of oxetanocins **174** and **175** in optically pure forms were prepared as sketched in Scheme 2.43. These two compounds were already cited in Chapter 1, Cycloaddition reactions for antiviral compounds, but here are obtained through a different synthetic approach where, again, the [2 + 2] cycloaddition reaction is the pivotal step.

The enantio- and diastereo-selective synthesis of homochiral **174** and **175** is based on the key feature of the reported synthetic route through the stereoselective and asymmetric [2 + 2] formation of a cyclobutane ring using a catalytic amount of a chiral titanium reagent starting from the oxazolidinone **168** and the 1, l-bis-(methylthio)ethylene **169** affording the four-membered ring derivative **170**. The following steps are the elimination of the oxazolidinone residue, required for enantioselection generation, and conversion of the ester moieties into protected hydroxymethylene groups as shown in the structures **171**–**173**. The mesylate derivative **173** was used to obtain the two target molecules **174** and **175** by replacement with suitable heterobases. Fifty percent inhibitory concentrations (IC_{50}) against herpes simplex virus types 1 and 2, human cytomegalovirus, hepatitis B virus, and HIV in cell cultures were 1.6, 1.6, 12, 0.024, and 0.03 μM for **174**, and 0.047, 0.08, 0.4, 0.86, and 0.3 μM for **175**, respectively. The authors, however, do not indicate the results regarding the anticancer activities of the synthesized molecules.[84]

Trimethylene-tethered thymine-β-lactam and thyminebis-β-lactam chimeras were prepared as novel classes of hybrid systems through selective mono- or bis-*N*-alkylation of thymine with cis-1-(3-bromopropyl)-β-lactams of type **176**. Their preparation includes initial imination of substituted benzaldehydes with 3-bromopropylamine, followed by a Staudinger [2 + 2]-cyclocondensation with a ketene that was prepared in situ from an alkoxyacetyl chloride (Scheme 2.44). In addition, acidic methanolysis of the β-lactam nucleus in these systems provided an entry into the class of thymine-β-amino-ester hybrids **177**.

A selection of the newly synthesized hybrid compounds **177** were assessed for their antiviral activity, cytotoxicity, and cytostatic activity, revealing a significant cytostatic effect of one of the derivatives against murine leukemia and human T-lymphocyte tumor cells. In particular, N^1,N^3-bis-substituted thymines showed good results with IC_{50} values of <100 μM for hybrids **177** bearing halogen-phenyl substituents, and the 1,3-bis{3-[cis-3-benzyloxy-4-(4-bromophenyl)-2-oxoazetidin-1-yl]propyl}thymine (**177**) was even found to possess an interesting cytostatic activity against leukemia and T-lymphocyte cells with IC_{50} values of 4.1 and 3.4 μM, respectively.

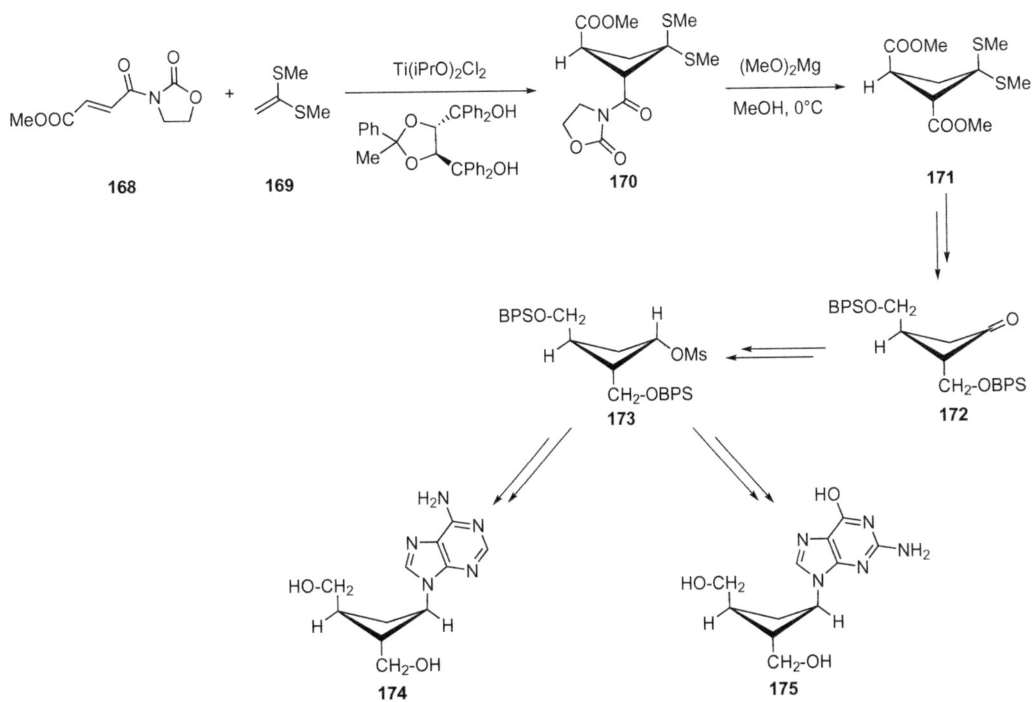

SCHEME 2.43 Synthesis of oxetanocins **174** and **175**.

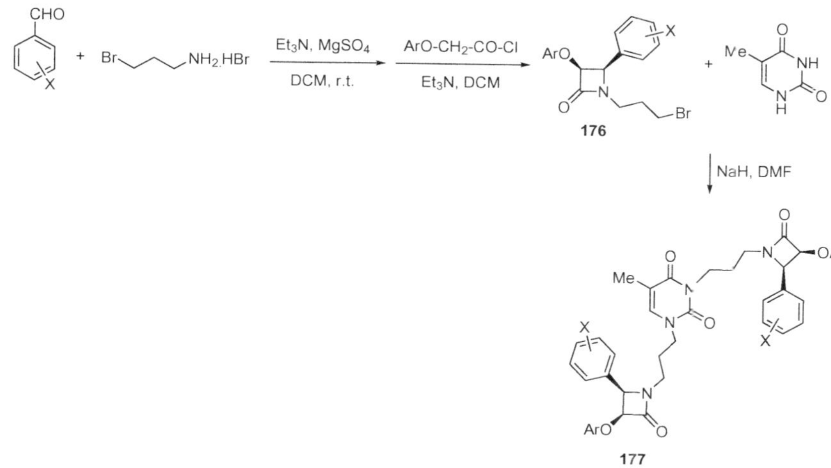

SCHEME 2.44 Synthesis of compound **177**.

The cytostatic activity against the HeLa cells was less pronounced (~20-fold) as compared to the cytostatic activity against the L1210 and CEM cells. Further studies should be performed to reveal the molecular mechanism for the observed cytostatic effects and to explore the potential applications of these compounds where the pivotal synthetic step is represented by the [2 + 2] cycloaddition reaction.[85]

Acknowledgment

Financial support by University of Pavia and MIUR (PRIN 2011, CUP: F11J12000210001) is gratefully acknowledged. We also thank "VIPCAT – Value Added Innovative Protocols for Catalytic Transformations" project (CUP: E46D17000110009) for valuable financial support.

References

1. Veronesi, U.; Betolini, F.; Curigliano, G. Oncologia. Scienze della Vita; Treccani: Roma, **2007,** 313–329.
2. Garr, A. N.; Luo, D.; Brown, N.; Cramer, C. J.; Buszek, K. R.; VanderVelde, D. *Org. Lett.* **2010,** *12*, 96–99 and references therein.
3. Thornton, P. D.; Brown, N.; Hill, D.; Neuenswander, B.; Lushington, G. H.; Santini, C.; Buszek, K. R. *ACS Comb. Sci.* **2011,** *13*, 443–448.
4. (a) Jackson, S. K.; Kerr, M. A. *J. Org. Chem.* **2007,** *72*, 1405–1411. (b) Jackson, S. K.; Banfield, S. C.; Kerr, M. A. *Org. Lett.* **2005,** *7*, 1215–1218. (c) Hitotsuyanagi, Y.; Fujiki, H.; Suganuma, M.; Aimi, N.; Sakai, S.-I.; Endo, Y.; Shudo, K.; Sugimura, T. *Chem. Pharm. Bull.* **1984,** *32*, 4233–4236. (d) Singh, S. B.; Ondeyka, J. G.; Jayasuriya, H.; Zink, D. L.; Ha, S. N.; Dahl-Roshak, A.; Greene, J.; Kim, J. A.; Smith, M. M.; Shoop, W.; Tkacz, J. S. *J. Nat. Prod.* **2004,** *67*, 1496–1506.

5. (a) Perchellet, J.-P. H.; Waters, A. M.; Perchellet, E. M.; Thorton, P. D.; Brown, N.; Hill, D.; Neuenswander, B.; Lushington, G. H.; Santini, C.; Chadrasoma, N.; Buszek, K. R. *Anticancer Res.* **2012,** *32*, 4671–4684. (b) Grosso, C.; Cardoso, A. L.; Rodrigues, M. J.; Marques, C.; Barreira, L.; Lemos, A.; Pinho e Melo, T. M. D. V. *Bioorg. Med. Chem.* **2018,** *25*, 1122–1131.
6. (a) Bentley, K. W. Morphine Alkaloids. In *The Alkaloids;* Manske, R. H. F., Ed.; Vol. 13; Academic Press: New York, 1971; pp 1–163. (b) Inubushi, Y.; Ibuka, T. The Hasubanan Alkaloids. In *The Alkaloids;* Manske? R. H. F., Ed.; Vol. 16; Academic Press: New York, 1977; pp 393–430. (c) Matsui, M. Hasubanan Alkaloids. In *The Alkaloids;* Brossi? A., Ed.; Vol. 33; Academic Press: New York, 1988; pp 307–347. (d) King, S. M.; Herzon, S. B. The Hasubanan and Acutumine Alkaloids. In *The Alkaloids;* Knölker? H.-J., Ed.; Vol. 73; Academic Press: London, 2014; pp 161–222.
7. Calandra, N. A.; King, S. M.; Herzon, S. B. *J. Org. Chem.* **2013,** *78*, 10031–10057.
8. Carrasco, E.; Álvarez, P. J.; Melguizo, C.; Prados, J.; Álvarez-Manzaneda, E.; Chahboun, R.; Messouri, I.; Vázquez-Vázquez, M. I.; Aránega, A.; Rodríguez-Serrano, F. *Eur. J. Med. Chem.* **2014,** *79*, 1–12.
9. (a) Ngoc, T. D.; Moons, N.; Kim, Y.; De Borggraeve, W.; Mashentseva, A.; Andrei, G.; Snoeck, R.; Balzarini, J.; Dehaen, W. *Bioorg. Med. Chem.* **2014,** *22*, 3292–3300. (b) Dangroo, N. A.; Singh, J.; Ratha, S. K.; Gupta, N.; Qayum, A.; Singh, S.; Sangwan, P. L. *Steroids* **2017,** *123*, 1–12.
10. Skiba, J.; Karpowicz, R.; Szabo, I.; Therrien, B.; Kowalski, K. *J. Organomet. Chem.* **2015,** *794*, 216–222.
11. (a) Stevens, J.; Blixt, O.; Tumpey, T. M.; Taubenberger, J. K.; Paulson, J. C.; Wilson, I. A. *Science* **2006,** *312*, 404–410. (b) Gagneux, P.; Cheriyan, M.; Hurtado-Ziola, N.; Brinkman van der, L.; Els, C. M.; Anderson, D.; McClure, H.; Varki, A.; Varki, N. M. *J. Biol. Chem.* **2003,** *278*, 48245–48250. (c) Daniels, M. A.; Hogquist, K. A.; Jameson, S. C. *Nat. Immunol.* **2002,** *3*, 903–910. (d) Bertozzi, C. R.; Kiessling, L. L. *Science* **2001,** *291*, 2357–2364.
12. (a) Daniels, M. A.; Hogquist, K. A.; Jameson, S. C. *Nat. Immunol.* **2002,** *3*, 903–910. (b) Martin, L. T.; Marth, J. D.; Varki, A.; Varki, N. M. *J. Biol. Chem.* **2002,** *277*, 32930–32938. (c) Blander, J. M.; Visintin, I.; Janeway, C. A., Jr; Medzhitov, R. *J. Immunol.* **1999,** *163*, 3746–3752.
13. Cheung, I. Y.; Vickers, A.; Cheung, N.-K. V. *Int. J. Cancer* **2006,** *119*, 152–159.
14. (a) Lee, L.; Chang, K.-H.; Valiyev, F.; Liu, H.-J.; Li, W.-S. *J. Chin. Chem. Soc.* **2006,** *53*, 1547–1555. (b) Pastuch-Gawolek, G.; Plesniak, M.; Komor, R.; Byczek-Wyrostek, A.; Erfurt, K.; Szeja, W. *Bioorg. Chem.* **2017,** *72*, 80–88.
15. (a) El Akri, K.; Bougrin, K.; Balzarini, J.; Faraj, A.; Benhida, R. *Bioorg. Med. Chem. Lett.* **2007,** *17*, 6656–6659. (b) Alaoui, S.; Dufies, M.; Driowya, M.; Demange, L.; Bougrin, K.; Robert, G.; Auberger, P.; Pagès, G.; Benhida, R. *Bioorg. Med. Chem. Lett.* **2017,** *27*, 1989–1992.
16. Goodman, J.; Hochstein, P. *Biochem. Biophys. Res. Commun.* **1977,** *77*, 797–803.
17. Wang, S.; Wang, Q.; Wang, Y.; Liu, L.; Wenga, X.; Li, G.; Zhang, X.; Zhou, X. *Bioorg. Med. Chem. Lett.* **2008,** *18*, 6505–6508.
18. Nicolaus, N.; Zapke, J.; Riesterer, P.; Neudörfl, J.-M.; Prokop, A.; Oschkinat, H.; Schmalz, H. G. *ChemMedChem* **2010,** *5*, 661–665.
19. Park, S. M.; Yang, H.; Park, S.-K.; Kim, H. M.; Kim, B. H. *Bioorg. Med. Chem. Lett.* **2010,** *20*, 5831–5834.
20. Montagu, A.; Roy, V.; Balzarini, J.; Snoeck, R.; Andrei, G.; Agrofoglio, L. A. *Eur. J. Med. Chem.* **2011,** *46*, 778–786.

21. Baraniak, D.; Kacprzak, K.; Celewicz, L. *Bioorg. Med. Chem. Lett.* **2011**, *21*, 723–726.
22. (a) Xia, Y.; Qu, F.; Peng, L. *Mini-Rev. Med. Chem.* **2010**, *10*, 806–821. (b) Liu, Y.; Xia, Y.; Li, W.; Cong, M.; Maggiani, A.; Leyssen, P.; Qu, F.; Neyts, J.; Peng, L. *Bioorg. Med. Chem. Lett.* **2010**, *20*, 3610–3613. (c) Wang, M.; Xia, Y.; Fan, Y.; Rocchi, P.; Qu, F.; Iovanna, J. L.; Peng, L. *Bioorg. Med. Chem. Lett.* **2010**, *20*, 5979–5983.
23. (a) Xia, Y.; Wang, M.; Demaria, O.; Tang, J.; Rocchi, P.; Qu, F.; Iovanna, J. L.; Alexopoulou, L.; Peng, L. *J. Med. Chem.* **2012**, *55*, 5642–5646. (b) Brandao, G. C.; Rocha Missias, F. C.; Arantes, L. M.; Soares, L. F.; Roy, K. K.; Doerksen, R. J.; Braga de Oliveira, A.; Rocha Pereira, G. *Eur. J. Med. Chem.* **2018**, *145*, 191–205.
24. Driowya, M.; Puissant, A.; Robert, G.; Auberger, P.; Benhida, R.; Bougrin, K. *Ultrason. Sonochem.* **2012**, *19*, 1132–1138.
25. Neidle, S.; Balasubramanian, S. *Quadruplex Nucleic Acids;* Royal Society of Chemistry: Cambridge, 2006.
26. (a) Stefan, L.; Guédin, A.; Amrane, S.; Smith, N.; Denat, F.; Mergny, J. L.; Monchaud, D. *Chem. Commun.* **2011**, *47*, 4992–4994. (b) Xu, H. J.; Stefan, L.; Haudecoeur, R.; Vuong, S.; Richard, P.; Denat, F.; Barbe, J. M.; Gros, C. P.; Monchaud, D. *Org. Biomol. Chem* **2012**, *10*, 5212–5218. (c) Haudecoeur, R.; Stefan, L.; Denat, F.; Monchaud, D. *J. Am Chem. Soc.* **2013**, *135*, 550–553.
27. (a) Dash, J.; Patil, A. J.; Das, R. N.; Dowdal, F.; Mann, S. *Soft Matter* **2011**, *7*, 8120–8126. (b) Das, R. N.; Kumar, Y. P.; Pagoti, S.; Patil, A. J.; Dash, J. *Chem. Eur. J.* **2012**, *18*, 6008–6014.
28. (a) Pavan Kumar, Y.; Bhowmik, S.; Das, R. N.; Bessi, I.; Paladhi, S.; Ghosh, R. Schwalbe, H.; Dash, J. *Chem. Eur. J.* **2013**, *19*, 11502–11506. (b) Panda, D.; Debnath, M. Mandal, S.; Bessi, I.; Schwalbe, H.; Dash, J. *Sci. Rep.* **2015**, *5*, 13183.
29. Duan, Y. C.; Zheng, Y.-C.; Li, X.-C.; Wang, M.-M.; Ye, X.-W.; Guan, Y.-Y.; Liu, G.-Z. Zheng, J.-X.; Liu, H.-M. *Eur. J. Med. Chem.* **2013**, *64*, 99–110.
30. (a) Liu, J.-F.; Sang, C.-Y.; Xu, X.-H.; Zhang, L.-L.; Yang, X.; Hui, L.; Zhang, J.-B.; Chen S.-W. *Eur. J. Med. Chem.* **2013**, *64*, 621–628. (b) Cui, F.; Mo, C. Z.; Su, S.; Chen, Y.; Ma, X.; Tang, H.; Wang, H.; Pan, Y.; Xu, Y. *Org. Lett.* **2018**, *20*, 925–929.
31. Wang, H.; Huang, W.; Orwenyo, J.; Banerjee, A.; Vasta, G. R.; Wang, L.-X. *Bioorg. Med. Chem.* **2013**, *21*, 2037–2044.
32. Su, Y.; Pan, S.; Li, Z.; Li, L.; Wu, X.; Hao, P.; Kwan Sze, S.; Yao, S. Q. *Sci. Rep.* **2015**, *5*, 7724.
33. Vasilyeva, S. V.; Kuznetsova, A. S.; Khalyavina, J. G.; Glazunova, V. A.; Shtil, A. A.; Gornostaev, L. M.; Silnikov, V. N. *Nucleosides Nucleotides Nucleic Acids* **2014**, *33*, 615–625.
34. Su, Y.-C.; Lo, Y.-L.; Hwang, C.-C.; Wang, L.-F.; Wu, M.-H.; Wang, E.-C.; Wang, Y.-M.; Wang, T.-P. *Org. Biomol. Chem.* **2014**, *12*, 6624–6633.
35. (a) Hannon, G. J. *Nature* **2002**, *418*, 244–251. (b) Fire, A.; Xu, S.; Montgomery, M. K.; Kostas, S. A.; Driver, S. E.; Mello, C. C. *Nature* **1998**, *391*, 806–811. (c) Pecot, C. V.; Calin, G. A.; Coleman, R. L.; Lopez-Berestein, G.; Sood, A. K. *Nat. Rev. Cancer* **2011**, *11*, 59–67. (d) Bennett, C. F.; Swayze, E. E. *Annu. Rev. Pharmacol. Toxicol.* **2010**, *50*, 259–293. (e) Kole, R.; Krainer, A. R.; Altman, S. *Nat. Rev. Drug Discov.* **2012**, *11*, 125–140. (f) Dias, N.; Stein, C. A. *Mol. Cancer Ther.* **2002**, *1*, 347–355.
36. Lee, R. G.; Crosby, J.; Baker, B. F.; Graham, M. J.; Crooke, R. M. *J. Cardiovasc. Transl. Res.* **2013**, *6*, 969–980.
37. Abendroth, F.; Seitz, O. *Angew. Chem. Int. Ed.* **2014**, *53*, 10504–10509.

38. Zilla, M. K.; Nayak, D.; Vishwakarma, R. A.; Sharma, P. R.; Goswami, A.; Ali, A. *Eur. J. Med. Chem.* **2014,** *77*, 47–55.
39. Khazir, J.; Hyder, I.; Gayatri, J. L.; Yandrati, L. P.; Nalla, N.; Chasoo, G.; Mahajan, A.; Saxena, A. K.; Alam, M. S.; Qazi, G. N.; Sampath Kumar, H. M. *Eur. J. Med. Chem.* **2014,** *82*, 255–262.
40. Kumar, D.; Mallick, S.; Vedasiromoni, J. R.; Pal, B. C. *Phytomedicine* **2010,** *17*, 431–435.
41. Chakraborty, B.; Dutta, D.; Mukherjee, S.; Das, S.; Maiti, N. C.; Das, P.; Chowdhury, C. *Eur. J. Med. Chem.* **2015,** *102*, 93–105.
42. Dang Thi, T. A.; Kim Tuyet, N. T.; Pham The, C.; Thanh Nguyen, H.; Ba Thi, C.; Thi Phuong, H.; Van Boi, L.; Van Nguyen, T.; D'hooghe, M. *Tetrahedron Lett.* **2015,** *56*, 218–224.
43. Wei, G.; Luan, W.; Wang, S.; Cui, S.; Li, F.; Liu, Y.; Liu, Y.; Cheng, M. *Org. Biomol. Chem.* **2015,** *13*, 1507–1514.
44. (a) Seligmann, J.; Twelves, C. *Future Med. Chem.* **2013,** *5*, 339–352. (b) Cortes, J.; Vidal, M. *Breast Cancer Res. Treat.* **2012,** *133*, 821–830. (c) Katsetos, C. D.; Dráber, P. *Curr. Pharm. Des.* **2012,** *18*, 2778–2792. (d) Stanton, R. A.; Gernert, K. M.; Nettles, J. H.; Aneja, R. *Med. Res. Rev.* **2011,** *31*, 443–481.
45. (a) Lu, Y.; Chen, J.; Xiao, M.; Li, W.; Miller, D. D. *Pharm. Res.* **2012,** *29*, 2943–2971. (b) Chen, J.; Liu, T.; Dong, X.; Hu, Y. *Mini-Rev. Med. Chem.* **2009,** *9*, 1174–1190. (c) Sivakumar, G. *Curr. Med. Chem.* **2013,** *20*, 892–898.
46. Sitnikov, N. S.; Sintsov, A. V.; Shchegravina, E. S.; Prokop, A.; Schmalz, H. G.; Fokin, V. V.; Fedorov, A. Y. *Russ. Chem. Bull. Int. Ed.* **2015,** *64*, 1362–1368.
47. Srinivasa Reddy, T.; Kulhari, H.; Ganga Reddy, V.; Subba Rao, A. V.; Bansal, V.; Kamal, A.; Shukla, R. *Org. Biomol. Chem.* **2015,** *13*, 10136–10149.
48. Swider, R.; Mastyk, M.; Zapico, J. M.; Coderch, C.; Panchuk, R.; Skorokhyd, N.; Schnitzler, A.; Niefind, K.; de Pascual-Teresa, B.; Ramos, A. *RSC Adv.* **2015,** *5*, 72482–72494.
49. Lu, H.; Zhou, Q.; Deshmukh, V.; Phull, H.; Ma, J.; Tardif, V.; Naik, R. R.; Bouvard, C.; Zhang, Y.; Choi, S.; Lawson, B. R.; Zhu, S.; Kim, C. H.; Schultz, P. G. *Angew. Chem. Int. Ed.* **2014,** *53*, 9841–9845.
50. Ramadoss, N.; Schulman, A. D.; Choi, S.; Rodgers, D. T.; Kazane, S. A.; Kim, C. H.; Lawson, B. R.; Young, T. S. *J. Am. Chem. Soc.* **2015,** *137*, 5288–5291.
51. Molnár, J.; Frank, E.; Minorics, R.; Kádár, Z.; Ocsovszki, I.; Schönecker, B.; Wölfling, J.; Zupkó, I. *PLoS One* **2015,** *10*, e0118104. Available from: https://doi.org/10.1371/journal.pone.0118104.
52. (a) Xavier, N. M.; Lucas, S. D.; Jorda, R.; Schwarz, S.; Loesche, A.; Csuk, R.; Conceição Oliveira, M. *Synlett* **2015,** *26*, 2663–2672. (b) Xavier, N. M.; Porcheron, A.; Batista, D.; Jorda, R.; Řezníčková, E.; Kryštof, V.; Conceição Oliveira, M. *Org. Biomol. Chem.* **2017,** *15*, 4667–4680.
53. Yu, B.; Qi, P.-P.; Shi, X.-J.; Huang, R.; Guo, H.; Zheng, Y.-C.; Yu, D.-Q.; Liu, H.-M. *Eur. J. Med. Chem.* **2016,** *117*, 241–255.
54. Baltus, C. B.; Jorda, R.; Marot, C.; Berka, K.; Bazgier, V.; Krystof, V.; Prie, G.; Viaud-Massuard, M. C. *Eur. J. Med. Chem.* **2016,** *108*, 701–719.
55. (a) Gazivoda Kraljevic, T.; Harej, A.; Sedic, M.; Kraljevic Pavelic, S.; Stepanic, V.; Drenjancevic, D.; Talapko, F.; Raic-Malic, S. *Eur. J. Med. Chem.* **2016,** *124*, 794–808. (b) Abdel-Wahab, B. F.; Mohamed, H. A.; Farhat, A. A. *Org. Commun.* **2014,** *7*, 1–27.
56. Klich, K.; Pyta, K.; Kubicka, M. M.; Ruszkowski, P.; Celewicz, L.; Gajecka, M.; Przybylski, P. *J. Med. Chem.* **2016,** *59*, 7963–7973.

57. Nairn, N. W.; Shanebeck, K. D.; Wang, A.; Graddis, T. J.; VanBrunt, M. P.; Thornton, K. C.; Grabstein, K. *Bioconj. Chem.* **2012,** *23*, 2087–2097.
58. Visweswara Sastry, K. N.; Rani Routhu, S.; Gupta Datta, S.; Nagesh, N.; Nagendra Babu, B.; Babu Nanubolu, J.; Ganesh Kumar, C.; Awatar Maurya, R.; Kamal, A. *Org. Biomol. Chem.* **2016,** *14*, 9294–9305.
59. Amdouni, H.; Robert, G.; Driowya, M.; Furstoss, N.; Métier, C.; Dubois, A.; Dufies, M.; Zerhouni, M.; Orange, F.; Lacas-Gervais, S.; Bougrin, K.; Martin, A. R.; Auberger, P.; Benhida, R. *J. Med. Chem.* **2017,** *60*, 1523–1533.
60. Seela, F.; Budow, S.; Peng, X. *Curr. Org. Chem.* **2012,** *16*, 161–223.
61. Campo, V. L.; Marchiori, M. F.; Ridrigues, L. C. *Glycoconj. J.* **2016,** *33*, 853–876.
62. Chingle, R.; Proulx, C.; Lubell, W. D. *Acc. Chem. Res.* **2017,** *50*, 1541–1556.
63. Fischer, R.; Druckovà, A.; Fisera, L.; Rybàr, A.; Hametner, C.; Cyranski, M. *Synlett* **2002,** 1113–1117.
64. Leggio, A.; Liguori, A.; Procopio, A.; Siciliano, C.; Sindona, G. *Nucleosides Nucleotides* **1997,** *16*, 1515–1518.
65. Chiacchio, U.; Corsaro, A.; Iannazzo, D.; Piperno, A.; Pistarà, V.; Rescifina, A.; Romeo, R.; Valveri, V.; Mastino, A.; Romeo, G. *J. Med. Chem.* **2003,** *46*, 3696–3702.
66. Romeo, R.; Giofré, S.; Iaria, D.; Sciortino, M. T.; Ronsisvalle, S.; Chiacchio, M. A.; Scala, A. *Eur. J. Org. Chem.* **2011,** *28*, 5690–5695.
67. Singh, G.; Kaur, A.; Sharma, V.; Suri, N.; Sharma, P. R.; Saxena, A. K.; Ishar, M. P. S. *MedChemComm.* **2013,** *4*, 972–978.
68. Singh, S.; Chopra, A.; Singh, G.; Saxena, A. K.; Ishar, M. P. S. *J. Pharm. Res.* **2013,** *7*, 337–341.
69. (a) Romeo, R.; Giofré, S. V.; Carnovale, C.; Campisi, A.; Parenti, R.; Bandini, L.; Chiacchio, M. A. *Bioorg. Med. Chem.* **2013,** *21*, 7929–7937. (b) Bkhaitan, M. M.; Mirza, A. Z.; Abdalla, A. N.; Shamshad, H.; Ul-Haq, Z.; Alarjah, M.; Piperno, A *Chem. Biol. Drug Des.* **2017,** 1–11. (c) Varrica, M. G.; Zagni, C.; Mineo, P. G.; Floresta, G.; Monciino, G.; Pistarà, V.; Abbadessa, A.; Nicosia, A.; Castlho, R. M.; Amata, E.; Rescifina, A. *Eur. J. Med. Chem.* **2018,** *143*, 583–590.
70. Romeo, R.; Giofré, S. V.; Carnovale, C.; Chiacchio, M. A.; Campisi, A.; Mancuso, R.; Cirmi, S.; Navarra, M. *Eur. J. Org. Chem.* **2014,** *25*, 5442–5447.
71. Romeo, R.; Carnovale, C.; Giofré, S. V.; Chiacchio, M. A.; Campisi, A.; Garozzo, A.; Amata, E.; Romeo, G.; Chiacchio, U. *Beilstein J. Org. Chem.* **2015,** *11*, 328–334.
72. Deng, L.; Shen, L.; Zhang, J.; Yang, B.; He, Q.; Hu, Y. *Can. J. Chem.* **2007,** *85*, 938–944.
73. Zeng, L.-F.; Zhang, H.-S.; Wang, Y.-H.; Sanchez, T.; Zheng, Y.-T.; Neamati, N.; Long, Y.-Q. *Bioorg. Med. Chem. Lett.* **2008,** *18*, 4521–4524.
74. Marastoni, M.; Scotti, A.; Trapella, C.; Ferretti, V.; Sforza, F.; Gavioli, R. *J. Pept. Sci.* **2014,** *20*, 258–265.
75. Wang, X.; Yang, L.; Xue, P.; Ruan, J. *Lett. Org. Chem.* **2017,** *14*, 39–42.
76. Strocchi, E.; Fornari, F.; Minguzzi, M.; Gramantieri, L.; Milazzo, M.; Rebuttini, V.; Breviglieri, S.; Camaggi, C. M.; Locatelli, E.; Bolondi, L.; Comes-Franchini, M. *Eur. J. Med. Chem.* **2012,** *48*, 391–401.
77. Thalassitis, A.; Katsori, A.-M.; Dimas, K.; Hadjipavlou-Litina, D.-J.; Pyleris, F.; Sakellaridis, N.; Litinas, K. E. *J. Enzyme Inhib. Med. Chem.* **2014,** *29*, 109–117.
78. (a) Tan, W.; Zhu, X.-T.; Zhang, S.; Xing, G.-J.; Zhua, R.-Y.; Shi, F. *RSC Adv.* **2013,** *3*, 10875–10886. (b) Filatov, A. S.; Knyazev, N. A.; Ryazantsev, M. N.; Suslonov, V. V.; Larina, A. G.; Molchanov, A. P.; Kostikov, R. R.; Boitsov, V. M.; Stepanov, A. V. *Org. Chem. Front.* **2018,** *5*, 595–605. Available from: https://doi.org/10.1039/c7qo00888k.

79. Kudryavtsev, K. V.; Yu, C. C.; Ivantcova, P. M.; Polshakov, V. I.; Churakov, A. V.; Brase, S.; Zefirov, N. S.; Guh, J.-H. *Chem. Asian J.* **2015,** *10*, 383–389.
80. Yu, B.; Sun, X.-N.; Shi, X.-J.; Qi, P.-P.; Zheng, Y.-C.; Yu, D.-Q.; Liu, H.-M. *Steroids* **2015,** *102*, 92–100.
81. Dong, H.; Song, S.; Li, J.; Xu, C.; Zhang, H.; Ouyang, L. *Bioorg. Med. Chem. Lett.* **2015,** *25*, 3585–3591.
82. Harwood, L. M.; Vickers, R. J. *Azomethine Ylides. Synthetic Applications of 1,3-Dipolar Cycloaddition Chemistry Toward Heterocycles and Natural Products;* Wiley and Sons, Inc.: Hoboken, NJ, 2003169–252.
83. Pyne, S. G.; Tang, M. *Curr. Org. Chem.* **2005,** *9*, 1393–1418.
84. Ichikawa, Y.; Narita, A.; Shiozawa, A.; Hayashi, Y.; Narasaka, K. *J. Chem. Soc. Chem. Commun.* **1989,** 1919–1921.
85. Piens, N.; De Vreese, R.; De Neve, N.; Van Hecke, K.; Balzarini, J.; De Kimpe, N.; D'hooghe, M. *Synthesis* **2014,** *46*, 2436–2444.

Chapter 3

Drug delivery and cycloaddition reactions

Daniele Merli
Department of Chemistry, University of Pavia, Pavia, Italy

General aspects

The synthesis of nano- and microstructured shuttles used for drug delivery purposes is an innovative research field that has led to the development of successful strategies for the management of otherwise difficult-to-treat diseases. This is particularly evident in the field of chemotherapy, where the introduction of nanovehicolated drugs has resulted in a reduction of the side effects (or at least a modification in the side effect profile) and an increase in the therapeutic index; sometimes a neat cross of the resistance of the neoplastic cells toward the classical chemotherapeutics is observed. In this sense, without the aim of being exhaustive, we can cite just two examples that are considered a commercial success: albumin-loaded paclitaxel (Abraxane) and liposomal doxorubicine (Caelix among others). These two antineoplastics are widely used in therapy, despite the higher cost with respect to the nonvehicolated drug, due to their improved pharmacokinetic and pharmacodynamic profile.

The profit of the described strategy (i.e., the development of drug delivery systems) is obviously not limited to the field of antineoplastic drugs, but has a wider significance: for example, the newly born field of gene therapy and gene delivery (along with the delivery of nucleic acids fragments, such as small interfering RNA (siRNA), and proteins' fragments interfering with genic expression of the cell) is constantly searching for a useful shuttle to deliver the active principle ingredient. Clearly, the major interest is devoted to the development of drug delivery systems capable of improving the prognosis of widely diffuse and scarcely responding diseases, such as some cancers, although efficient drug delivery systems could be seen with favor in the case of less invasive drugs (e.g., antalgics) or more benign pathologies, and also in the cosmetic and cosmeceutic fields.

For these reasons, efficient synthetic methods should be perfected to obtain the drug delivery systems and eventually for the conjugation of the

Modern Applications of Cycloaddition Chemistry. DOI: https://doi.org/10.1016/B978-0-12-815273-7.00003-4

drug with the shuttle. In this regard, it should be pointed out that, for a drug delivery system being introduced in therapy, the synthetic procedure should be high-yielding, as green as possible, easily scalable, and tolerant toward the functional groups present in the selected drug.

So, the use of synthetic strategies based on cycloaddition reaction is particularly benign, as it is completely bioorthogonal with respect to other synthetic strategies that can be used in the construction of the final drug-loaded shuttle, and it is tolerant toward different functional groups and in some case high yielding. Among these click processes, Cu(I)-catalyzed azide-alkyne cycloaddition (CuAAC) has attracted the most attention based on its high orthogonality, reliability, and experimental simplicity for nonspecialists. A renewed interest in the use of efficient classical transformations has been also observed (e.g., thiolene coupling, Michael addition, Diels–Alder).[1] Clearly, click chemistry itself is highly desirable as it is enantioselective, including simple reaction conditions (ideally tolerant to oxygen and water) with no use of chromatographic methods[2]; of course, for use as a drug delivery system, the final product obtained must be stable under physiological conditions. Despite this, it must be considered that despite all the exempla reported in the following and that involve the synthesis of polymeric materials, it often requires a very dispendious purification step that requires prolonged dialysis and a further lyophylization step. The dialytic step is particularly time-consuming, as it requires long time (one or more days), but of course is independent from the use of the click chemistry procedure, as it is required whenever these polymers are synthesized; this aspect is only reported here and not further discussed in every procedure reported in this chapter, as the procedure is similar in all cases.

In the case of copper-promoted click chemistry, a further purification step is often required; in fact, copper can interfere with cell proliferation and cell viability, so that it must be removed from the synthesized vehicle, for example, by using chelating resins.

Contamination of the obtained products by different metals or by organic impurities is often a problem with all the synthetic procedures, and can be ascribed to the use of suboptimal purity reagents, catalyst, and solvents. This is an absolutely fundamental issue to be considered, as biological test results can be completely affected by contaminated products; moreover, the degree of contamination can vary from batch to batch of the prepared material, so fluctuating results in viability and/or biological responses often found their explanation in the contamination of the drug delivery system. Chemists often neglect this aspect, and biologists are unconscious of this possibility: a perfect mix to have biased results.

Sterility is also a crucial point when the drug delivery systems are used for in vitro/in vivo studies, although this aspect is curiously rarely considered. In fact, bacterial or fungal contamination of biocompatible vectors is often reported in laboratory practice, and results of the tests can be profoundly altered by this type of contamination. Sometimes the bacterial proliferation is clearly detectable during the test, sometimes it is not, but in all cases it results in time- and money-wasting procedures, as it is mandatory to reprepare the experiments.

One desired characteristic that can be attained following the reported strategy is to produce stimuli-responsive vehicles that can discharge their load (i.e., the drug) upon a physiological or external stimulus [e.g., light, infrared (IR) irradiation, presence of reducing agents][3] or that act as tunable and/or sustained release systems,[4] so that a targeted release only near the organ/tissue needing the pharmaceutical agent is reached; eventually, the pharmacokinetics of the parent drug in this way can be advantageously modified (i.e., a minor number of administration is needed if a sustained release product is used, and moreover the plasma concentration is more constant).

Regarding the chemical/pharmacological aspects of the proposed vehicles, of particular importance is the loading efficiency (i.e., the amount of drug that can be loaded onto the vehicle), as the quantity of the drug loaded must be significant to prevent a huge amount of adduct to be administered. Moreover, the vehicle should be nontoxic and biodegradable or at least eliminable from the body, not accumulating in it. At this proposal, inorganic-based vehicles pose serious problems in terms of toxicity and accumulation, in particular if based on nanometric silica or similar structures. Of course, an easy synthetic procedure is mandatory.

In the following, a selection of the more recently published synthetic strategies based on cycloaddition reactions is reported. The classification is based on the therapeutic indication of the drug loaded more than on the synthetic strategies of the shuttle. The list is far from being complete, as a huge amount of publication is produced every year in this field. The preference is accorded to delivery systems that are nanostructured, and whose impact in this research field seems to be higher.

Nucleic acids delivery

Nonviral-based vectors are promising transporting systems for the intracellular delivery of therapeutic gene sequences, and they can deeply influence the success of a gene therapy. However, the associated drawbacks like low transfection rate, significant toxicity, and poor targetability require further improvement. Here, the most recent proposals are reported.

Inorganic structures

Inorganic structures such as carbon nanotubes (CNTs) are traditionally modified by cycloaddition reactions, and the functionalization of CNTs with DNA molecules can be used in gene transporters. The nucleic acid molecule can be loaded onto the structure by π-stacking between the aromatic bases and the CNTs' surface. The interaction between DNA and CNTs is between anionic groups of DNA, with cationic groups on CNTs upon chemical modification of their surface, resulting in superior biostability. The combination of CNT with DNA can enhance the expression of modified DNA and provide a mechanism for cellular uptake, as reported in literature.[5] Despite this, the poor biocompatibility of CNTs, their potential toxicity (i.e., only slightly reduced by their functionalization), the poor excretion of the vast majority of the CNT derivatives that lead to a bioaccumulation in the organism, strongly preclude the possibility of seriously using these carbon structures as drug delivery systems, and in general the use of CNTs in biomedical applications is expected to decrease in the coming years due to these concerns.

Organic structures

Sugar-based structures

Cyclodextins (CDs) allow the covalent attachment of a precise number of biorecognizable ligands in precise positions of the CD core. The commercial availability of the native CDs in three different sizes, their axial symmetry, and the possibility of position- and face-selective functionalization allow strict control of the valency and spatial orientation of the recognition motifs (glycotopes) in low, medium, high, and hyperbranched glycoclusters, including glycodendrimer-CD hybrids. Click-type ligation chemistries, including CuAAC, thiol-ene coupling or thiourea-forming reactions, have been implemented to warrant full homogeneity of the adducts. Nanometric glycoassemblies have the capability of adapting the intersaccharide distances and orientations in the presence of a receptor partner and are able to mimic the fluidity of biological membranes.[6] A major requirement for those channels is the design of a sufficiently diverse ensemble of compounds for parallel evaluation of their capabilities to condense DNA into transfection nanoparticles (NPs) where the gene material is protected from the environment, and in this sense CD offers the same advantages and poses the same problems as calixarenes.[7]

For this purpose, CDs having amphiphilic characteristics have been prepared by implementing two click methodologies, namely cuprous cation catalyzed azide-alkyne cycloaddition (CuAAC) and thiourea forming reaction. The use of solid-supported Cu(I) catalysts was found to be well suited for multiple reaction sequences, while facilitating purification of the adduct. More in detail, a modification strategy example is reported in Scheme 3.1. After the esterification of the hydroxylic group of the hepta-(C-6)-azide **1** to produce heptakis(6-azido-6-deoxy-2,3-di-*O*-alkanoyl)cyclomaltoheptaose **2**, this last compound was reacted with bis[2-(*N*-tert-butoxycarbonylamino) ethyl]propargylamine to afford an 1,4-substituted triazole cluster **3** in yields ranging from 79% to 95% (depending on the alkanoyl chain length). Finally, trifluoroacetic acid-promoted hydrolysis of the carbamate groups provided

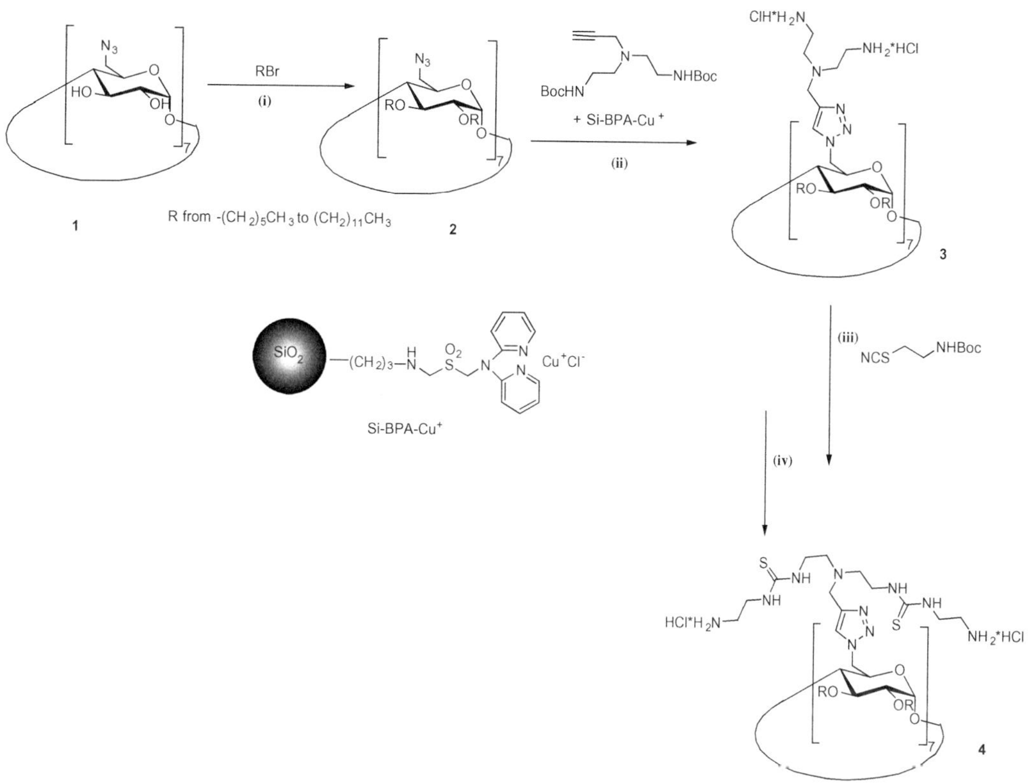

SCHEME 3.1 Synthesis of the click triazole-thiourea adduct using the silica-supported heterogeneous catalysts Si-BPA-Cu^+. (i) TBAI, *tert*-BuOH: H_2O 3:1, reflux, 8 h, yield: 45%–50%; (ii) (1) CuAAC (Cu-supported catalyst): *tert*-BuOH:H_2O 2:1, reflux, 12 h, yield: c. 80%–90%, then (2) TFA:DCM 1:1,2 h, r.t., followed by (3) freeze-drying from a solution of 0.1N aq. HCl, yield (2 + 3): 95%–99%; (iii) DCM, Et_3N, r.t, 16 h, yield: 50%; (iv) (1) TFA:DCM 1:1, 2 h, r.t., then (2) freeze-drying from a solution of 0.1N aq. HCl, yield (1 + 2): 99%. *DCM*, Dichloromethane; *TBAI*, tetrabutylammonium iodide; *TFA*, trifluoroacetic acid; *Et_3N*, triethylamine.

the corresponding polycationic click cluster, which was reacted with a linear thiocyanate (2-(*N*-tert-butoxyaminocarbonyl)ethyl isothiocyanate) to obtain **4** after deprotection of the Boc groups. The strategy is compatible with molecular diversity–oriented approaches, which has been exploited to generate a small library of click polycationic amphiphilic CDs for assessing the influence of structural modifications in the ability to complex, compact, and efficiently load plasmidic DNA (pDNA) into cells and promote transfection.[8]

Another research group underlined that β-CD and calix[4]arene (CA4) vectors with facial amphiphilic character and an hydrophobic domain, formed by hexanoyl or hexyl chains, self-assemble in the presence of pDNA through a process that involves the formation of βCD or CA4 bilayers between the oligonucleotide filaments. Spherical transfectious NPs are thus obtained following a synthetic strategy that employs cycloaddition chemistry at the core of the synthetic procedure. Biological transfection evaluation evidenced the importance of having primary amino groups in the vector to warrant high levels of transfection, while the optimal platform (βCD or CA4) strongly depends on the cell type.[7]

This strategy was also adopted in a slightly modified way, synthesizing hepta-guanidino-β-cyclodextrin (G-CD), its hepta-PEG conjugate (G-CD-PEG), and the corresponding anisamide terminated PEG conjugate (G-CD-PEG-AA) and comparing them as delivery vectors for siRNA to prostate cancer cells and tumors in vivo, in which the anisamide targets the sigma receptor. Results were promising both in vitro and in vivo.[9]

A multifunctional polymeric carrier for gene and drug codelivery, a new cyclodextrin derivative containing poly(L-lysine) dendrons **7**, was prepared by the click conjugation of per-6-azido-β-cyclodextrin **6** with propargyl focal point poly(L-lysine) dendron **5** of third generation, in the presence of copper sulfate and sodium ascorbate (Scheme 3.2). It was found that such a conjugate (**7**) could form colloidally stable nanocomplexes with plasmid DNA in an aqueous system and exhibited high gene transfection efficiency. Moreover, it could load efficiently methotrexate (an antineoplastic) and showed a sustained release behavior. Different from commonly used amphiphilic copolymers with cationic character, the as-obtained shuttle may be used directly for the combinatorial delivery of nucleic acid and lipophilic anticancer drugs without a complicated micellization process.[10]

Remaining in the field of supramolecular assemblies involving dendrones and to better understand their possible role in drug delivery, acetylated-per-azido-β-cyclodextrin was reacted with series of alkyne dendrons (containing 1–5 ethyleneamine units) using copper-catalyzed 1,3-dipolar cycloaddition, to form a series of click clusters bearing 1,2,3-triazole linkers. The macromolecules bind and compact pDNA into spherical NPs in the size range of 80–130 nm. The polycations protect DNA against nuclease degradation, while the cellular uptake profiles demonstrate that all the click clusters efficiently deliver DNA into HeLa and H9c2 cell lines.[11]

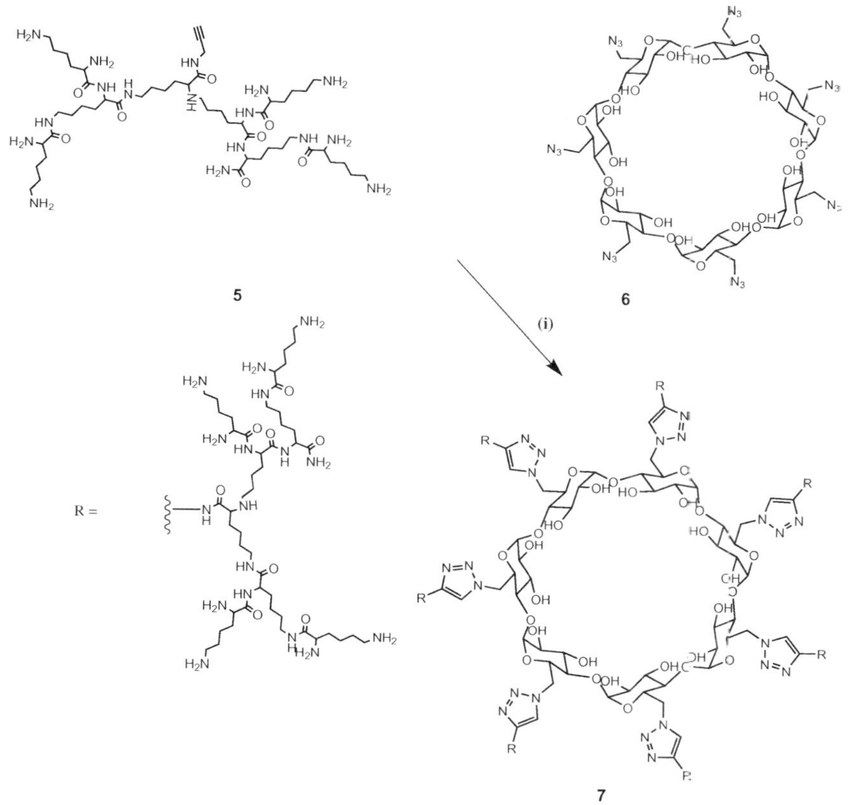

SCHEME 3.2 Synthesis of the dendrone-CD adduct. $CuSO_4$, sodium ascorbate, DMSO; 40°C. 48 h, yield: 65%. *CD*, Cyclodextrin.

In another study involving dendrons, six analogs of a trehalosepentaethylenehexamine glycopolymer (compound **11** as depicted in Scheme 3.3 for a chosen example) that contain, among others, alkynyl-oligoethyleneamine **9** and azido trehalose **8** were synthesized, and evaluated the effects of polymer end group on the ability of these systems to bind, compact, and deliver pDNA to cultured HeLa cells. The polymers were synthesized in one-pot azide-alkyne cycloaddition reactions with an adaptation of the Carothers equation for step-growth polymerization to produce a series of polymers with similar dispersivity in molecular weight. An excess of end-capping monomer (**10**) was added at the end of the polymerization to maximize functionalization efficiency. The polymers were all found to bind and compact pDNA; the study shows that very subtle structural changes in polymer chemistry, such as end groups, can yield very significant differences in the biological delivery efficiency and transgene expression.[12]

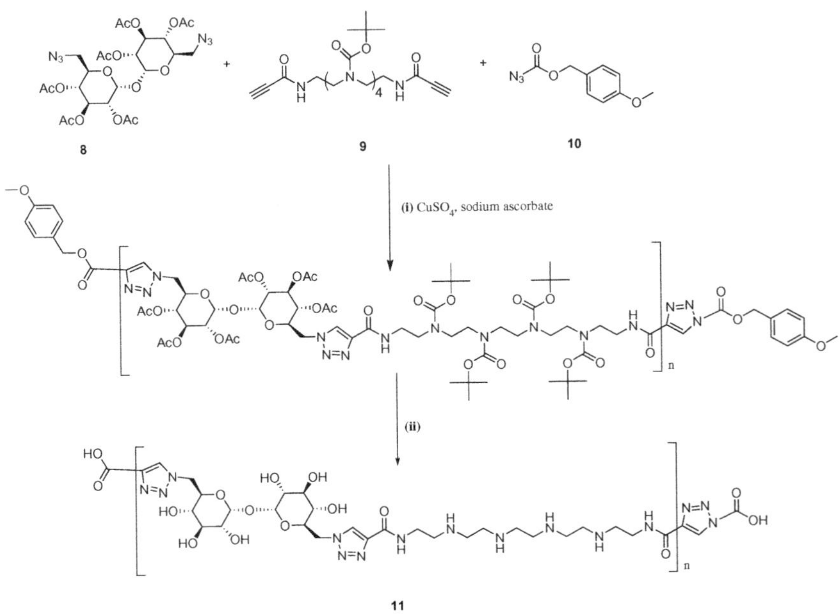

SCHEME 3.3 Synthesis of a chosen example of the dendrone-threalose adducts by one-pot azide-alkyne cycloaddition reactions. (i) Copper sulfate, sodium ascorbate, DMF, 70°C, 24 h; (ii) (1) NaOMe, MeOH, 18 h, r.t. (to remove acetyl groups/4-methoxybenzyloxycarbonyl groups) then (2) HCl 4N, dioxane, 18 h (to remove Boc groups). Yield (i + ii) 13%. *NaOMe*, Sodium methoxide.

As well known, dendrones can be made with very different building blocks; among them, poly(amidoamine) dendrons look very promising, as they can, for example, be conjugated onto biocompatible dextran by a CuAAC.

For resultant dextran conjugates with various generations of poly(amidoamine) dendrons, their characteristics for the complexation with plasmid DNA were investigated and their complexes with plasmid DNA were studied with respect to their transfection efficiency in human embryonic kidney (HEK293) cell lines. In the case of a higher generation of poly(amidoamine) dendrons, dextran conjugate was found to have much lower cytotoxicity and better gene delivery capability when compared to branched polyethylenimine (25 kDa), a commonly used gene vector.[13]

Nonsugar structures

Lipid formulations are considered a good alternative to viral vectors for DNA cell transfection, and among them a lipid shuttle was obtained by mixing two distinct cationic surfactants: namely, tocopheryl succinate–based cationic lipid and 1,12-dodecane-based bolaamphiphile. The adduct proved to be internalized by different cell types following the clathrin-mediated endocytosis.

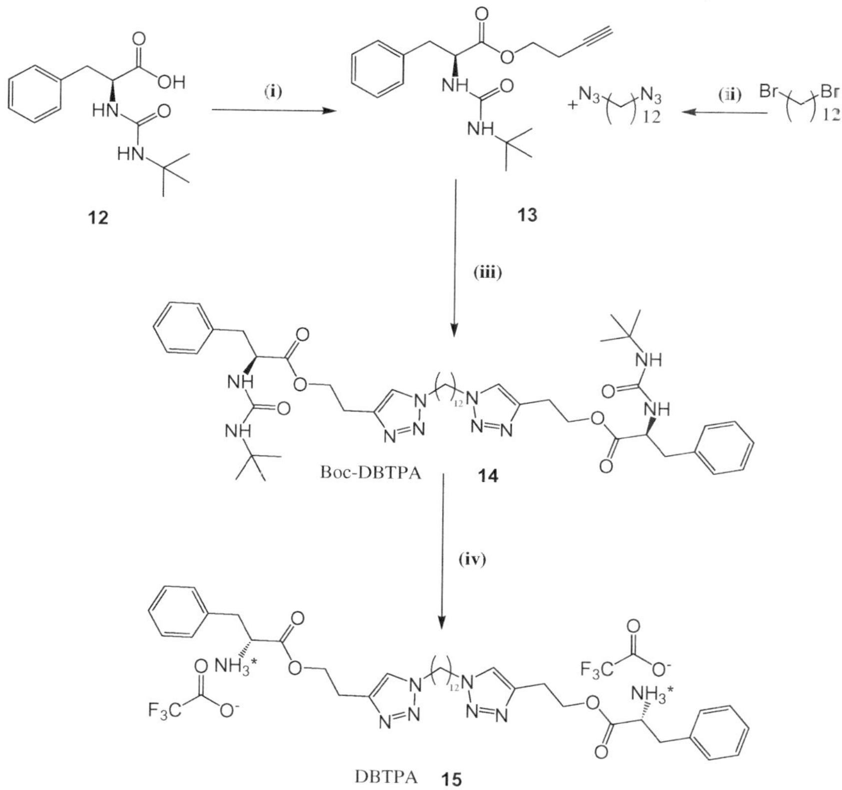

SCHEME 3.4 Reagents and conditions: (i) but-3-*yn*-1-ol, DCC, DCM, DMAP, r.t., 24 h, yield: 92%; (ii) NaN_3, acetonitrile, 80°C, 16 h, yield: 96%; (iii) $CuSO_4$, sodium ascorbate, THF:H_2O: tert-BuOH 2:2:1, r.t, 24 h, yield: 92%; (iv) dry DCM:TFA 2:1, r.t., 12 h, yield: quantitative.

These aggregations showed unilamellar vesicle morphology and could transfect efficiently different varieties of cell lines, competing with the gold standard, Lipofectamine 3000 (L3K).

In details, dodecane-based cationic symmetrical bolaamphiphile DBTPA (**15**) was synthesized by following the steps mentioned in Scheme 3.4, which include the synthesis of the acetylene derivative **13**, starting from Boc-L-phenylalanine **12** and esterifying it with 3-butyn-1-ol with the aid of *N,N*-dicyclohexylcarbodiimide (DCC). Compound **13** was subjected to the cycloaddition reaction with azide functional groups present on either side of 1,12-diazidododecane by following a standard click chemistry protocol ($CuSO_4$, sodium ascorbate, THF:H_2O:*tert*-BuOH 2:2:1, r.t., 24 h), obtaining Boc − DBTPA (**14**). The protected amines were protonated by removing the Boc on acidic conditions to afford bolalipid DBTPA (**15**), having two protonated amines with trifluoroacetate as counter-ion in hydrophilic part and dodecane in hydrophobic region.[14]

Always about lipidic shuttles, poly(α-hydroxy acids) (PAHAs) are a class of biodegradable and biocompatible polymers that are widely used in numerous bioapplications. One drawback of these conventional polymers, however, is their lack of side-chain functionalities, which makes it difficult to conjugate active moieties to PAHA or to fine-tune the properties of PAHA-derived materials through side-chain modifications. However, it is highly challenging to introduce functional groups into conventional PAHAs derived from ring-opening polymerization (ROP) of lactides and glycolides to yield functional PAHAs with favorable properties, and amino acids and their *O*-carboxyanhydrides (OCAs) derivatives, a class of five-membered cyclic anhydrides are excellent resources for functional polymers because of their low cost, availability, and structural as well as stereochemical diversity. The ready availability of starting materials from renewable resources and the facile postmodification strategies such as azide-alkyne cycloaddition and the thiol-yne click reaction have enabled the production of a multitude of PAHAs with controlled molecular weight, narrow polydispersity, high terminal group fidelities, and structural diversities that are amenable for self-assembly and bioapplications.[15]

Another approach for gene delivery makes use of a polymer able of self-assembly in water, that effectively resists premature serum degradation and is able of releasing its nucleic acid cargoes upon a pH change, and is based on polyethylene glycol moiety instead of lipidic chains. In fact, some authors described a pH-sensitive polymer, poly(oligo(ethylene glycol) monomethyl ether methacrylate)-*co*-poly(2-(dimethylamino)ethyl methacrylate)-block-poly(propargyl methacrylate-graft-propyl-(4-methoxy-benzylidene)-amine) (p(PMA)-*b*-p(OEGMA-DMAEMA)-R (**20** in Scheme 3.5) for in vitro and in vivo gene transfer.

To synthesize the polymers, poly(trimethylsilyl)propargyl methacrylate (p(TMS-PMA)), **17**, was first prepared via RAFT polymerization of (trimethylsilyl)propargyl methacrylate **16** with cyanopentanoic acid dithio-benzoate, followed by consecutive RAFT polymerization with methyl ether methacrylate (OEGMA) and 2-(dimethylamino)ethyl methacrylate (DMAEMA) to obtain the block copolymer, p(TMSPMA)-*b*-p(OEGMA-DMAEMA) **18**. The trimethylsilyl protecting groups were then removed to obtain pPMA-*b*-p (OEGMA-DMAEMA) **19**. The benzoic imine moiety on the side chain of pPMA-*b*-p(OEGMA-DMAEMA) to obtain pPMA-*b*-p(OEGMA-DMAEMA)-R **20** was introduced by click chemistry, using CuBr as a catalyst (40°C, 24 h, c. 80% yield).

In physiological condition, the hydrophobization of p(OEGMA-DMAEMA) polycations by p(PMA) significantly enhanced the stability of its polyplexes counterpart. In endosomes, the polymer undergoes an acid-triggered hydrophilic transition through the cleavage of benzoic imines, thus

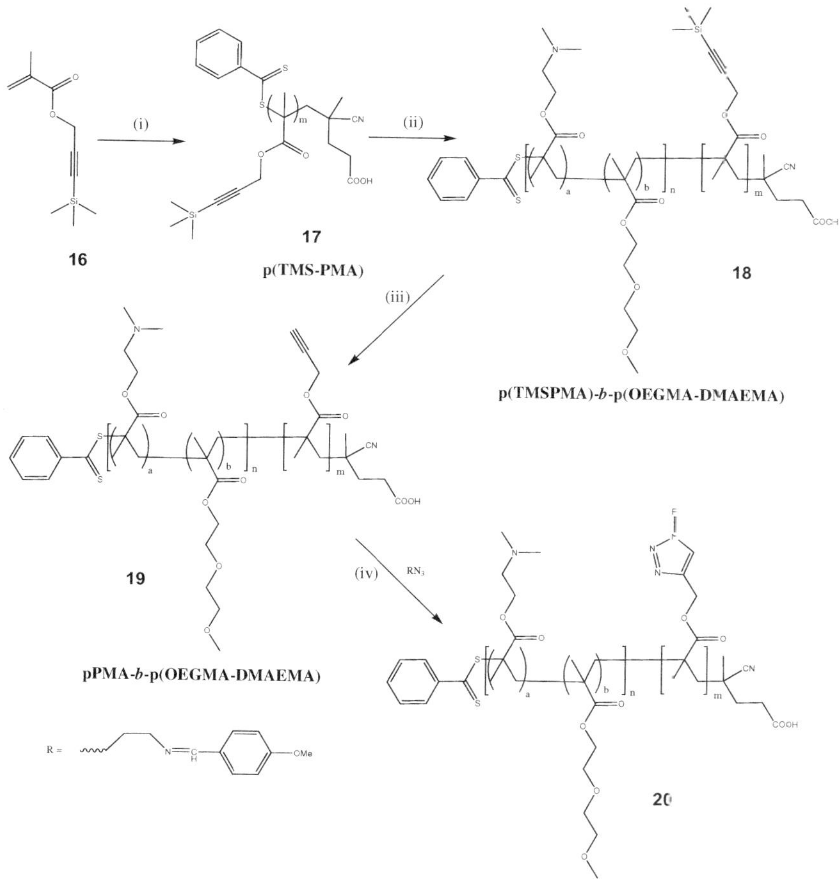

SCHEME 3.5 Scheme of synthetic routes of p(PMA-PMBA)-*b*-p(OEGMA-DMAEMA) **20**. (i) CPABD, AIBN, toluene; 60°C, 8 h, yield: 75%; (ii) OEGMA + DMAEMA, AIBN dioxane; 60°C, 20 h, yield: 70%; (iii) TBAF, acetic acid −15°C 30 min, then r.t. 12 h, yield: 60%; (iv) CuBr, DMF, 40°C, 24 h, yield: 80%. *AIBN*, Azobisisobutyronitrile; *CPABD*, cyanopentanoic acid dithio-benzoate; *DMAEMA*, 2-(Dimethylamino)ethyl methacrylate; *OEGMA*, methyl ether methacrylate; *TBAF*, tetrabutylammonium fluoride.

allowing the vector to quickly release nucleic acid cargo due to the loss of hydrophobic functionalization.[16]

Structure–activity relationships of a series of linear cationic click polymers with various structures that remember PEG units for their structures, were also investigated for in vitro gene delivery and in vivo gene transfer. The experimental results revealed that the minimal structure variation could result in a crucial effect on DNA-binding ability, buffering capacity, and the cellular delivery capacity of polymer, all of which brought about the effects on their transfection efficiencies. The polymers were synthesized from diazide monomer containing bis-ethylenediamine unit and dialykene monomer containing bis-ethylene glycol unit, and showed high in vitro and in vivo transfection efficiency.[17]

Although these positively charged polymers are described as endowed with cytotoxic action, some authors explored the possibility of using cationic polymer lacking significant toxicity due to its tridimensional arrangement; specifically, PEIs are promising gene delivery vectors due to their inherent ability to condense genetic material and successfully affect its transfection. A set of cyclic PEIs that showed reduced toxicity relative to the linear or branched PEI and were highly efficient in gene transfection was synthetized.[18] The low toxicity of cyclic cathionic polymers with respect to the linear counterpart was also underlined by another study, in which the architecture of polycation gene carriers has been shown to affect both their transfection efficiency and cytotoxicity. The work reports the synthesis of cyclic polycations and their use for gene transfer to mammalian cells. Cyclic poly-((2-dimethylamino) ethylmethacrylate) (pDMAEMA) homopolymers of various molecular weights were synthesized by intrachain click cyclization of α-alkyne-ω-azide heterodifunctional linear precursors prepared by atom transfer radical polymerization (ATRP), following a synthetic route similar to that illustrated in Scheme 3.5. Cyclic polymers formed more compact particles with plasmid DNA compared to linear analogs. In general, cyclic polymers complexed and delivered nucleic acids with efficiencies similar to their linear counterparts, despite their reduced toxicity.[19]

Cycloaddition reactions could be used also for the preparation of nanostructured gels, and the development of these nanogels (as nanoscale multifunctional polymer-based matrixes) is really interesting, due to the fact that the size of an object strongly influences the interactions with living cells; in fact, NPs could be subjected to active or passive cellular uptake only within a defined range. For this reason, some authors proposed two methods to synthesize rhodamine-modified nanogels in order to produce nanostructures that could be traced during the cellular interactions and internalization and that are also suitable as carriers for drugs or genes. The approach is complementary of what was stated earlier, as mixed structures composed of polyethylene glycol (PEG) units and PEI units were created. In detail, PEG was bis-activated by reaction with 1,1′-carbonyldiimidazole, while PEI was reacted with propargyl bromide obtaining the alkyne modified polymer (Fig. 3.1). The nanogel, able to compact DNA due to the positive charges given by PEI, was prepared using the modified emulsification-evaporation method, while rhodamine bearing an azide group was reacted with alkyne modified PEI via a click chemistry reaction catalyzed by copper iodide and sodium ascorbate, before (path A in Fig. 3.1) or after gelation (path B in Fig. 3.1). In both cases, the reaction took 24 h at 50°C (water as solvent; yield not reported).[20]

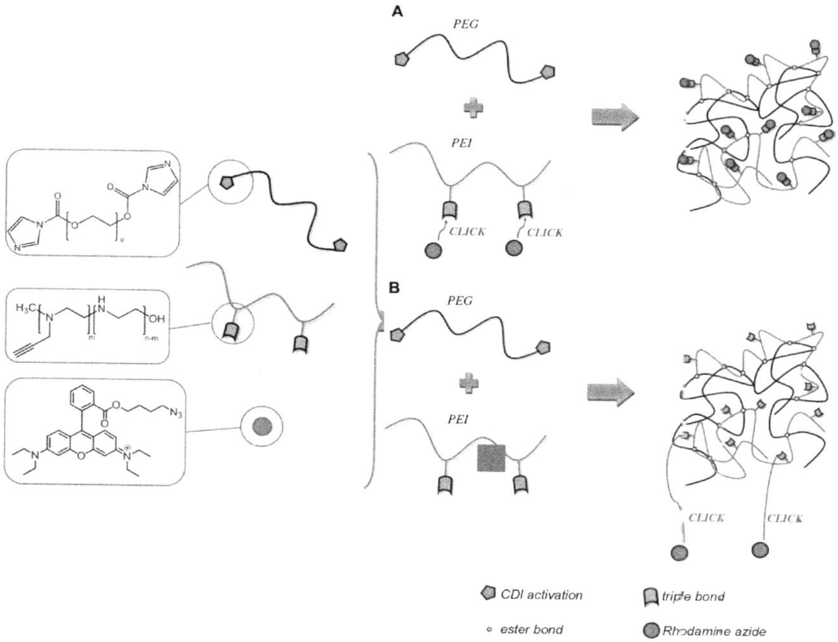

FIGURE 3.1 Representation of the two procedures to synthesized rhodamine-modified nanogels PEG-PEI. *PEG*, Polyethylene glycol; *PEI*, poly(ethylene imine). *Mauri, E.; Moroni, I.; Magagnin, L.; Masi, M.; Sacchetti, A.; Rossi, F.* React. Funct. Polym. ***2016***, 105, *35–44*.

Small interfering RNA delivery

A particular interest is reported in recent times toward the delivery to living organisms of the so-called siRNA, which are RNA fragments that can modulate gene expression by causing gene silencing or inducing transcription when properly introduced in the cell nucleolus. This is a promising treatment not only for cancer, but for a plethora of diseases in which the expression of a gene is pathologically altered. Despite this, the difficulty of efficiently transporting the siRNA sequence in the nucleolus and the complex relationship that can prevent the drug to efficiently modulate gene expression has resulted in no siRNA-based drug approved as of 2018. A similar gene therapy approach based on an antisense oligodesoxyribonucleotide (oblimersen) produced a similar failure and disappointed hopes after some FDA clinical trials, teaching us that there is a long and winding road from chemical success to clinical success. Moreover, despite the broad applicability of the Huisgen cycloaddition reaction, the click functionalization of RNAs with peptides still remains a challenge. In another work, some authors describe a straightforward method for the click functionalization of siRNAs with peptides of different sizes and complexities. In fact, some peptides have pronounced affinities for selected neoplastic cell lines due to overexpression of their receptor on them, and can be used as shuttles to deliver into these cells particular siRNA sequences. More specifically, it is well known that some integrins (specially, the $\alpha_v\beta_3$ subfamily) recognize the tripeptide motif RGD (arginyl-glycyl-aspartic acid) with high affinity and selectivity, offering the possibility of using RGD-containing peptides to deliver drugs into specific cancer cell types overexpressing them. Another candidate for tumor-drug delivery is octreotide, a potent synthetic cyclooctapeptide agonist of the endocrine hormone somatostatin that displays high affinity for the somatostatin subtype-2 receptor, which is very prevalent in tumor cells. Similarly, several attempts to overcome the limitations associated with antibody-based anti-HER2 therapies have led to the development of several small antigen-mimetic peptides of great therapeutic potential. Among them, a peptide carrier based on a cyclic anti-HER2 peptide (AHNP) linked to a Tat cell-penetrating peptide (Tat-AHNP) has been reported to enter HER2 + breast cancer cells with high efficiency. For this purpose, a series of these promising peptidic carriers based on the just reported principles for the selective siRNA delivery into HER2 + breast cancer cell lines has been reported, as schematically illustrated in Fig. 3.2.

In detail, azido-modified peptides, whose aminoacidic structures are reported in Fig. 3.2, are reacted with a 5′-alkynyl-modified sense siRNA strand targeting the endogenous gene HER2 (i.e., able of silencing this precise gene). To achieve the RNA 5′-end conjugation, a deoxyuridine phosphoramidite with an octadiyne handle at C5 (5oU) was used during RNA synthesis. The click reactions take place in presence of copper sulfate and sodium ascorbate: conversion was quantitative within 90 min and the conjugate was obtained as a single product in 35%–45% isolated yield, depending on the adduct considered.[21]

Apart from peptides, nonpeptidic charged shuttles also were tested for siRNA delivery, using a novel collaborative assembly approach via a combination of an electrostatically driven physical assembly and a facile click reaction-mediated chemical assembly, which showed some advantages over the conventional approach that is only based on the physical assembly. This strategy exploits the cationic residual charges of a lipid-based complex to load siRNA with a high capacity. The direct chemical modification via click chemistry of a model polyanion, hyaluronic acid (HA), that was used to fabricate the complex partially shielded the positive charge of complex without affecting the siRNA binding, in this way reducing the toxicity and enhancing the blood stability of the complex.

The obtained lipid-polymer hybrid NP (RSC-HA) using collaborative assembly presented great in vivo stability. After endocytosis into the cells, the protection of RSC-HA on siRNA turned off, while the release of siRNA induced by the intracellular signals produced an efficient gene silencing.[22]

Similarly, the use of a redox-marked siRNA delivery system was also explored, ideally allowing the targeting of cells with a reductive environment

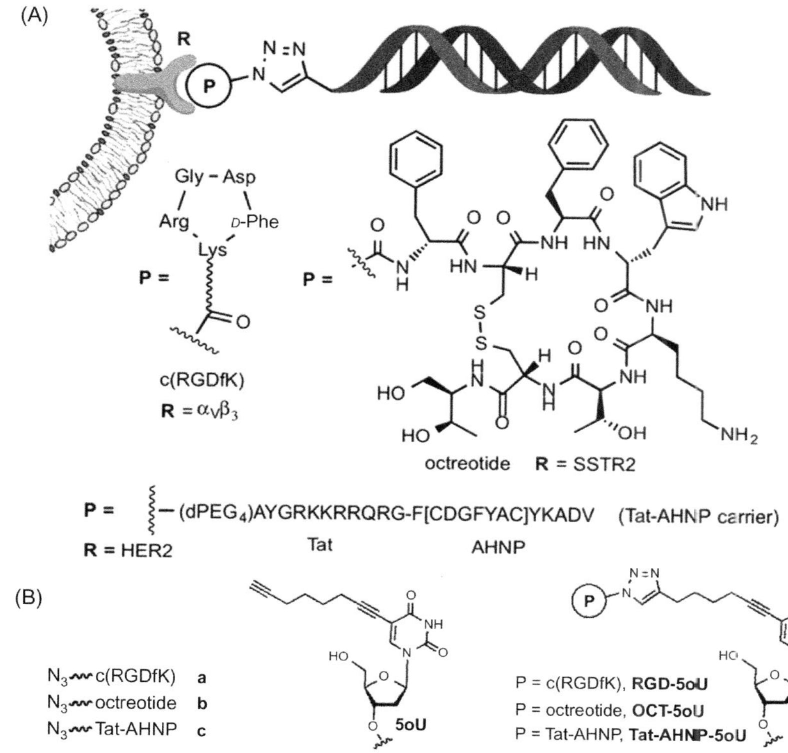

FIGURE 3.2 (A) siRNA–peptide conjugates used in the study. (B) Detail of clickable 5-alkynyl-sense RNA: the deoxyuridine with an octadiyne handle at C5 (5oU); azido peptides (a,b,c) and siRNA–peptide (P) click conjugates targeting the endogenous gene HER2 + are also reported. R = receptor type to which the corresponding peptide binds. The peptide sequences are reported with one-letter notation (i.e., each amino acid corresponds to one letter). For other abbreviations see the text. *siRNA*, small interfering RNA. *Reproduced from Gandioso, A.; Massaguer, A.; Villegas, N.; Salvans, C.; Sanchez, D.; Brun-Heath, I.; Marchan, V.; Orozco, M.; Terrazas, M.* Chem. Comm. ***2017***, 53, *2870–2873, with permission of The Royal Society of Chemistry.*

(as neoplastic cells often are); for example, a ferrocenium-capped amphiphilic pillar[5]arene (FCAP) was synthesized and self-assembled to cationic vesicles in aqueous solution.

As can be seen in Fig. 3.3, to synthesize FCAP **24**, **21** was first functionalized with sodium azide. Azide-modified pillar[5]arene **22** was subjected to CuAAC (in the presence of copper sulfate and sodium ascorbate) to form pillar[5]arene **23** in a yield of 88%. Oxidation of the ferrocenyl groups on **23** by $FeCl_3$ gave FCAP **24** stoichiometrically, accompanying a polarity shift from hydrophobicity of **23** to amphipathy of FCAP, which could lead to the formation of regular aggregates in water by self-assembling. The cationic

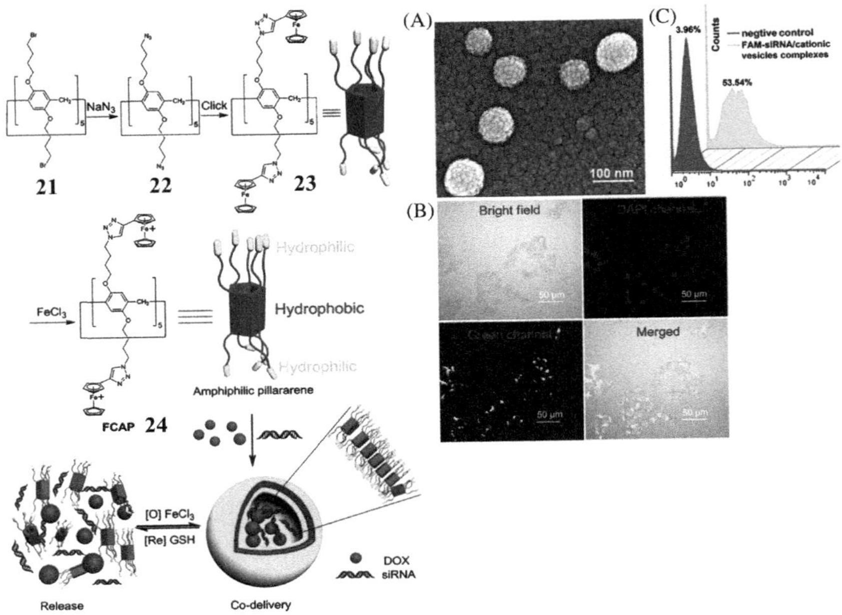

FIGURE 3.3 (Left side) Illustration of the synthesis of FCAP, formation of cationic vesicles, and their redox-responsive drug/siRNA release. (right side) (A) SEM image of FAM-siRNA/cationic vesicle complexes. (B) Flow cytometry analyses of HeLa cells after incubation with FAM-siRNA/cationic vesicle complexes. (C) CLSM images of HeLa cells, after being treated with FAM-siRNA/cationic vesicles complexes; scale bar 50 μm. *siRNA*, Small interfering RNA. *Reproduced from Chang, Y.; Yang, K.; Wei, P.; Huang, S.; Pei, Y.; Zhao, W.; Pei, Z.* Angew. Chem. Int. Ed. ***2014***, *53, 13126–13130, with permission of John Wiley and Sons.*

vesicles, displaying low cytotoxicity and significant redox-responsive behavior due to the redox equilibrium between ferrocenium cations and ferrocenyl groups, allow building an ideal glutathione (GSH)-responsive drug/siRNA codelivery system for rapid drug release and gene transfection in cancer cells in which higher GSH concentration exists. The vesicle can be loaded with doxorubicin (DOX) and/or siRNA, as depicted in Fig. 3.3. For siRNA loading, freshly prepared unloaded vesicles were incubated with an siRNA (20 μM) solution. The mixture was mixed and allowed to stand for 20 min before use.[23]

As targeting agents, apart from redox-responsive probes, vitamin derivatives were sometimes used, due to the siRNA transfection ability of B vitamin-based cationic clickable amphiphiles (VBs). The VBs derived from vitamins B2, B3, B5, B6, and B7 formed nanoassembled low molecular weight hydrogelators (LMWGs,). In Scheme 3.6, the synthesis of vitamin B7 LMWG is illustrated. The azide derivative of a cationic lipid (**25**) is

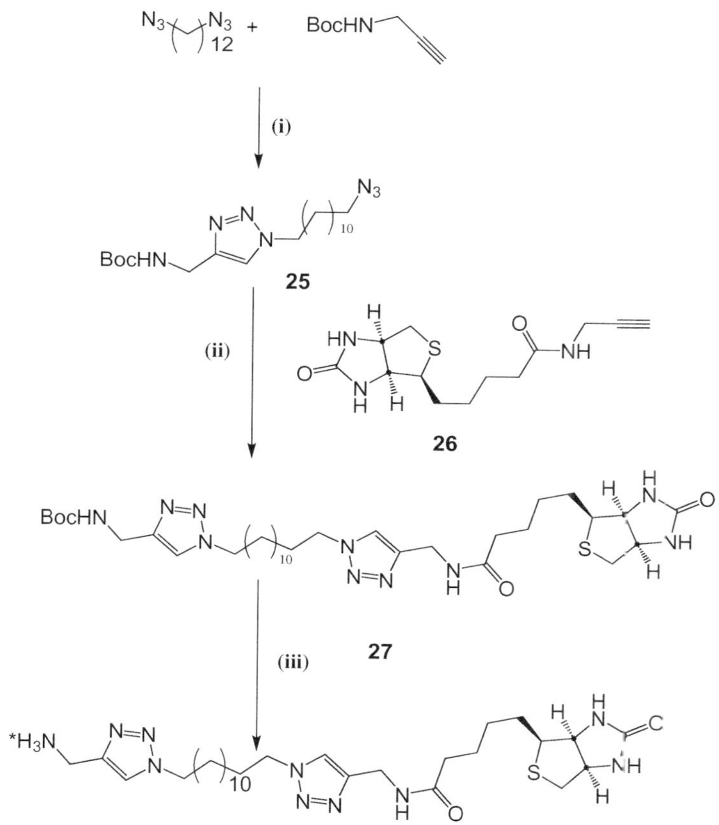

28

SCHEME 3.6 (i) and (ii): $CuSO_4$, sodium ascorbate, THF:*tert*-BuOH:H_2O 2:2:1, 45°C, 48 h (i: yield 96%; ii: yield 71%); (iii) 4M HCl in dioxane, r.t., 8 h, yield: 97%.

synthesized by cycloaddition catalyzed by Cu(II) and ascorbate of dodecandiazide and Boc-protected propargylamine. Coupling of the acetylenated vitamin B7 derivative **26** with **25**, through click chemistry, provided the precursors VB7 **27**; subsequent deprotection of the Boc protecting group provided the desired compounds **28**.

The LMWGs VB2, VB6, and VB7 (derived from vitamins B2, B6, and B7, respectively) facilitated delivery of siRNA, efficiently silencing gene expression specifically into cancer cell lines; in addition, the LMWGs derived from vitamins B3, B5, and B6 were biocompatible. The gene silencing efficiency of vascular endothelial growth factor siRNA delivered by LMWGs was dependent on the nature of the vitamin headgroup, and, interestingly, on the hydrogelation properties of the VBs.[24]

Also vitamin E (VE) derivatives were tested, and a novel type of RNA duplex-bindable molecule with an oligodiaminosaccharide structure was synthesized. These 2,6-diamino-2,6-dideoxy-(1-4)-*b*-D-galactopyranose oligomers (oligodiaminogalactoses; ODAGals) conjugated with α-tocopherol (VE) or a VE analog were designed as novel siRNA-bindable molecules that can be utilized to target liver cells.[25]

Other approaches based on amphiphilic polymers make use of carbosilane, generation 1 to 3 (Scheme 3.7, **30**), and PEG-based dendrons (**29**) functionalized at the periphery with NHBoc groups and at the focal point with azide and alkyne moieties, respectively. The coupling of these two types of dendrons via click chemistry led to the formation of new hybrid dendrimer (**31**) with two distinct moieties: the hydrophobic carbosilane and the hydrophilic PEG-based dendron. The protected dendrimers were transformed into cationic ammonium dendrimers. These amphiphilic dendrimers were studied as vectors for gene therapy against HIV in peripheral blood mononuclear cells. The presence of the PEG moiety afforded lower toxicities and evidenced a weaker interaction between dendrimers and siRNA when compared to the homodendrimer analogs.[26]

Delivery of antineoplastic agents

Much research in the field of therapy of neoplastic disease with cytotoxic agents has focused more on developing new delivery systems to enhance chemotherapy efficacy and minimize side effects than in the search for new therapeutic cytotoxic agents. This is true in the case of compounds having untargeted antineoplastic properties (antimetabolites, alkylating agents, DNA intercalant agents, etc.), because of course there is a continuous search for targeted therapies based on small molecules (e.g., tyrosine kinase inhibitors, such as gefitinib, imatinib, and sunitinib) or on high molecular weight biological compounds (e.g., antibodies, such as adalimumab, trastuzumab, and bevacizumab).

Of all the chemotherapeutic agents, the most studied in search for targeted delivery systems is DOX. The reason for this choice by several authors can rely on the fact that it was already proven that the targeted delivery version of this drug (by liposomes) has a dramatically different pharmacokinetic and toxicity profile, hoping that a further manipulation of the shuttle used for the delivery could result in an even greater success. Moreover, DOX is active against a wide spectrum of tumors, ranging from Hodgkin lymphoma to sarcomas to small cell lung cancer. Of great (synthetic) importance, the quinonic groups of DOX can be very easily conjugated by amino-group bearing moieties by pH-reversible Schiff base formation, that make the use of this specific compound attractive.

However, this approach has produced a plethora of drug delivery systems that were optimized for DOX, but no one much better than the other as

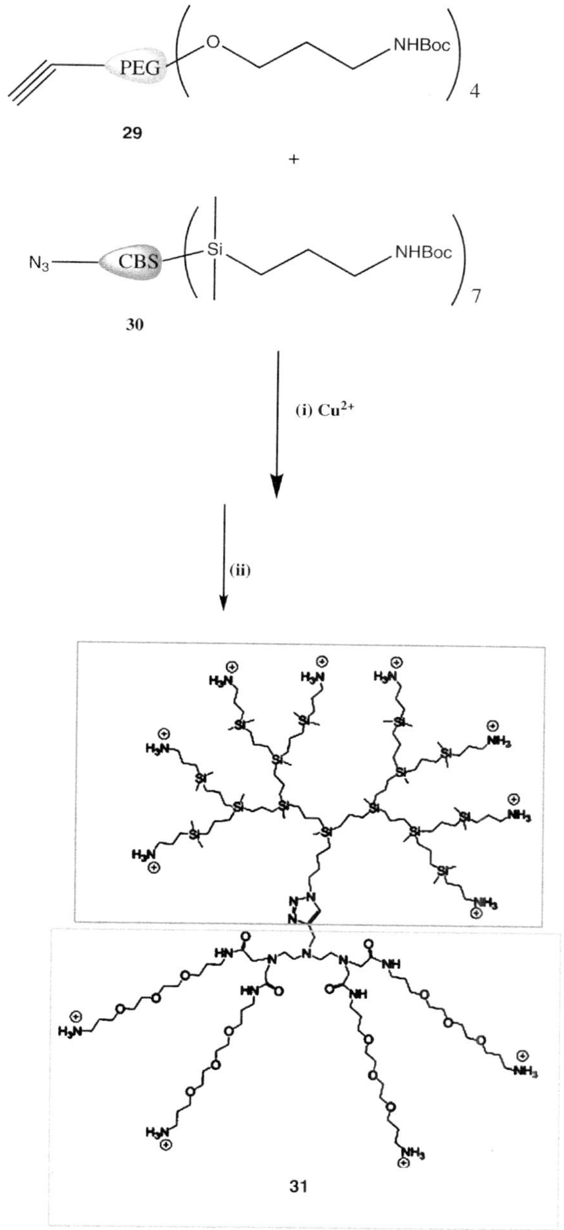

SCHEME 3.7 Synthesis of cationic amphiphilic dendrimers (generation 3) from carbosilane (CBS) dendrons and PEG-based dendron. (i) $CuSO_4$, sodium ascorbate, NEt_3, THF:H_2O (6:2v:v), 16 h, 60°C, yield: 23%; (ii) HCl, DCM:H_2O, 3 h, r. t., yield: 89%.

shuttles, while for a huge amount of other useful therapeutic agents an efficient delivery system was never exploited.

Shifting from antineoplastic to antibiotic drug, for example, the problem is particularly heard: while the argument is only seldom considered by scientists, the problem of multidrug resistant strain of pathogenic bacteria is a world emergency, and the availability of delivery systems able to change the pharmacokinetic and pharmacodynamic properties of known molecules would be a great step forward. Nevertheless, scientists sometimes think erroneously that cancer is a more noble cause to fight, and a great number of the published study related to drug delivery involves this last disease (or more precisely this wide family of diseases), neglecting other important medical conditions.

In the following, of the really enormous number of articles devoted to the drug delivery of DOX (and to a lesser extent to paclitaxel), only the ones considered more innovative, at least from the chemical point of view, are reported, taking into account the cycloaddition reactions as a key step to obtain the vehicle or to conjugate the drug to it.

Doxorubicin delivery

The drug delivery of DOX was attempted using an inorganic/organic hybrid, whose evident drawback is that the silica NP at the core of the shuttle is nonbiodegradable, possibly nonexcreted by the organism once adsorbed and potentially toxic due to its nanoscale structure. As depicted in Fig. 3.4, bis (clickable) mesoporous silica nanospheres (c. 100 nm) were obtained by the cocondensation of tetraethyl orthosilicate (TEOS) with variable amounts (2%–5% each) of two clickable organosilanes (Prec-Alk and AzPTES) in the presence of the cathionic surfactant CTAB (hexadecyltrimethylammonium bromide). Such NPs could be easily functionalized with two independent functions using the CuAAC reaction to transform them into nanomachines bearing cancer cell targeting ligands with the ability to deliver drugs on demand. The active targeting was made possible after anchoring aminopropargyl folic acid by CuAAC click reaction, whereas the controlled delivery was performed by clicked 4′-azido-azobenzene fragments. Indeed, the azobenzene groups are able to obstruct the pores of the NPs in the dark whereas upon irradiation in the UV or in the blue range, their *trans*-to-cis photoisomerization enable the delivery of the cargo molecules. The added value of the folic acid ligand was evidenced by the difference of cell-killing cytotoxicity toward MCF-7cells induced by doxorubicin-loaded NPs under blue irradiation, depending on whether the particles featured the clicked folic acid ligand or not, as depicted in Fig. 3.4.[27] This can reasonably ascribed to the fact that folic acid enable NPs' internalization by the cell, acting as a targeting in DOX delivery. Blue light is partially permeable to soft tissue, so that an external irradiation of the breast cancer with light in this wavelength region is *theoretically* feasible.

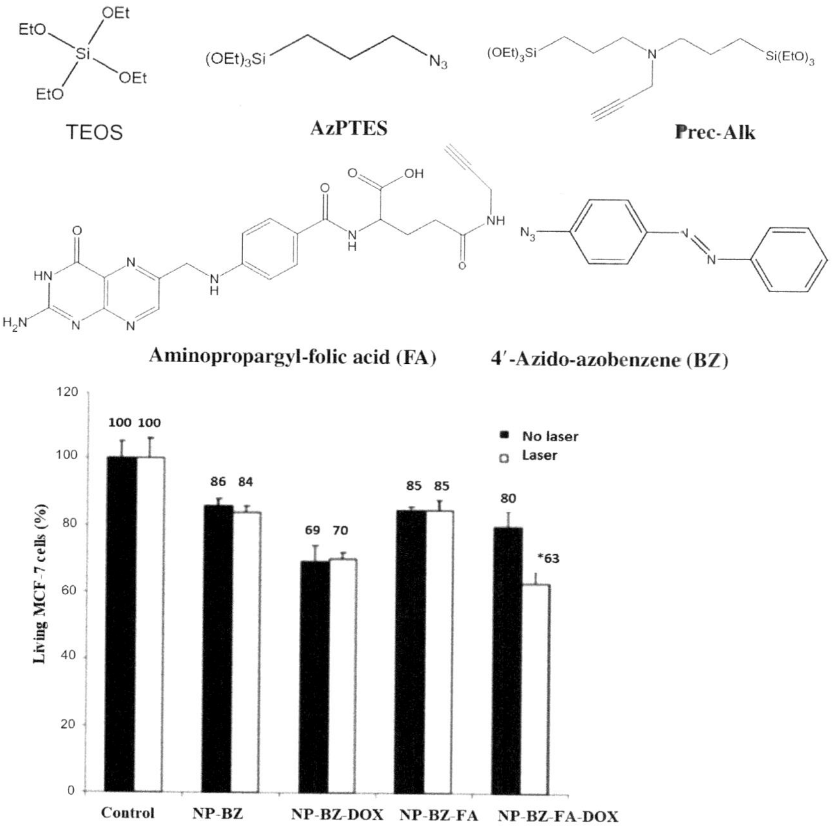

FIGURE 3.4 (A) Sol–gel precursors (up) and clickable molecules (down). *AzPTES*, (3-Azidopropyl)triethoxysilane; *Prec-Alk*, bis(triethoxysilylpropyl)amine propargyl; *TEOS*, tetraethyl orthosilicate. (B) Cell-killing properties of the NP2-derived nanomachines. MCF-7 cells were incubated with the corresponding NPs (2 μg/mL) for 6 h then rinsed, irradiated at 400–420 nm, and left for 48 h before MTT assay was performed. $^*P < .05$ versus control. *BZ*, 4′-azido-azobenzene; *FA*, folic acid; *NP*, nanoparticle. *Reproduced from Noureddine, A.; Gary-Bobo, M.; Lichon, L.; Garcia, M.; Zink, J. I.; Wong Chi Man, M.; Cattoen, X.* Chemistry ***2016*** *22, 9624–9630, with permission of John Wiley and Sons.*

Although it is an interesting chemical study, there is really a small possibility that such a designed vector could pass in vivo testing and arrive at clinical trials, not only due to the scarcely biocompatible inorganic part of the shuttle, but also due to the presence of azobenzene (i.e., so toxic to be considered a forbidden pharmacophore) and also because up to now targeting drug delivery systems with folic acid has not produced the desired high selectivity of internalization in neoplastic cells when in vivo experiments were performed.

Mesoporous silica was used also by other authors, which realized the synthesis of a zinc porphyrin derivative possessing eight triethoxysilyl groups through a CuAAC-click reaction. This porphyrin was covalently entrapped in ethenylene-bridged mesoporous organosilica NPs, which allowed DOX delivery and two-photon imaging of breast cancer cells.[28]

Similarly, enzyme-responsive, hybrid, magnetic silica NPs have been employed for multifunctional applications in selective drug delivery and intracellular tumor imaging. In one study, DOX-conjugated, enzyme-cleavable peptide precursors were covalently tethered onto the surface of uniform silica-coated magnetic NPs through click chemistry. The silica magnetic NP complex was synthesized by the Stöber method. Hydrophilic iron oxide particles were first prepared in an aqueous phase. Silane-containing amine functional groups were then used to coat the iron oxide particles to form core–shell magnetic silica structures (indicated as gray spheres in Scheme 3.8). This core–shell silicon-coated structure exhibited the advantages of stabilizing the NP suspension in biological systems and enabling easy surface modification. In detail, as can be seen in Scheme 3.8 starting from the Fmoc protected phenylalanine and Fmoc protected lysine, the alkenyl compound **32** was synthesized, and its phenolic hydroxyl group was activated with *bis*(*p*-nitrophenyl) carbonate (bis-PNP carbonate) to obtain **33**. Subsequently, it was reacted with DOX HCl to obtain **34**, and the 4-methyltrityl (Mtt) protecting group was removed with trifluoroacetic acid (**35**). This last DOX–peptide modified with an alkynyl group was reacted with an azidooligo(ethylene)glycol linker coupled to the silica-coated magnetic NPs through [2 + 3] cycloaddition in the presence of copper sulfate and ascorbic acid to produce uniformly drug-loaded functional NPs, with a drug-loading content of DOX to NPs about 5%.

This enzyme-responsive NP conjugate demonstrated efficient DOX release upon specific enzyme interactions in vitro and allows imaging of the tumor cells with high cathepsin B expression due to the possibility of imaging the magnetic core of the NPs.[29]

A more biocompatible approach makes use of an enzyme-responsive DOX-Janus dendron conjugate-based nanoscale system (Fig. 3.5). The tumor-associated enzyme matrix metalloprotease-2 sensitive linker of the proline-valine-glycine-leucine-isoleucine-glycine (Pro-Val-Gly-Leu-Ile-Gly, PVGLIG) oligopeptide was synthesized and conjugated to the Janus dendron (Boc-G2L-G3L-OMe) modified with methoxypoly(ethylene glycol) (mPEG) via a two-step highly efficient CuAAC reaction.

Janus mPEGylated dendron-PVGLIG-DOX conjugate self-assembled into compact NPs in the presence of DOX; the adduct included 4.0% of DOX. Compared to free DOX, the nanoscale system substantially reduced the side effects accompanied by a similar antitumor efficacy in a murine model.[30]

A really innovative and elegant approach is based on a class of materials called sunflower polymers, synthesized by controlled radical polymerization of hydrophilic petals from a cyclic multimacroinitiator core. This synthesis strategy afforded control over the size of the polymer nanostructures based

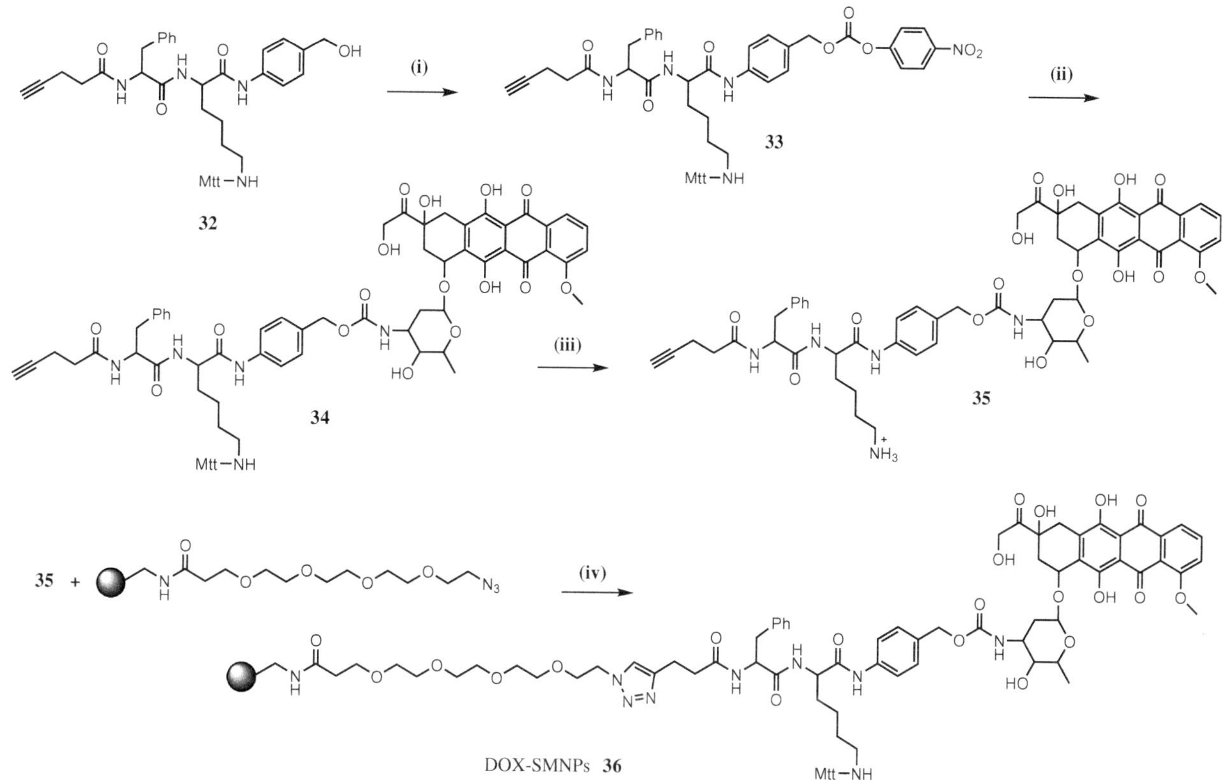

SCHEME 3.8 Synthesis of the DOX SMNP substrate. (i) bis(*p*-nitrophenyl) carbonate (bis-PNP carbonate), DIPEA, r.t., 2 days, yield: 91%; (ii) DOX · HCl, Et_3N, DMF, r.t., 18 h; then (iii) 0.5% TFA, DCM, 4 h, r.t. Yield (ii + iii): 41%; (iv) H_2O: *tert*-BuOH 1:1, ascorbic acid, copper sulfate, r.t, 24 h. *Fmoc*, Fluorenylmethyloxycarbonyl; *Mtt*, 4-methyltrityl; *N,N*-DIPEA, diisopropylethylamin; *TFA*, trifluoroacetic acid.

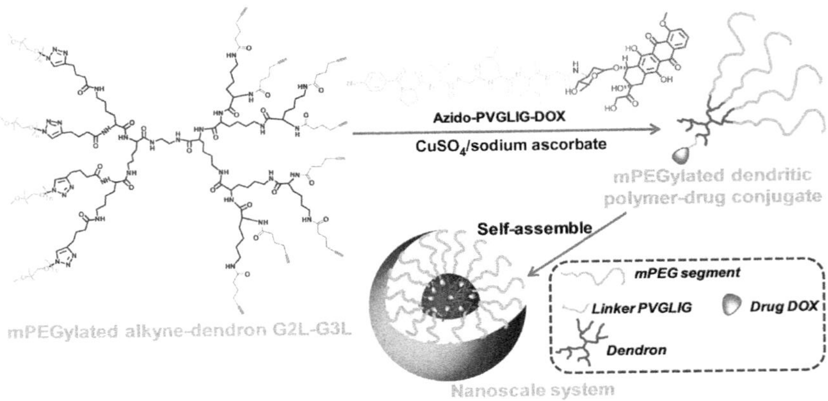

FIGURE 3.5 Illustration of the preparation of the mPEGylated dendritic polymer–drug conjugate and its self-assembly into the nanoscale drug delivery system (nanoscale system). *PVGLIG*, Proline-valine-glycine-leucine-isoleucine-glycine. *Reproduced from Li, N.; Guo, C.; Duan, Z.; Yu, L.; Luo, K.; Lu, J.; Gu, Z. J.* Mater. Chem. B: Mater. Biol. Med. ***2016***, 4, *3760–3769, with permission of The Royal Society of Chemistry.*

on their petal polymerization time. Particle size can be further tuned by varying the dimension of the cyclic core in addition to that of the petals. Click chemistry was at the core of the synthetic strategies, as it allows the cyclization of the backbone, containing an alkyne moiety and an azido group to form the petals. The reaction takes place under argon atmosphere at 100°C in dimethylformamide (DMF), with a catalyst composed of PMDETA and CuBr (48 h reaction time, yield: 80%). Folate itself is conjugated by click reaction to the backbone of the polymer, by using an alkyne-bearing folic acid derivative and azido groups introduced in the polymer (PMDETA, CuBr, r.t., 24 h, yield: 86%). DOX was linked via idrazone bond. Folate-targeted, DOX-conjugated sunflower polymers undergo receptor-mediated uptake into cancer cell and pH-triggered drug release leading to substantial cytotoxicity.[31]

A theranostic polymeric nanometric device charged with DOX was also proposed. A galactose targeted pH responsive amphiphilic multiblock copolymer conjugated with both the drug and a near-IR (NIR) fluorescence probe has been designed and prepared by a four-step process (Scheme 3.9): (1) ROP of *N*-carboxy anhydride (NCA) monomers of lysine or arginine using propargylamine as initiator to obtain **38**; (2) reversible addition-fragmentation chain transfer (RAFT) polymerization of oligo(ethylene glycol) methacrylate (OEGMA) and galactose monomer (MalpGP, 6-*O*-methacryloyl-1,2:3,4-di-*O*-isopropylidene-D-galactopyranose) coupling by the azido modified RAFT agent **36** to obtain **37** and subsequently **39**; (3) combining the obtained two polymeric segments **38** and **39** by click reaction (**40**) and deprotection/activation to obtain **41**; lastly,

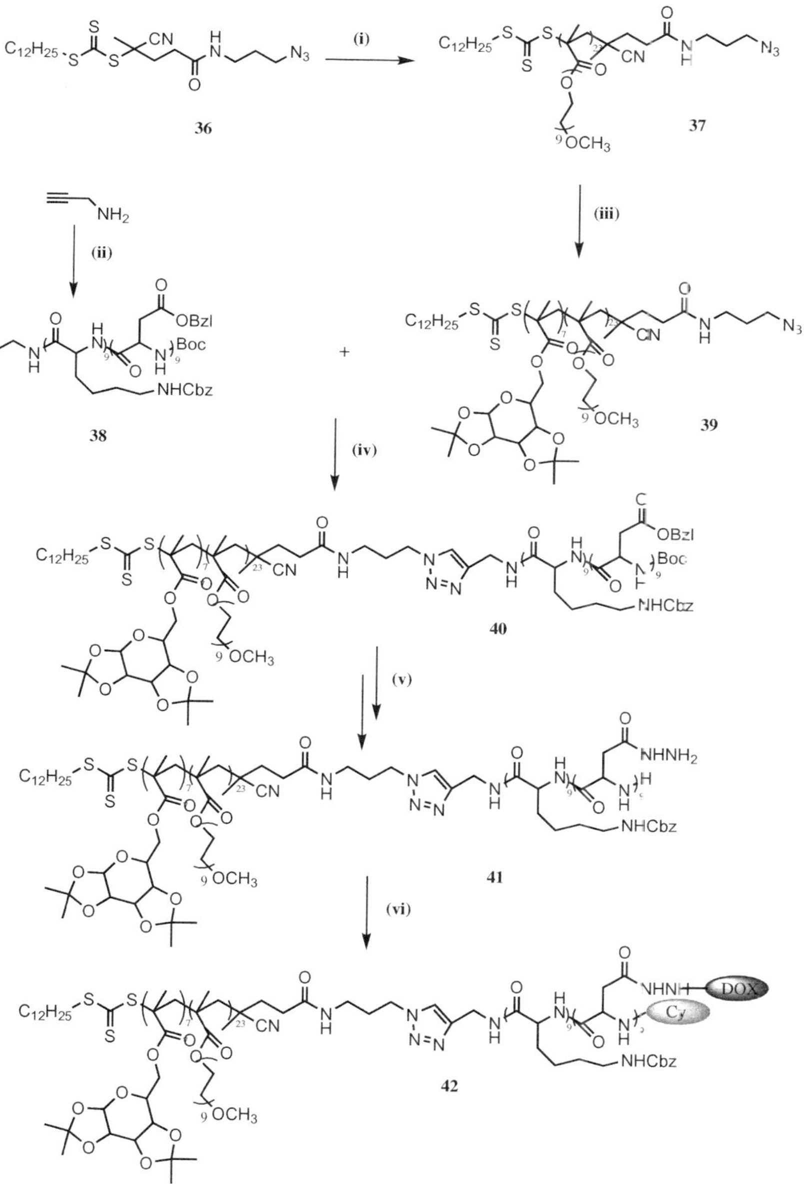

SCHEME 3.9 Synthesis of the targeting, amphiphilic, near-IR and pH-responsive triblock copolymer via radical and ring-open polymerization.(i) OEGMA, AIBN, 80°C, 12 h, yield: 99%; (ii) propargylamine, DMF, Lys-NCA or Asp-NCA, 0°C, 3 days, then Boc2O, r.t., 12 h, yield:95%; (iii) MalpGP, AIBN, DMF, 80°C, 24 h, yield: 99%; (iv) CuBr, DMF, PMEDTA; 30°C, 3 days, yield: 92%; (v) 80% formic acid, r.t., 48 h, yield: 86%, then hydrazine dichloride, DIEA. DMF, 45°C, 36 h, yield: 94%; (vi) DMF, DOX*HCl, 48 h, 45°C, yield: 90%, then NIR Aminocyanine Dye, HOSu, DMF, DCC,24 h, r.t., yield 89%. *MalpGP*, 6-*O*-methacryloyl-1,2:3,4-di-*O*-isopropylidene-D-galactopyranose; *PMEDTA*, *N,N,N′,N′,N′*-pentamethyldiethylene triamine; *Boc*, *tert*-butyloxycarbonyl; *Cbz*, carboxybenzyl; *Bzl*, benzyl; *NCA*, *N*-carboxy anhydride; *Lys*, lysine; *Asp*, aspartate; *DIEA*, diisopropylethylamine; *HOSu*, *N*-hydroxysuccinimide; *DCC*, *N,N*-dicyclohexylcarbodiimide; *Cy*, cyanine dye, whose structure is reported as inset in Scheme 3.9.

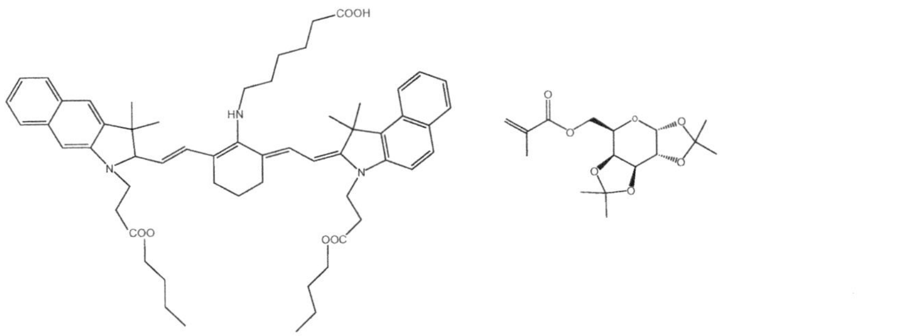

Cy, Near Infrared Aminocyanine Dye

MalpGP, 6-O-methacryloyl-1,2:3,4-di-O-isopropylidene-D-galactopyranose

SCHEME 3.9 *(Continued).*

(4) NIR copolymer prodrug **42** was synthesized by chemical linkage of both cyanine dye (whose structure is reported in Scheme 3.9B) and DOX to the block copolymer **41** via amide bond and hydrazine bond, respectively. The obtained copolymer has strong fluorescence in the NIR region, and a pH sensitive drug release behavior. The galactose targeted polymeric prodrug shows a fast and enhanced endocytosis due to the specific interaction for HepG2 cells, indicating that the polymer is a candidate for theranosis of liver cancer.[32]

Another approach was inspired by virus behavior, and led to the synthesis of chemically defined molecular constructs that like viruses achieve targeted delivery through a series of coordinated functions. An efficient click chemistry technique was used to synthesize aptamer-polymer hybrids (APHs), coupling cell-targeting aptamers to block copolymers that secure a therapeutic payload in an inactive state. The azido-derivatized aptamer was linked to the polymer backbone bearing an alkyne group based on a novel tricarboxylate ligand 5,5′,5′-[2,2′,2″-nitrilotris(methylene)tris(1*H*-benzimidazole-2,1-diyl) tripentanoic acid tripotassium salt (BimC4A)3, which stabilizes Cu(I) during the CuAAC. In comparison to traditional CuAAC ligands, the use of (BimC4A)3 does not cause irreversible damage to the aptamers, leads to higher yield, and offers superior water solubility. The reaction takes place at room temperature in 12 h approx.

Upon recognizing the targeted cell-surface marker, the APH enters the host cell via endocytosis, at which point the payload is triggered to be released into the cytoplasm. After visualizing this process with a chemically linked coumarin dye, the targeted killing of tumor cells with linked DOX was demonstrated.[33]

Another research group introduced a modified peptide-decorated polymeric NP for cancer cell targeting, which can deliver DOX to several kinds of cancer cells. Specifically, a nucleolin-targeting NP was employed with a matrix based on a copolymer of acrylamide (AAm) and 2-carboxyethyl acrylate (CEA). The negatively charged co(CEA-AAm) NP was conjugated with a nucleolin-targeting F3 peptide using a highly efficient and specific copper(I) catalyzed azide-alkyne click reaction, resulting in about seven peptides/NPs in the final product (copper sulfate, L-ascorbic acid, r.t., 12 h). DOX was conjugated via postloading of the preformed NPs (i.e., particle solution in water and DOX solution in water were mixed together and kept stirring overnight; the DOX-loaded NP solution was purified by a centrifuge filter). F3 peptide binds to angiogenic tumor vasculatures and other nucleolin overexpressing tumor cells. Attaching F3 peptide onto the NP increases the NP uptake by the nucleolin-expressing glioma cell line 9 L and the breast cancer cell line MCF-7. A high loading and slow release of doxorubicin were achieved.[34]

F3 peptide is not the only one that has attracted the interest of the researchers due to its tumor-targeting ability; for example, peptide dendrimer-based NPs have presented significant potential as candidates for a drug delivery system. In a study, the tetra-peptide sequence Gly-Phe-Leu-Gly (GFLG) was explored as an enzyme-responsive linker to connect DOX to the periphery of mPEGylated peptide dendrimer. The enzyme-responsive mPEGylated peptide dendrimer-GFLG-doxorubicin conjugate (dendrimer-GFLG-DOX) was proposed as a chemotherapeutic drug delivery nanocarrier, prepared via a two-step highly efficient CuAAC reaction of alkyne dendrimer and azido-mPEG (copper sulfate, ascorbic acid, DMF: water 3:1, r.t., 3 days, yield: 68%). The dendrimer-GFLG-DOX was capable of self-assembling into NP, and compared to the free drug DOX, it demonstrated higher accumulation and retention within SKOV-3 ovarian tumor tissue, resulting in a significant antitumor activity.[35]

In a similar manner, NPs with high drug loading and pH-responsivity were prepared by nanoprecipitation of a hydrophobic polymer-drug conjugate (PDC). The PDC, polylactide-graft-doxorubicin (PLA-*g*-DOX, Scheme 3.10), was synthesized by azide-alkyne click reaction to transform acetylene-functionalized PLA **43** into PLA-graft-aldehyde (PLA-*g*-ALD **45**) by reacting with 6-azidohexyl-4-formylbenzoate **44** in the presence of copper sulfate and sodium ascorbate at room temperature, followed by DOX (**46**) conjugation to form acid-sensitive Schiff base linkage between drug moieties and polymer scaffold (PLA-*g*-DOX, **47**). The DOX loading amount in PLA-*g*-DOX was determined to be 32 wt.%. NPs prepared under the optimal conditions exhibited well-defined spherical morphology; due to the Schiff base conjugation linkage in PLA-*g*-DOX, acid-sensitive drug release behavior of the NPs was observed. In vitro studies against MCF-7 breast cancer cells showed that the NPs can be readily taken up and result in enhanced therapeutic efficiency as compared to DOX.[36]

Daunorubicin

Daunorubicin is an anthracycline antineoplastic agent strictly correlated to DOX. In a study devoted to cancer therapy, the synthesis and biological

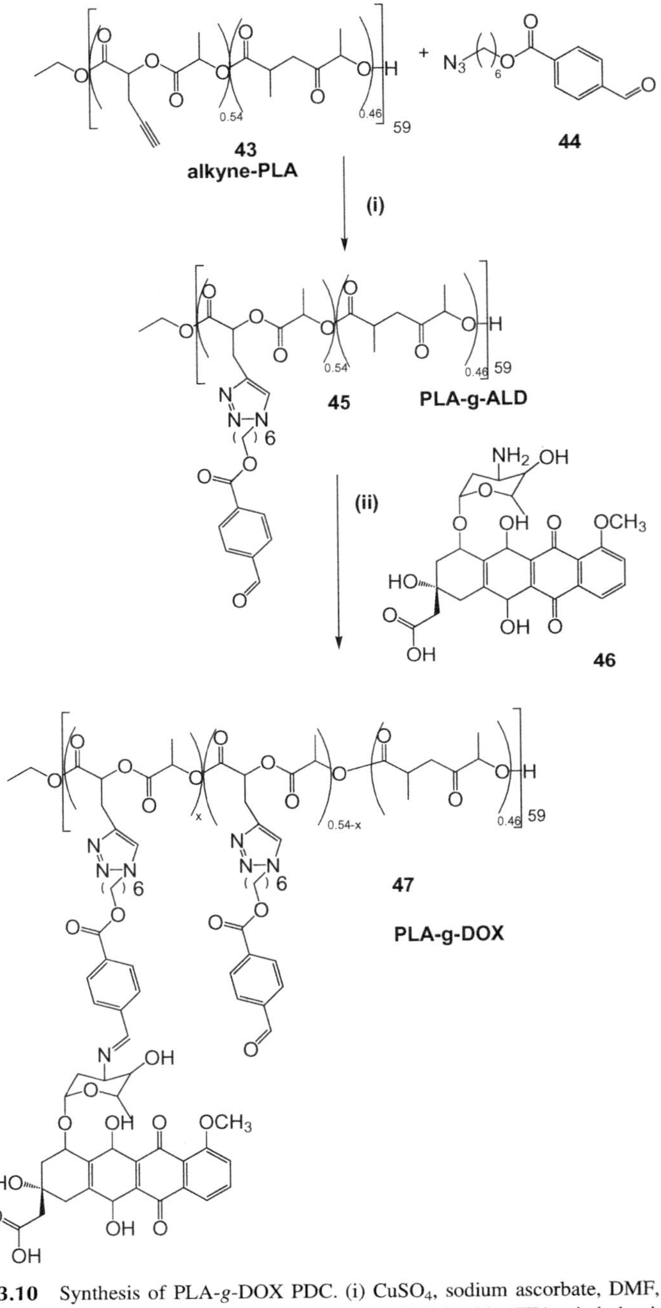

SCHEME 3.10 Synthesis of PLA-*g*-DOX PDC. (i) $CuSO_4$, sodium ascorbate, DMF, r.t., 24 h; yield: 81%; (ii) TEA, DMSO, r.t., 24 h, 86% yield. *PLA*, Polylactide; *TEA*, triethylamine.

evaluation of a cell-penetrating peptide (CPP) that is partly cyclized via triazole bridge was reported: the choice was performed as cyclized peptides have recently gained interest for their metabolic stability and intracellular uptake properties. A CPP based on human calcitonin was used in the study, and side chain cyclization was achieved via copper catalyzed alkyne-azide click reaction; the reaction takes place in ammonium ascorbate buffer in 30 min (reaction conditions however are poorly defined). Furthermore, using this novel peptide as drug transporter for daunorubicin was performed, and efficient uptake in breast cancer MCF-7 cells was evidenced.[3]

Antimicrotubule agents (paclitaxel) delivery

Paclitaxel (PTX) is a taxoid antineoplastic that interferes with the normal function of microtubule growth. It is used in the treatment of lung, ovarian, and breast cancer, although its poor pharmacokinetic properties (e.g., reduced solubility in water due to its hydrophobicity) have stimulated the search for drug delivery systems that can overcome these problems.

In particular, as depicted in Fig. 3.6, a nanometric polymeric vehicle for PTX was developed by linking to the Janus PEGylated peptide dendrimer via an enzyme-sensitive linker glycyl-phenyl-alanyl-leucyl-glycine (GFLG) tetrapeptide by efficient click reaction, resulting in Janus dendritic prodrug with 20.9% PTX content. To the dendrone, cyanine 5.5 (Cy 5.5), an NIR-fluorescence-emitting dye, was linked with the same click strategies to follow the fate of the NP once put in contact with the biological tissues. It was observed that the prodrug self-assembled into nanoscale particles. In addition to high stability during circulation, the NPs were able to release the drug. The NPs accumulated quite selectively in the 4T1 breast cancer cells line, and exhibited significantly enhanced antitumor efficiency. Moreover, in a murine model, the NPs reduced the occurrence of side effects.[37]

In another study, the selective uptake of the PTX loading NPs exploits the integrins as targeting molecules, since these proteins are associated with tumor cell survival and progression, and their expression has been shown to be increased in tumors. Thus as illustrated in Scheme 3.11, conjugates of the tripeptide integrin ligand Arg-Gly-Asp (RGD) **49** and the double clickable alkyne **48** (or strictly correlated compounds) esterified with PTX were prepared. A bivalent derivative RGD-PTX (**50**) was selected for in vivo studies in an ovarian carcinoma cell model xenografted in immunodeficient mice, showing an antitumor activity similar to free PTX, but with a more favorable toxicity profile.[38]

Another research group developed another delivery strategy, synthesizing PTX-loaded reduction-responsive core-cross-linked micelles in aqueous media via click chemistry. An amphiphilic block copolymer with multiple pendant azide groups was first synthesized through the controlled ring-opening copolymerizaton of ε-caprolactone and 5,5-dibromomethyl

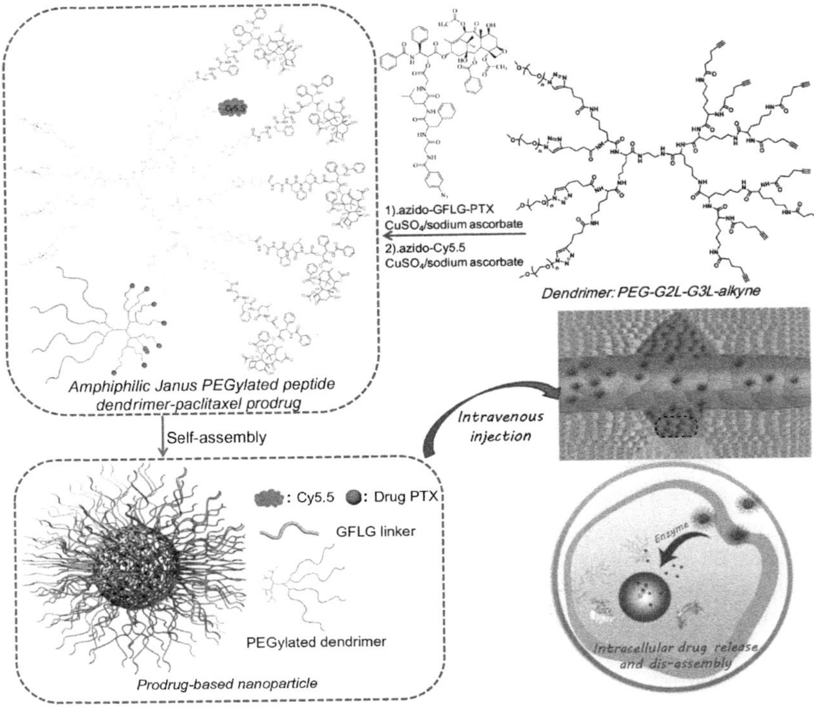

FIGURE 3.6 Structure and synthesis schemas of the Janus dendritic prodrug and the illustration of its self-assembled nanoparticles. The nanoparticles accumulated to tumor after circulation and released the free drug through intracellular enzymolysis after endocytosis. *Azido-GFLG-PTX*, Azido-glycyl-phenylalanyl-leucyl-glycine-paclitaxel conjugate; *Cy 5.5*, Cyanine 5.5; *PEG*, poly(ethylene glycol). *Reprinted with permission from Li, N.; Cai, H.; Jiang, L.; Hu, J.; Bains, A.; Hu, J.; Gong, Q.; Luo, K.; Gu, Z.* ACS Appl. Mater. Interfaces ***2017***, 9, *6865–6877. Copyright 2017 American Chemical Society.*

trimethylene carbonate in the presence of methoxypoly(ethylene glycol) (mPEG) as a macroinitiator, followed by azidation. This amphiphilic block copolymer could self-assemble into micelles and PTX could be encapsulated into the micellar core to form PTX-loaded micelles, which were core-cross-linked in situ by propargyl dithiopropionate via click chemistry (sodium ascorbate, copper sulfate, phosphate buffer as solvent). The mixture was stirred under argon for 24 h; drug loading content and drug loading efficiency were 4.43% and 59.8%, respectively). The in vitro release studies revealed the minimized release of PTX under physiological conditions, whereas a burst release of PTX was observed in response to reductive conditions (splitting of the disulfide bonds); because neoplastic cells are generally anoxic, according to the authors the drug delivery systems would be quiet selective as a cytotoxic agent for cancer cells.[39]

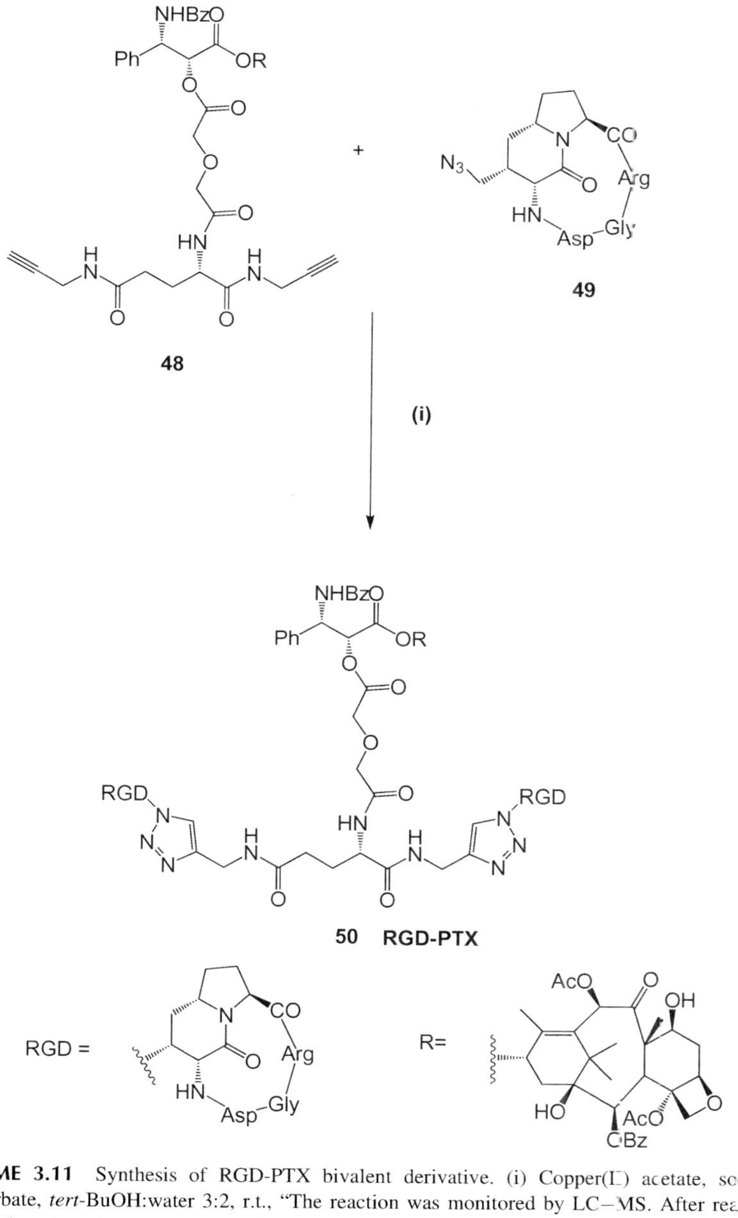

SCHEME 3.11 Synthesis of RGD-PTX bivalent derivative. (i) Copper(II) acetate, sodium L-ascorbate, *tert*-BuOH:water 3:2, r.t., "The reaction was monitored by LC–MS. After reaction completion, the solvent was removed by lyophilisation" [sic] yield: 57%.

PTX was also conveyed using sulfur-containing nanoshuttles, synthesized using a hyperbranched polyester polymer containing sulfur-pendants (HBPE-S) in the branching points and applying a nanoformulation technique based on solvent diffusion. The HBPE-S polymer is composed of spherical-shaped, aliphatic three-dimensional (3D) architecture with carboxylic acid groups on the surface. The presence of sulfur pendants in the polymeric cavities demonstrated an important role in the effective encapsulation of bismuth-DOTA (1,4,7,10-Tetraazacyclododecane-1,4,7,10-tetraacetic acid) complexes and, as a consequence, good X-ray contrast images were obtained from these bismuth-containing NPs. In addition, the HBPE-S polymer's spherical structure with amphiphilic cavities allow for encapsulation of PTX (encapsulation efficiency 73%) and can be decorated with the tumor targeting molecule folic acid, showing cytotoxicity against a lung carcinoma cell line.[40] Azide functional folic acid was attached to the triple bond functionalized HBPE-S NPs working in sodium bicarbonate buffer (pH = 8.4) as solvent, in the presence of CuI; the reaction was continued for 10 h at room temperature. The amount of folic acid bound was not reported.

Ester-based polymers more readily cleavable by enzymatic activity inside the cells (i.e., phosphate esters) can be prepared using a quiet different approach and obtaining biodegradable NPs that were also used in in vivo studies (in mice) showing promising results (Fig. 3.7). Polyphosphoester-based polymeric micelles (PTX loaded micelles in Fig. 3.7) and shell cross-linked knedel-like NPs (SCKs; PTX loaded SCKs in Fig. 3.7) in fact have been designed from amphiphilic PEGilated block-graft terpolymers (PEBP-*b*-PBYP-*g*-PEG) by supramolecular assembly, followed by shell cross-linking via thiol − yne click chemistry with hexa(ethylene glycol)dithiol (0.2 equiv to alkyne groups) in the presence of 2,2-dimethoxy-2-phenylacetophenone and UV irradiation at 365 nm. The NP could incorporate up to 10% PTX.

Drug release studies indicated that PTX released from these nanostructures was defined through a structure−function relationship, showing in vitro cytotoxic activity against osteosarcoma cell lines known to metastasize to the lungs (CCH-OS-O and SJSA). Evaluation of these materials in vivo has provided an understanding of the effects of NP structure−function relationships and related biodistribution and pharmacokinetics data.[41]

Another very beaten path, as already reported and just seen, is the use of folic acid as a targeting agent for neoplastic cells, as these last are known to have in some cases an overexpression of folic acid receptor and an accelerated metabolism for this vitamin (i.e., the rationale for the use and designed synthesis as antineoplastic of folic acid antagonist, such as methotrexate and raltitrexed). Although targeting of drug delivery systems with folic acid and analogs has not yet resulted in a drug so effective as to justify clinical trials, the synthesis of a pH-sensitive Boltorn H40-based nanogel by click reaction through miniemulsion polymerization was reported. The nanogels were synthesized by cross-coupling H40-poly(ε-caprolactone) (H40-PCL) dendrimers

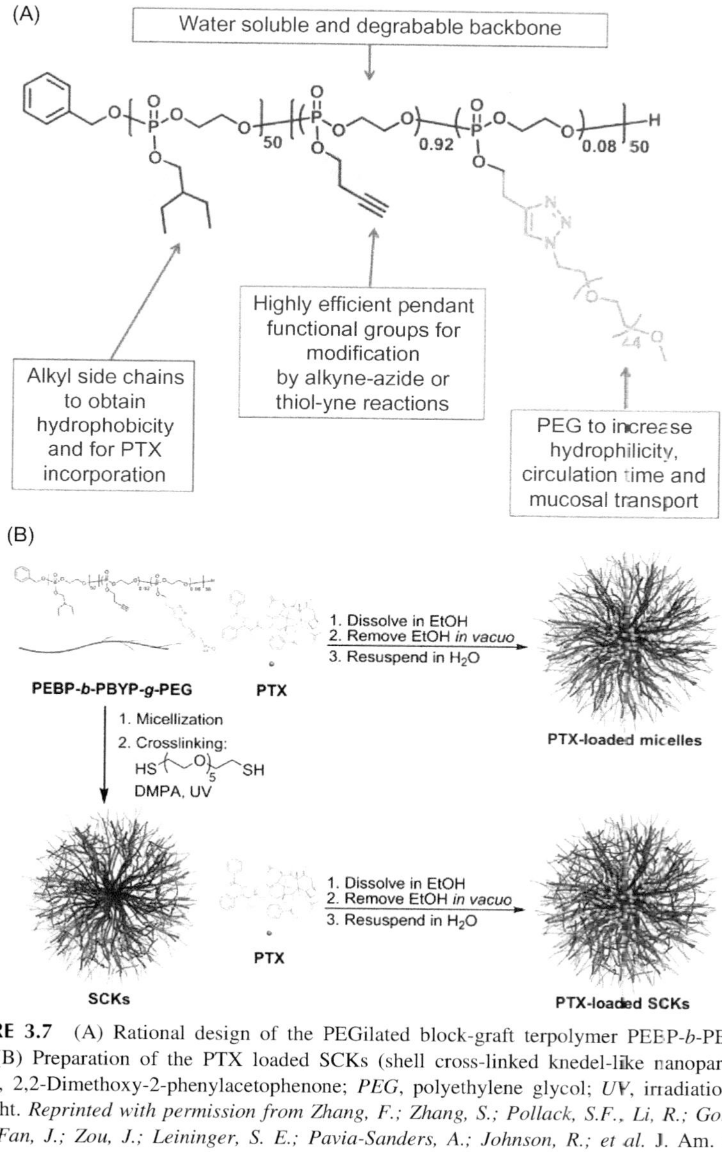

FIGURE 3.7 (A) Rational design of the PEGilated block-graft terpolymer PEEP-*b*-PBYP-*g*-PEG. (B) Preparation of the PTX loaded SCKs (shell cross-linked knedel-like nanoparticles). *DMPA*, 2,2-Dimethoxy-2-phenylacetophenone; *PEG*, polyethylene glycol; *UV*, irradiation with UV light. *Reprinted with permission from Zhang, F.; Zhang, S.; Pollack, S.F., Li, R.; Gonzalez, A.M.; Fan, J.; Zou, J.; Leininger, S. E.; Pavia-Sanders, A.; Johnson, R.; et al.* J. Am. Chem. Soc. ***2015***, 137, *2056–2066. Copyright 2015 American Chemical Society.*

as cores and poly(vinylpyridine) (PVP) as a pH sensitive cross-linker; the reaction takes place under argon in dioxane, using PMEDETA and CuCl as catalyst (60°C for 24 h). The reaction solution was exposed to air to stop the polymerization. The PEG pendant group was introduced at the middle of

PVP chains as hydrophilic segment to act as a hydrophilic shell at the final nanogel for better dispersity. Folic acid was conjugated at the end of some of PEG through the free carboxyl group. PTX could be loaded, exploiting its hydrophobic characteristic and the hydrophobic nature of the NP's core; drug loading content and encapsulation efficiency of nanogels were determined to be ~16 wt% and ~63%, respectively. These nanogel swell and release the drug loaded in a pH-sensitive manner.[42]

Another drug delivery system was prepared by the CuAAC reaction between an azide functional di-block copolymer and an alkyne-functional PTX. The azide-functional di-block copolymer, poly(ethylene glycol) (PEG)-*b*-P(OEGEEMA-*co*-AzPMA), was synthesized via the atom transfer radical polymerization of oligo(ethylene glycol) ethyl ether methacrylate (OEGEEMA) and 3-azidopropyl methacrylate (AzPMA) using PEG-Br as macroinitiator and CuBr/*N*,*N*,*N*′,*N*″,*N*″-pentamethyldiethylenetriamine (PMDETA) as a catalytic system. The alkyne-functional PTX was covalently linked to the copolymer via a click reaction (CuBr, PMDETA, DMF; r.t., 24 h,), and the loading content of PTX could be easily tuned by varying the feeding ratio (drug loading content modular from 13% to 33%). The drug-loaded copolymers could self-assemble into micelles in aqueous solution. Moreover, the drug release behavior of PEG-*b*-P(OEGEEMA-*co*-AzPMA-PTX) was pH dependent, and the PTX-polymer conjugate could efficiently deliver PTX into HeLa and SKOV-3 cells leading to cytotoxic activity.[43]

An interesting approach (depicted in Fig. 3.8) evidencing once more the role of stepwise orthogonal click chemistry, involved the strain-promoted azide-alkyne cycloaddition (SPAAC) and microwave-assisted CuAAC to assemble a derivative of PTX (azido-mPTX, **51**) and a targeting ligand (trivalent galactoside, T-Gal, in its azido modified form: azido-T-Gal, **52**) on a fluorescent silicon oxide amino-terminated NP by using dialkyne linker as a bridge (Dialkyne@Cy_3SiO_2NP). In the fabrication of the final adduct TGal-PTX@Cy_3SiO_2NP, high loading of the sugar onto the modified silica NPs serves a targeting function and also overcomes the low water solubility of PTX. The presence of a fluorescent probe, a targeting ligand, and an anticancer drug allows for real-time imaging, specific cancer-cell targeting, and evidencing the cell-killing effect.[44]

An older but interesting study reported the synthesis of micellar NPs based on linear PEG block dendritic cholic acid (CA) copolymers (telodendrimers) for the targeted delivery of PTX. The micellar NPs have been decorated with a high-affinity cyclic peptide against α-3 integrin receptor (named OA02) to improve the tumor-targeting specificity; this receptor is overexpressed on the surface of ovarian cancer cells. Click chemistry was used to conjugate alkyne-containing OA02 peptide to the azide group at the distal terminus of the PEG chain in a PEG^{5k}-CA_8 telodendrimer (micelle-forming unit); the Dde-protected amino group on the proximal lysine in the obtained OA02-PEG^{5k}-CA8 telodendrimer can be removed and used for the

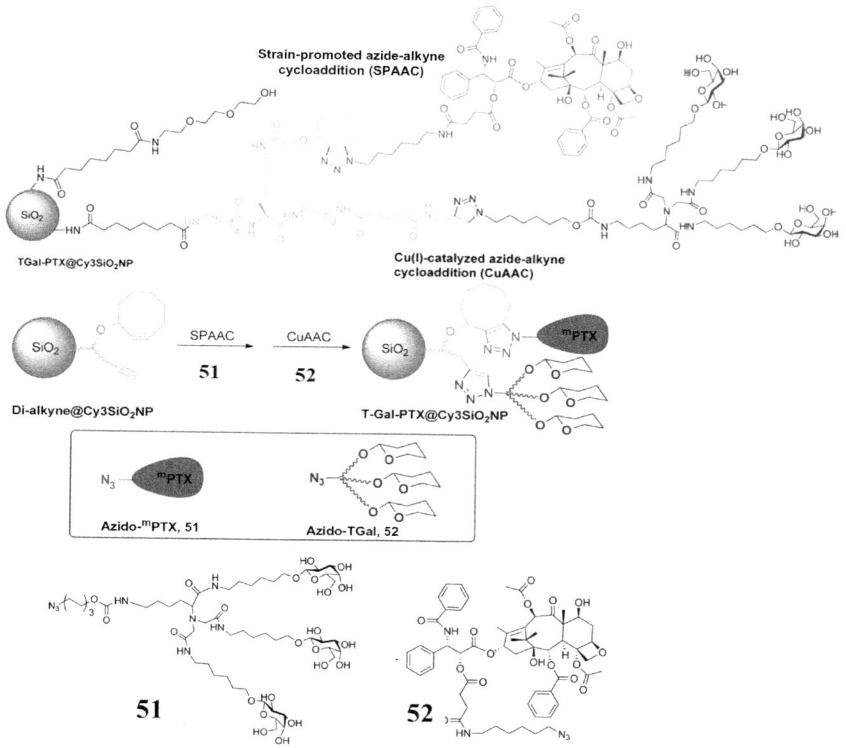

FIGURE 3.8 General strategy for SOCC on fluorescent silicon oxide NPs. The detailed chemical structures of **51** and **52** are also shown. *T-Gal*, trivalent galactoside. Other abbreviations are explained in the text. *NPs*, Nanoparticles; *SOCC*, stepwise orthogonal click chemistry. *Reprinted with permission from Lai, C. -H.; Chang, T. -C.; Chuang, Y. -J.; Tzou, D. -L.; Lin, C. -C* Bioconjug. Chem. ***2013***, 24, *1698–1709. Copyright 2013 American Chemical Society.*

conjugation of fluorescence dyes such as fluorescein isothiocyanate or Cy 5.5 for cell and animal imaging. Micelles are formed in water solution by reacting CA, PTX (eventually a fluorescent dye), and OA02-PEG^{5k}-CA$_8$.

OA02 peptide enhanced the uptake efficiency of PEG^{5k}-CA$_8$ NPs in SKOV-3 and ES-2 ovarian cancer cells via receptor-mediated endocytosis. The in vivo (mice) biodistribution study showed OA02 peptide greatly facilitated tumor localization in SKOV3-luc tumor bearing mice and exhibited antitumor efficacy with low systemic toxicity profile.[45]

A very interesting approach is to develop biodegradable and adhesive coating material for fabricating PTX-containing eluting stents. One of the developed items is a mixed stent coated with PTX and a blend of two polymers also containing PTX; these last provide a sustained release of the drug while the free drug provides a "burst" initial release (PBS, pH 7.4). The blend of the two polymers is composed of PTX-endcapped poly(lactic acid)-*b*-polyisobutylene (PTX-PLA-*b*-PIB) di-block copolymer, which possesses favorable biocompatibility, and a second polymer that is a mixture of PIB-*b*-PLA di-block copolymer and PTX. The PTX-PLA-*b*-PIB was prepared through a combination of ROP and CuAAC click reaction (PMDETA, DCM, CuBr, 25°C, 24 h, yield: 61%). The solutions of PTX-PLA-*b*-PIB and the PIB-*b*-PLA/PTX mixture were separately coated onto the bare metal stents to form the PTX-containing stents. The results showed that the coating formed by a blend of PTX-PLA-*b*-PIB, PIB-*b*-PLA, and PTX yielded a release of the cytotoxic drug that was better sustained than those of the individual polymers.[46]

Closely related to PTX, cabazitaxel (CTX) has recently gained interest and a drug delivery system was synthesized based on an amphiphilic triblock copolymer composed of a hydrophilic PEG block, a central poly(acrylic acid) (PAA) block, and a hydrophobic poly(methyl methacrylate) (PMMA) block; click coupling of the intermediate *tert*-butyl protected PMMA-N_3 and PEG-alkyne was performed in presence of CuBr, *N,N,N′,N″,N″*-pentamethyldiethylenetriamine (PMDETA) in DMF (48 h, r.t.), while *tert*-butyl groups were removed with trifluoroacetic acid to obtain PEG-*b*-PMMA.

The polymerization of PAA and PEG-*b*-PMMA, as depicted in Fig. 3.9, was performed using atom transfer radical polymerization technique. The self-assembly of the so-obtained PEG-*b*-PAA-*b*-PMMA copolymers in aqueous solutions formed nanosized micelles able to encapsulate the hydrophobic guest molecule CTX. 2,2b′-(propane-2,2-diylbis(oxy))-diethanamine was used to react with the carboxylic acid groups of the central PAA block forming acid-labile, shell cross-linked micelles (SCLM). The effect of CTX-loaded non-cross-linked micelles (NSCLM) and SCLM-50 on the viability of PC-3 and C4-2B cancer cells was investigated. Results show that SCLM released CTX in a pH-dependent manner (greater release at lower pH) and are cytotoxic in vitro toward a prostate cancer cell line. Both NP types have an encapsulation efficiency around 82% and loading capacity around 7.5%.[47]

Topoisomerase I inhibitors delivery (camptothecins derivatives)

Camptothecins (CPTs) interact with topoisomerase I, an enzyme that regulates DNA topology during cell replications. CPTs cause double-strand DNA damage during the S-phase of cell replication (being a cell cycle phase-specific cytotoxic agent), inducing apoptosis in the cells. As neoplastic cells have an accelerated replication rate, the drugs are somewhat specific for neoplastic cells, although not in a complete manner. Several semisynthetic analogs of CPT were synthesized and are used in clinic (irinotecan, topotecan) to overcome problems associated with the pristine molecule. Several drug delivery systems were also tested to enhance the specificity of CPTs in killing cancerous cells, reducing the side effects.

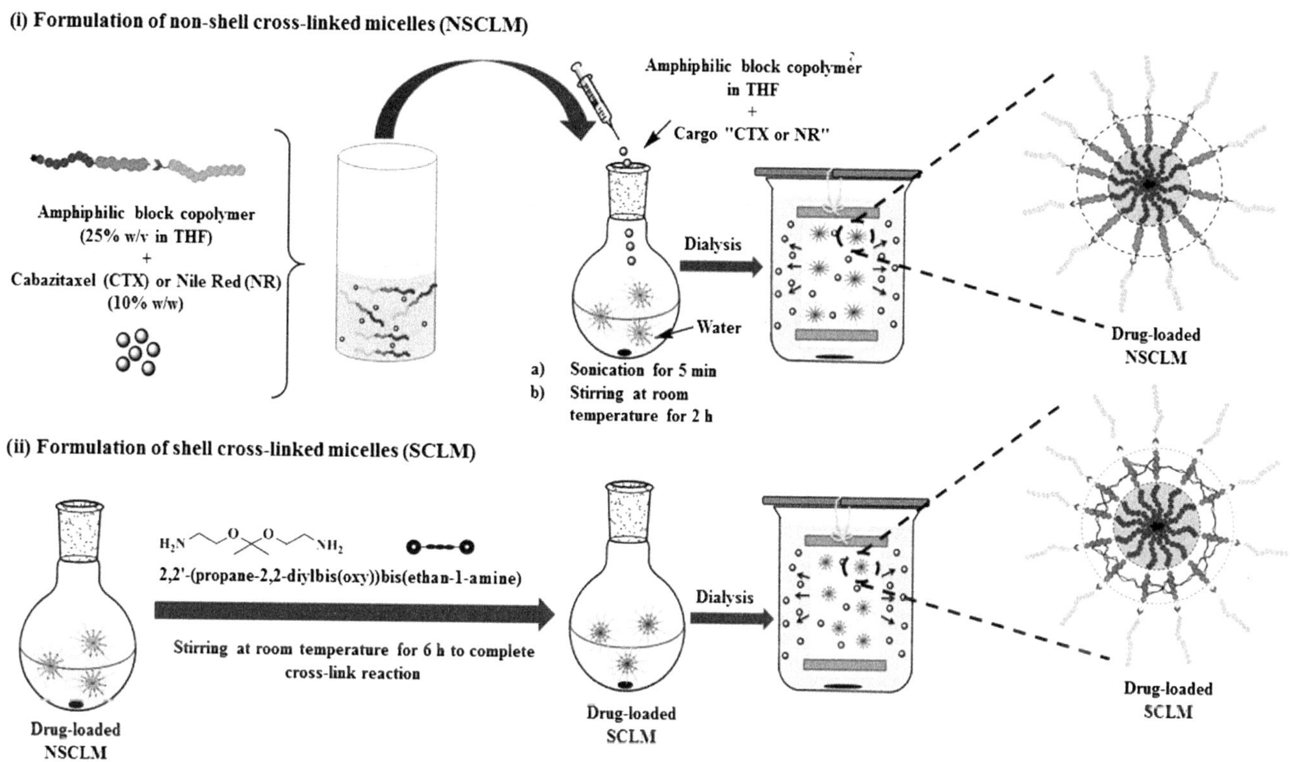

FIGURE 3.9 (A) Scheme for the synthesis of PEG-*b*-PAA-*b*-PMMA copolymers. (B) A schematic drawing showing the formulation of PEG-*b*-PAA-*b*-PMMA polymers into NR- and CTX-loaded NSCLM and SCLM micelles. *Reprinted with permission from Formulation of Acid-Sensitive Micelles for Delivery of Cabazitaxel into Prostate Cancer Cells By Aydin, O.; Youssef, I.; Yuksel Durmaz, Y.; Tiruchinapally, G.; ElSayed, M. E. H.* Mol. Pharmaceutics ***2016***, 13, *1413–1429. Copyright 2016 American Chemical Society.*

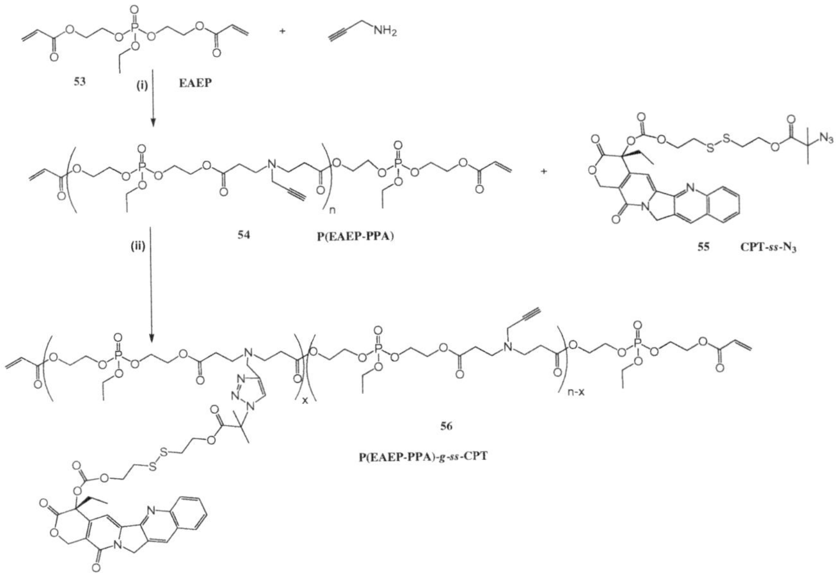

SCHEME 3.12 Synthetic routes of reduction-responsive polymeric prodrug P(EAEP-PPA)-*g*-*ss*-CPT. (i) DCM, 60°C, 6 days, yield: 50%. (ii) $CuSO_4$, sodium ascorbate, DMF, 35°C, 24 h, yield: 57%.

Among them, polyphosphoesters (PPEs) are considered potential candidates for the synthesis of biocompatible and biodegradable drug shuttles. Various synthetic methods have been employed in the preparation of linear PPEs such as the one-step Michael addition polymerization, which can be used to synthesize **54** (P(EAEP-PPA)) starting from ethyl-bis(2-(acryloxy)ethyl) phosphate (EAEP) **53** and 2-propynylamine (Scheme 3.12). Subsequently a CPT derivative containing disulfide bonds and azido groups (CPT-*ss*-N_3, **55**) was linked onto the side chain of the PPE **54** through CuAAC click chemistry to yield a reduction-responsive polymeric prodrug P(EAEP-PPA)-*g*-*ss*-CPT **56**.

The amphiphilic prodrug could self-assemble into micelles in aqueous solutions; the micelles possess a favorable stability in physiological conditions and can be degraded in a reductive medium, as usually observed in anoxic neoplastic cells (nonneoplastic cells, as already stated, tend to have an oxygen-reach, oxidative environment, different from neoplastic cells). The prodrug micelles can efficiently release CPT into 4T1 or HepG2 cells to inhibit the cell proliferation.[48]

In a very similar strategy (Scheme 3.13), the synthesis of a reductive-cleavable polymer-CPT prodrug, in which the polymer backbone consists of a biodegradable di-block polyphosphoester (PBYP-*b*-PEEP, **58**) prepared in a sequential manner via the one-pot sequential ROP using benzyl alcohol as the initiator and stannous octoate as the catalyst, was proposed. After

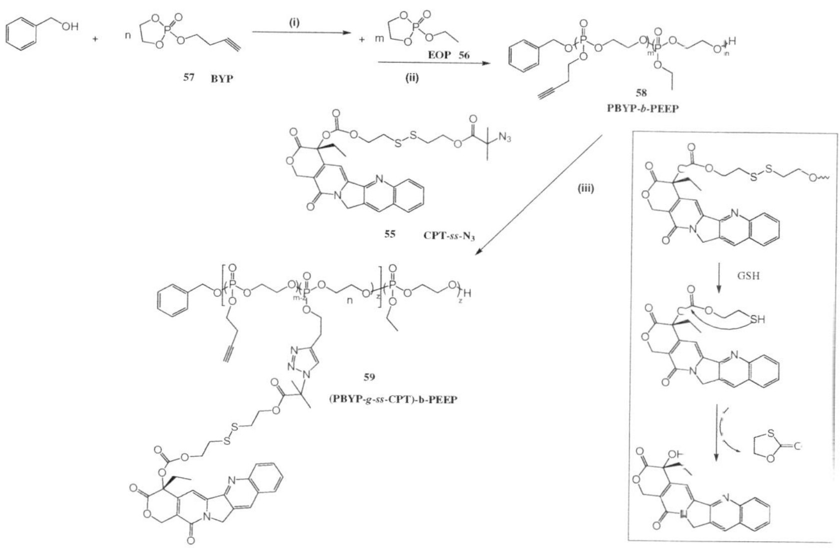

SCHEME 3.13 Synthetic routes of reduction-responsive polymeric prodrug (PBYP-*g*-*ss*-CPT)-*b*-PEEP; (inset) mechanism of CPT release from (PBYP-*g*-*ss*-CPT)-*b*-PEEP in the presence of glutathione (GSH). (i) Stannous octoate, DCM, 25°C, 1 h; (ii) 25°C, 3 h, yield: 79%; (iii) CuBr, 2,2′-dipyridyl, DCM, 25°C, 24 h, yield: 60%.

complete conversion of the first monomer 2-(but-3-*yn*-1-yloxy)-2-oxo-1,3,2-dioxaphospholane (BYP), **57**, the second monomer 2-Ethoxy-2-oxo-1,3,2-dioxaphospholane (EOP) **56** was added into the reaction mixture.

A modified CPT (CPT-*ss*-N_3, **55**) is linked onto the pendant alkynes of PBYP via CuAAC click reaction to yield the polymeric prodrug, (PBYP-*g*-*ss*-CPT)-*b*-PEEP **59**. The resulting prodrug could self-assemble into uniform prodrug micelles in aqueous solution. Since the releasable disulfide carbonate between the CPT and the PPE would be disrupted under an intracellular reducing environment, the disassociation of prodrug micelles could result in a rapid release of the CPT parent drug. The in vitro drug release studies showed that the CPT release from prodrug micelles was proceeded in a GSH-dependent manner and could efficiently inhibit the proliferation of HepG2 cells (inset in Scheme 3.13).[49]

The same approach, consisting of a reduction-dependent release of the drug loaded, was also exploited using a dithiomaleimide-based CPT-containing NPs.[50]

Accordingly, CPT-conjugated, core cross-linked (CCL) micelles that are subject to redox-responsive cleavage of the built-in disulfide bonds, result in disruption of the micellar structure and rapid release of CPT. CCL micelles were prepared via coprecipitation of disulfide-containing CPT-poly-(tyrosine-(alkynyl)-OCA) conjugate and monomethoxy poly-(ethylene glycol)-*b*-poly-(tyrosine-(alkynyl)-OCA), followed by cross-linking of the micellar core via azide-alkyne click chemistry, using copper(II) chloride and sodium ascorbate in water (r.t., 24 h, yield not reported). CCL micelles are stable under physiological conditions, while they underwent rapid dissociation in reductive environment, resulting in burst release of CPT and exhibiting cytotoxicity against human breast cancer cells in vitro. Drug loading of 13.4% was achieved.[51]

Finally, and with a different approach, a clickable and cleavable lipid analog was chosen to maximize the efficiency of bioconjugation and triggered release. This compound contains a dibenzocyclooctyne (DBCO) group for biorthogonal derivatization: **63** can be used to prepare liposomes (94% L-α-phosphatidylcholine (PC), 6% lipid analog **63**; round structures in Fig. 3.10), to which an azide-tagged analog of CPT (CPT-N3) can be linked by click chemistry (no copper catalyst is used; the doubt remains that at least part of the CPT-N3 is not chemically linked to the liposome structure but just inside the liposome or incorporated in its walls). The ester attachment of the azide tag acted as a caging group for release by intracellular esterases. This resulted in a dose-dependent cytotoxic action of the drug delivery system loaded with CPT.

In detail, the synthetic route to clickable lipid **63** (Scheme 3.14) entailed coupling of alkyne-bearing 2-nitrobenzene **60** to azidolipid scaffold **61** via click chemistry (copper sulfate + sodium ascorbate), followed by reaction with carboxylic acid **62** in presence of the coupling reagents *O*-(benzotriazol-1-yl)-*N*,*N*,*N*′,*N*′-tetramethyluronium hexafluorophosphate (HBTU) and hydroxybenzotriazole (HOBt).[52]

Antimetabolites delivery

Several different antimetabolites were subjected by investigations regarding the synthesis of drug delivery systems exploiting the chemistry of cycloaddition reactions.

Of them, methotrexate is of particular interest, as its use is not limited nowadays in the field of chemotherapy, but is also used as an immunomodulatory in inflammatory diseases in which the autoactivation of the immune system plays a great role (e.g., rheumatoid arthritis, Crohn's disease, sarcoidosis). The purpose of the study was to conjugate and evaluate methotrexate with C60-fullerenes and multiwalled CNTs (MWCNTs) for drug delivery to cancer cells, and also to compare these two systems. C60-fullerenes (**64**, Scheme 3.15) and MWCNTs were functionalized by 1,3-dipolar cycloaddition using glycine and paraformaldehyde (Scheme 3.15 shows the process for fullerenes). The reactive species, azomethine ylides, are formed by condensation of α-amino acid (glycine) and an aldehyde (paraformaldehyde) and lead to compound **65** in the case of fullerene. The results revealed the presence of ≈0.036 milliequivalents of amino groups per mg of fullerene.

Methotrexate was esterified as diethyl ester (**66**) and conjugated to the functionalized carbon-based carriers (**67**). The conjugates were characterized

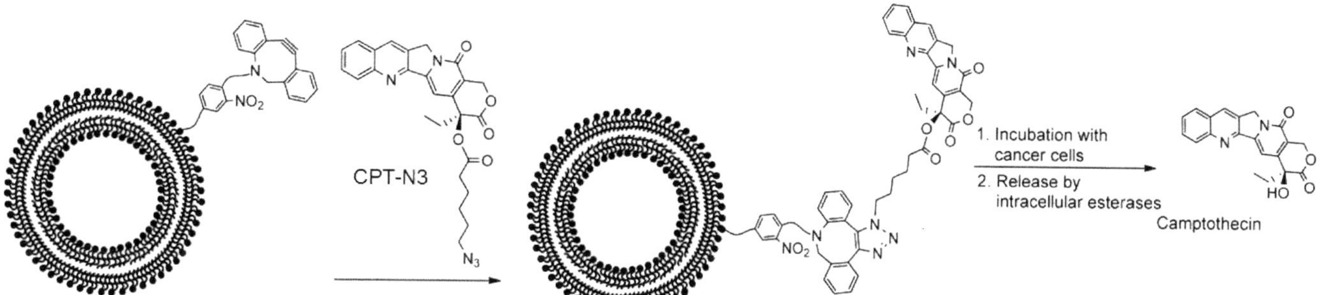

FIGURE 3.10 Delivery of camptothecin to cancer cells. L-α-PC liposomes (round structures) doped with clickable lipid **63** (see Scheme 3.14) were reacted with azido-camptothecin CPT-N_3 through click chemistry (in the absence of any catalyst). The liposomes were then incubated with cancer cells to deliver CPT through ester hydrolysis within cells. *CPT*, Camptothecin; *PC*, phosphatidylcholine. *Reprinted with permission from Alam, S.; Alves, D. S.; Whitehead, S. A.; Bayer, A. M.; McNitt, C. D.; Popik, V. V.; Barrera, F. N.; Best, M. D.* Bioconjug. Chem. ***2015***, 26, *1021–1031. Copyright 2015 American Chemical Society.*

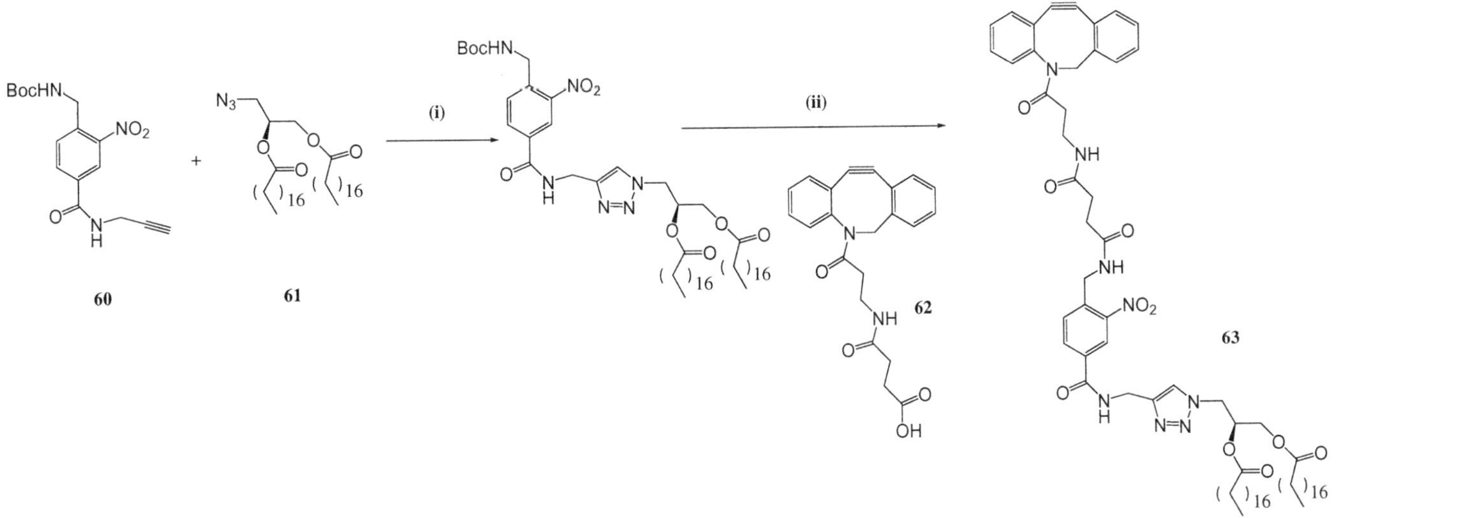

SCHEME 3.14 Synthetic route to clickable lipid delivery agent **63**. (i) sodium ascorbate, $CuSO_4$, THF:H_2O 1:1, r.t., 12 h, yield: 50%; (ii) (1) TFA:DCM 1;1, r.t.,5 h, then (2) HBTU, HOBt, DIEA, DCM, 16 h, r.t., yield (steps 1 + 2): 38%. *DIEA*, Diisopropylethylamine; *HBTU*, *O*-(benzotriazol-1-yl)-*N*,*N*,*N′*,*N′*-tetramethyluronium hexafluorophosphate; *HOBt*, hydroxybenzotriazole; *TFA*, trifluoroacetic acid.

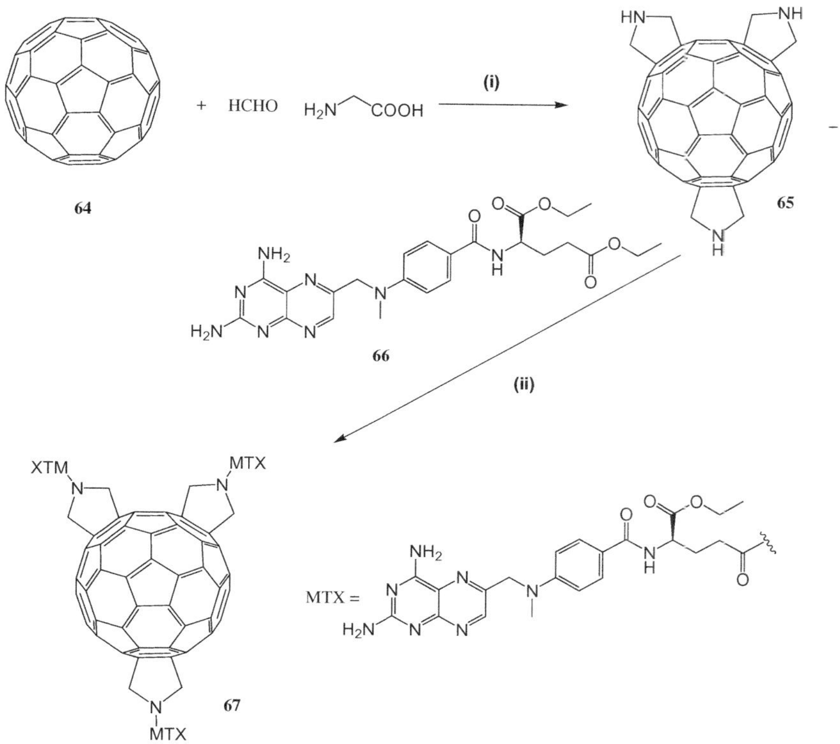

SCHEME 3.15 Amination step for C60-fullerene by 1,3-dipolar cycloaddition and conjugation with methotrexate diethyl ester. (i) DMF, reflux (130°C), 120 h; (ii) DMF, reflux, 4 h.

for micromeritics and drug conjugation. The developed systems released more drug at the supposed pH of cancer cells (acidic) to that of the pH of plasma, although strong in vivo evidence is completely lacking. C60-fullerenes were observed to be better than MWCNTs in the delivery. The underlined chemistry is interesting, although proposing such expensive (fullerenes) and poorly biocompatible carbon backbones makes the study extensibly difficult in vivo, and the ways in which some experiments are conducted is doubtful (e.g., the calculation of drug loading is disputable).[53]

Antibiotic drug delivery systems

Electroconductive hydrogels are composed of three-dimensionally structured hydrogels and conducting molecules with reversible redox properties. These hybrid materials can be utilized as electrically stimulated biomaterials in drug delivery systems. In a study schematized in Fig. 3.11, CNT-incorporated polyvinyl alcohol (PVA)-based hydrogels were synthesized by

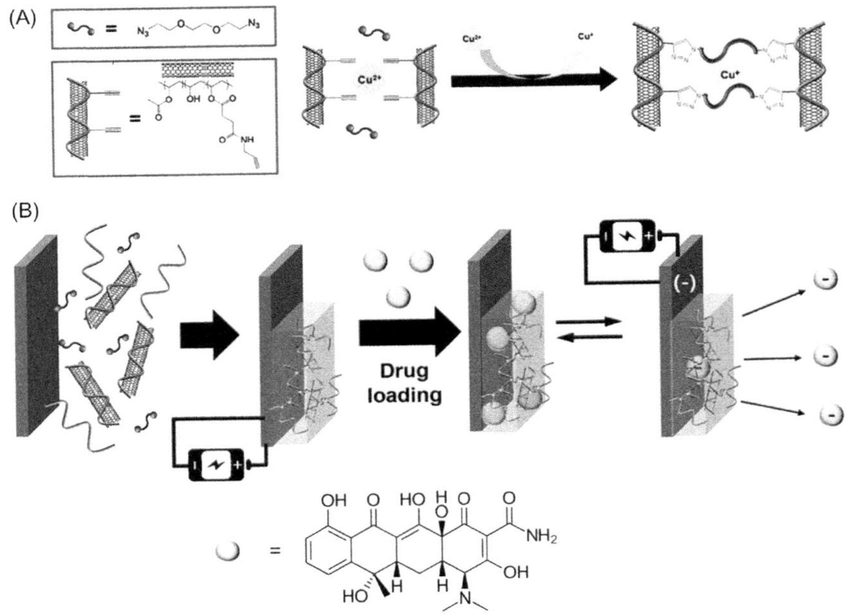

FIGURE 3.11 (A)In the presence of Cu(I) generated by an electric potential, CNT-wrapping alkyne-functionalized PVA chains are cross-linked with bis-azide via Cu(I) catalyzed alkyne–azide cycloaddition. (B) Schematic illustration of electroconductive hydrogel films synthesized by the electrically controlled click reaction. CNT-embedded hydrogel films were synthesized onto the electrode at the reduction potential of Cu(II). These hydrogel films can be used in an electrostimulated drug delivery system (tetracycline, as depicted in this figure). *CNT*, Carbon nanotube; *PVA*, polyvinyl alcohol. *Reproduced from Choi, E. J.; Shin, J.; Khaleel, Z. H.; Cha, I.; Yun, S. -H.; Cho, S. -W.; Song, C.* Polym. Chem. ***2015***, 6, *4473–4478, with permission of The Royal Society of Chemistry.*

an electro-click reaction, which was controlled by an electrochemically generated Cu(I) catalyst. When the reduction potential of Cu(II) ions was applied, PVA-based hydrogels (from alkyne-functionalized PVA) were deposited onto indium-tin oxide-coated glass electrodes, reacting with 1,2-bis(2-azidothoxy)ethane. This electro-click reaction has been exploited in creating multifunctional polymer materials since further stabilization of the easily oxidized Cu(I) during film deposition is not necessary, and spatial and temporal control of the reaction is also possible.

When the hydrogels contained CNTs, thicker films were deposited because the embedded CNTs provided a larger electrochemically active area. The electrostimulated drug release behavior with electro-click conductive hydrogels using tetracycline as a model drug was investigated. The drug is loaded on the gel by dipping it in a tetracycline solution; the releasing behavior of this amphoteric substance appeared to be largely dependent on

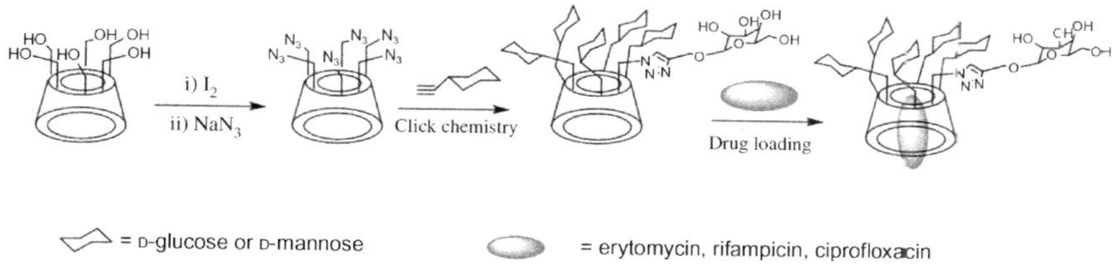

FIGURE 3.12 Schematic illustration of the preparation procedures for CD-MAN (or CD-GLU) and CD-MAN-antibiotic (or CD-GLU-antibiotic) complex. CD-N3: Heptakis-(6-deoxy-6-azido)-β-cyclodextrin. Click chemistry conditions: mannose-alkyne + *N,N,N′,N″,N″*-pentamethyl diethylenetriamine (PMDETA) + CuBr (60°C, 24 h). Yield not reported. *CD*, Cyclodextrin; *GLU*, glucose; *MAN*, mannose.

the pH of the solutions, and it was found that such a behavior could be controlled by electrical stimuli.[54] This research is particularly attractive both for the underlying chemistry and for the effort dedicated to the drug delivery of an antibiotic compound; it is a hope that nonbiodegradable, potentially toxic CNTs could be substituted by a more benign conductive material.

The use of "Trojan Horse" nanocarriers for antibiotics to enhance the activity of antibiotics against susceptible and resistant bacteria is investigated. Antibiotic carriers (CD-MAN and CD-GLU) were prepared from β-cyclodextrin grafted with sugar molecules (D-mannose and D-glucose, respectively; Fig. 3.12) via azide-alkyne click reaction. Heptakis(6-deoxy-6-azido)-β-cyclodextrin (CD-N_3), mannose-alkyne, and (*N,N,N′,N″,N″*-pentamethyldiethylenetriamine) PMDETA were reacted in the presence of CuBr to obtain the antibiotic carriers (CD-MAN and CD-GLU). The sugar molecules serve as a chemoattractant enticing the bacteria to take in higher amounts of the antibiotic, resulting in rapid killing of the bacteria.

Three types of antibiotics—erythromycin, rifampicin, and ciprofloxacin—are used as model drugs and loaded into the carriers. The minimum inhibitory concentration of the antibiotics in the CD-MAN-antibiotic and CD-GLU-antibiotic complexes for a number of gram-negative and gram-positive bacteria was investigated. Interestingly, *Staphylococcus aureus* strains, including the methicillin-resistant strains, are reduced by a factor ranging from 3 to >100. The CD-MAN-antibiotic complex is also able to prolong the stability of the loaded antibiotic and inhibit development of intrinsic antibiotic resistance in the bacteria. These sugar-modified nanocarriers are noncytotoxic.[55]

Antiviral drug delivery systems

An innovative system for the delivery of analogs of the antiretroviral zidovudine (AZT) was proposed using the nonbiocompatible silicon dioxide NPs as a platform. For this purpose, SiO_2-drug conjugates were prepared using the click reaction between AZT-triphosphates (AZT*TP) and premodified NPs containing the alkyne groups. This reaction was conducted in DMSO: water: triethylammonium acetate buffer (pH 7), ascorbic acid, copper sulfate, r.t., 2 h, yield: 75%. The nanocomposites penetrate into cells and can inhibit the reproduction of POX and herpes viruses. The loading capacity was 0.32 μmol AZT/mg adduct.[56]

The same substance was the object of other research. In detail, a synthetic strategy has been developed to create tailor-made poly(ethylene glycol) and poly(propylene glycol) copolymers functionalized with alkynyl groups that are introduced into the copolymer chains by the addition of glycidyl propargyl ether (GPE) as comonomer along with propylene oxide and ethylene oxide. The alkynyl-decorated copolymer obtained (PEG-PPG-GPE, **68** in Scheme 3.16) is used as a starter material to synthesize a drug-copolymer conjugate (**70**) by copper catalyzed ($CuSO_4$ + sodium ascorbate) azide-alkyne cycloaddition (CuAAC) reaction (click reaction) with AZT (**69**), possessing an azide group. This way, hydrophobic drugs can be transformed into water-sol compounds by improving its absorbability and bioavailability by joining them to tailor-made PEG–PPG–GPE copolymers with low polydispersity and precise molecular weight. Up to now, biological studies of this adduct are lacking.[57]

Miscellaneous drugs

Albendazole

Although used as an antihelminthic, curiously albendazole (**78**, Scheme 3.17) was considered a model for a cytotoxic drug in an ovarian carcinoma cell line (OVCAR-3), based on studies that have preconized its antitumor effects[58] Poly(2-hydroxyethyl acrylate-s-adamantyl methyl acrylate) (**73**, P (HEA_{17}-s-$AdMA_7$) and a block copolymer based on poly(*N*-iso-propyl-acrylamide) (PNIPAAm) and a block with a statistical distribution of poly (2-hydroxyethyl acrylate) (PHEA) and repeating unit with carrying β-cyclodextrin (PNIPAAm 80-*b*-P (HEA12-s-PMA-β-CD 18), **77**, Scheme 3.17) were prepared separately via reversible addition-fragmentation chain transfer (RAFT) polymerization and click reaction.

In detail, **73** is prepared from 2-hydroxyethyl acrylate (HEA) **71** and 1-adamantylmethyl acrylate (AdMA) **72** by RAFT polymerization in presence of 3-benzylsulfanylthiocarbonylsulfanyl propionic acid (3-BSPA). The RAFT polymerization of **71** and trimethylsilylpropargyl methacrylate (TMSPMA) **74** in presence of PNIPAAm macro-RAFT agent (NIPAAm = *N*-isopropyl acrylamide) lead to **75**, that was deprotected (desilylation) and reacted by CuAAC with 6I-azido-6I-deoxy-β-cyclodextrin (β-cyclodextrin azide, **76**) in presence of copper (II) sulfate/ascorbic acid (Huisgen azide-alkyne 1,3-dipolar cycloaddition) to obtain **77**. Copolymerization with HEA was deemed necessary because the reaction with β-cyclodextrin azide may otherwise be subject to steric congestion.

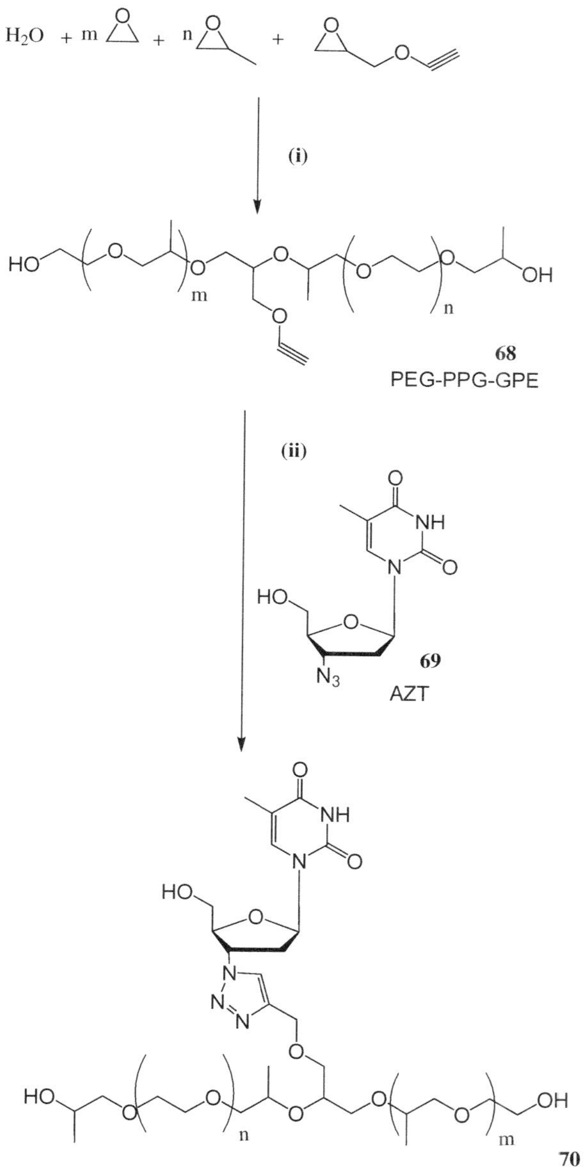

SCHEME 3.16 One-pot synthesis of PEG–PPG–GPE copolymer and click reaction between polymer terminal alkyne groups and the azide group of AZT. (i) KOH, 3 bar, 120°C, 5 min; (ii) copper sulfate, sodium ascorbate, THF:EtOH:H_2O 1:2:2., 67°C, 1 h. Yields not reported.

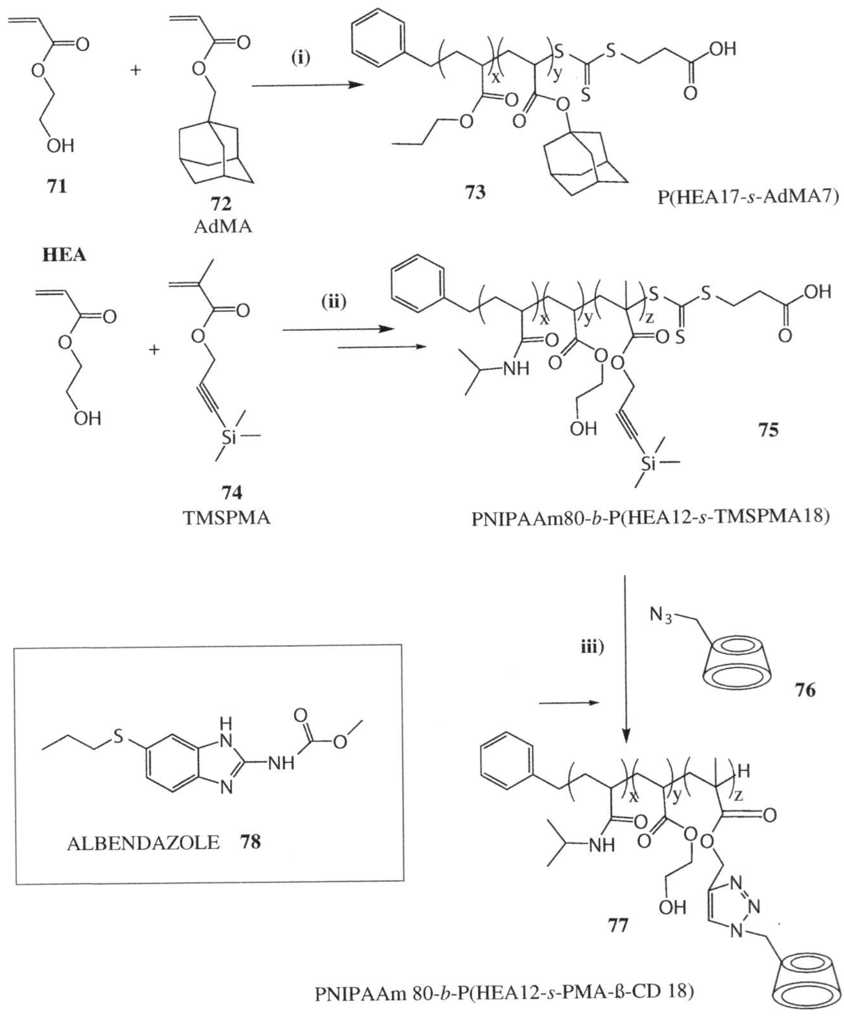

SCHEME 3.17 Schematic synthetic approach to block copolymers with pendant β-cyclodextrin. (i): 3-BSPA, AIBN, toluene, 60°C; 8 h (conversion: 60%); (ii) (1) PNIPAAm macro-RAFT agent, AIBN, DMAc, 60°C, 10 h (conversion: 75%), then (2): TBAF, acetic acid, THF, r.t., 48 h (yield not reported); (iii): CuAAC, DMF, 100°C for 96 h, reported 100% clicking efficiency. *AIBN*, 2,2′-Azobisisobutyronitrile; *DMAc*, *N,N*-dimethylacetamide; *TBAF*, tetrabutylammonium fluoride; *3-BSPA*, 3-benzylsulfanylthiocarbonylsulfanyl propionic acid. For other abbreviation, see the text.

Addition of (**73**) to the block copolymer **77** led to the formation of cross-linked nanometric micelle to which the drug (albendazole) can assemble; the micelles are stabilized via host–guest complexation between β-cyclodextrin and adamantane in **73**.

Cell viability studies using OVCAR-3 cell lines show a higher toxicity of the shell cross-linked micelle compared with the free block copolymer.[59]

Steroids: prednisolone

Prednisolone is a corticosteroid widely used for its antiinflammatory properties, acting on several bodies' districts. For this reason, the pharmacological effects are burdened by dangerous side effects, particularly when high-dose regimens are protracted over time. Therefore it is desirable to have a drug delivery system that can selectively release the drug in selected tissues, avoiding systemic exposure. In this regard, a drug delivery system for prednisolone 21-acetate (PA) based on dual-responsive CCL micelles was prepared by alkyne-azide click chemistry. Poly(ethylene oxide)-*b*-poly(glycidyl methacrylate), PEO-*b*-PGMA, was first synthesized by RAFT polymerization, and the PGMA parts were subsequently functionalized with azido groups to obtain **79**; Fig. 3.13. The CCL micelles of the PEO-*b*-PGMA-N3 and covalently entrapped PA were prepared using the alkyne-functionalized hydrazone-containing PA **80** and disulfide-containing cross-linking agent **81** (obtaining CCL-I, **83**) or sulfur-free cross-linking agent **82** (obtaining CCL-I, **84**) in the presence of Cu (II) and sodium ascorbate. The CCL micelles **83** and **84** showed an 83% loading efficiency. The CCL micelles are stable under physiological conditions, while decross-linking through the cleavage of disulfide groups took place rapidly in reductive circumstance (e.g., in presence of 1,4-dithio-D,L-threitol, DTT) in the case of **83**, while as expected no decross-link is found with **84**. In addition, the pH-sensitive hydrolysis of the hydrazone groups in PA derivative in the micellar core of (decross-linked) **83** presented a burst release of the drug at pH 5, in marked contrast to little release at pH 7.4, making the structure dual-stimuli responsive.[60]

Proteins: antibody, peptides, vaccines

Endosomolytic peptides are often coupled to drug delivery systems to enhance endosomal escape, which is crucial for the delivery of macromolecular drugs that are vulnerable to degradation in the endolysosomal pathway. Among others, melittin (**89**, Scheme 3.18) is a 26-amino-acid peptide derived from bee venom that has a very high membranolytic activity. However, such lytic peptides have similar activity against red blood cells and other nontarget cell membranes. So, it is highly desirable to control the membrane-disrupting capacity of this peptide in time and space by physically constraining onto a nanocarrier surface in such a way that they only become activated when delivered inside a proper cellular compartment. So, by carefully choosing the conjugation site and using a pH-activated analog of the melittin peptide, lytic activity toward a model membrane is completely inhibited at physiological pH.

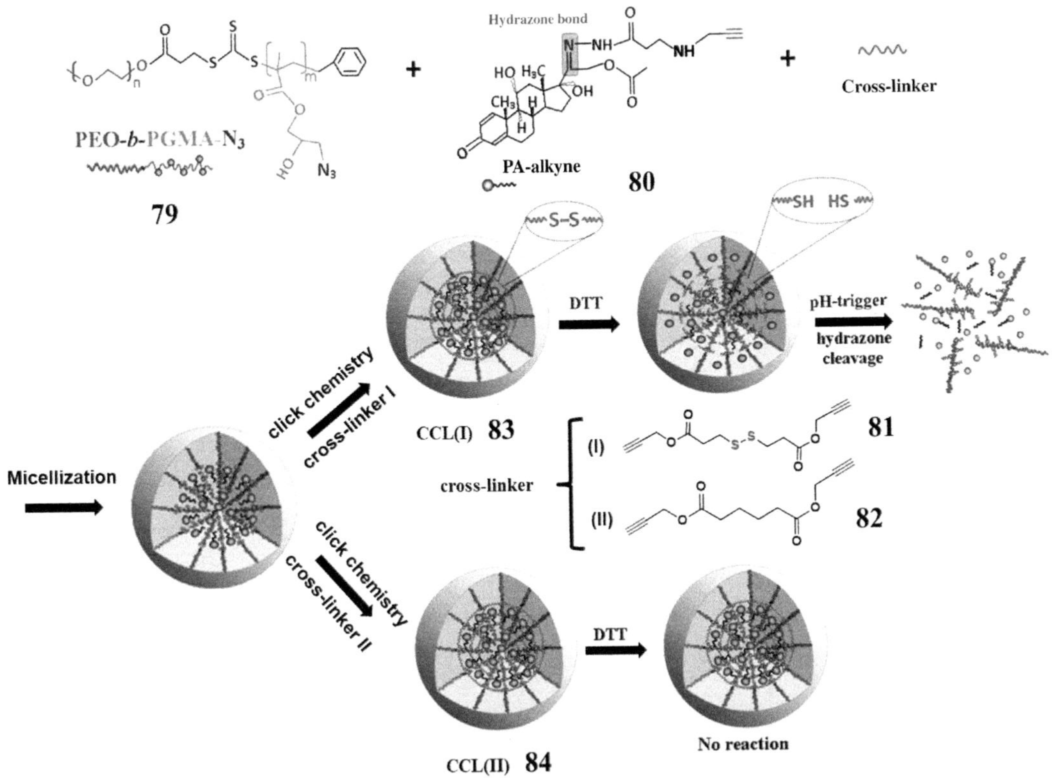

FIGURE 3.13 The preparation of CCL micelles of PEO-*b*-PGMA-N3 block copolymers, followed by degradation and drug release in the presence of DTT (1,4-dithio-D,L-threitol). *CCL*, Core cross-linked; *PA*, prednisolone 21-acetate. *Cao, X. T.; Kim, Y. H.; Park, J. M.; Lim, K. T.* Eur. Polym. J. ***2016***, 78, *264–273*.

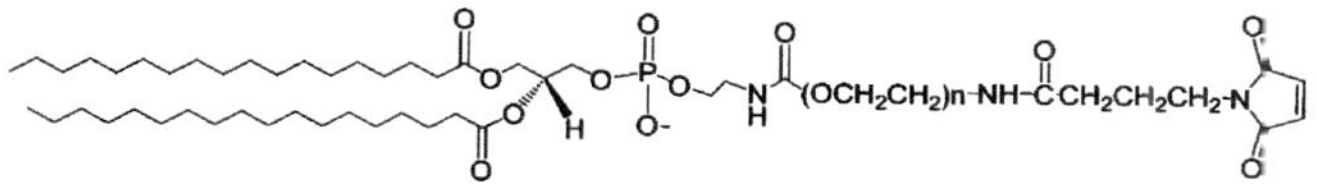

DSPE-PEG(2000)-maleimide **85**

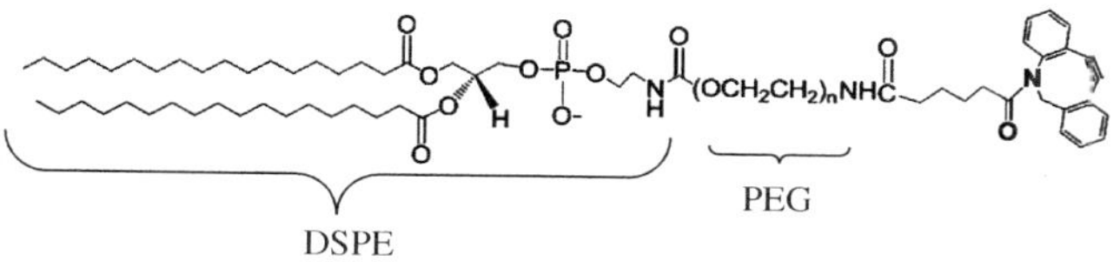

DSPE-PEG(2000)-dibenzocyclooctyne **86**

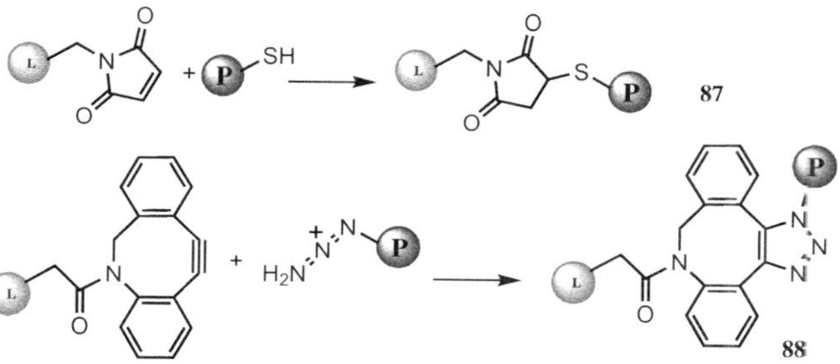

87

88

Melittin: GIGAVLKVLTTGLPALISWIKRKRQQ **89**

Azido-Melittin: 

90

SCHEME 3.18 Synthesis of the melittin-modified liposomes and structure of the precursors. In the case of the synthesis of **87**, the −SH residue of a cysteine in melittin structure is involved in the reaction. Melittin aminoacidic sequence is reported (one-letter notation). *L*, Liposome; *P*, peptide (melittin).

In detail, the melitin-containing liposome was prepared by using two different click chemistry strategies, starting from commercially available clickable liposome-forming units based on 1,2-distearoyl-*sn*-glycero-3-phosphoethanolamine (DSPE).

In detail, the alkyne DSPE-PEG(2000)-DBCO **86** (Scheme 3.18) or the maleimide-derivative DSPE-PEG(2000)-maleimide **85** were used to prepare carrier liposomes composed of 1,2-dipalmitoyl-*sn*-glycero-3-phosphatidylcholine (DPPC), 1,2-distearoyl-*sn*-glycero-3-phospho ethanolamine-*N*-[methoxy(polyethyleneglycol)-2000] (DSPE-PEG2000), and cholesterol (equaling 6% of surface PEGylation) (plus **85** or **86**). In the case of liposome **87**, a click chemistry reaction between the −SH residue of cysteine of melitin **89** and maeimide of **85** is expected (DMSO was used as solvent). In the case of liposome **88**, a click chemistry reaction (catalyst-free) between the triple bond of the dibenzocyclooctyl moiety of **86** and an azido-derivative of melitin **90** takes part.[61]

Antibody-drug conjugates (ADCs) are promising alternatives to naked antibodies for selective drug-delivery applications and treatment of diseases such as cancer. Construction of ADCs relies upon site-selective, efficient, and mild conjugation technologies. The choice of a chemical linker is crucial, and hydrophilic bifunctional chemical linkers based on carbohydrates would be a useful class of derivatization agents for the construction of linker-drug conjugates and ADCs. The synthesis of carbohydrate-based derivatization agents (**92**, Scheme 3.19) that allows the synthesis of glycolinker-drug conjugates featuring the tubulin inhibitor monomethyl auristatin E (MMA E, **91**) has been described; the conjugate **93** can be linked to an anti-EGFR antibody like cetuximab (CTX, **94**). This can target specifically neoplastic cell lines in which the epidermal growth factor receptor (EFGR) is overexpressed, leading to a double cytotoxic action: one attributable to the drug, and one due to the antibody itself. The synthesis requires the introduction in **94** of a clickable azodibenzocyclooctyne

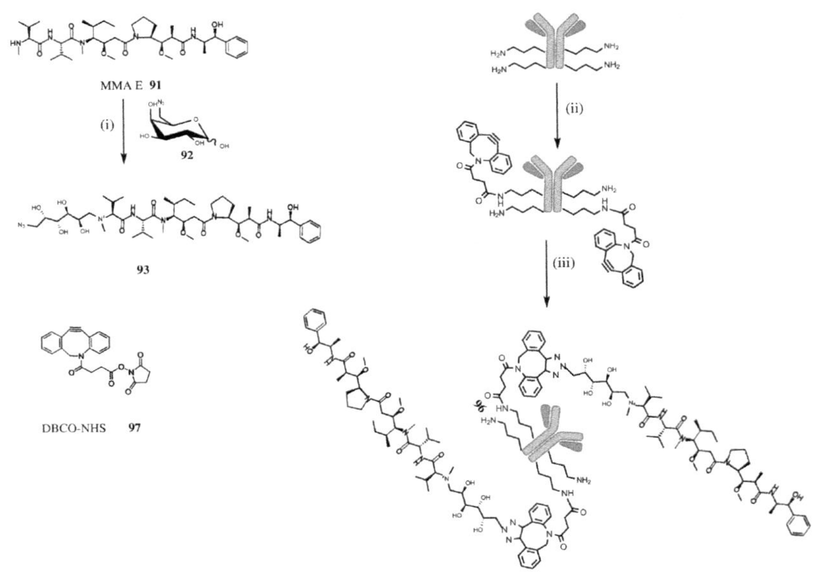

SCHEME 3.19 (A) (Upper part) Synthesis of MMA E–glycolinker conjugates. Reagents and conditions: (i) $NaCNBH_3$, DIPEA, DMSO:H_2O 80:20, 6 days, 60°C, yield: 34%. (Lower part) Structure of DBCO-NHS ester, **97**. (B) Synthesis of Cetuximab–glycolinker–MMA E conjugate **96**; (ii) DBCO-NHS, PBS, DMSO, 2 h, r.t.; (iii) PBS, DMSO, 12 h, r.t., label-to-antibody ratio: 10. *MMA E*, Monomethyl auristatin E; *PBS*, phosphate buffered saline; *DIPEA*, *N,N*-diisopropylethylamine.

Double mutant Exendin:
HGEGTFTSDLSKQXEEEAVRXFIEWLKNGGPSSGAPPPS **98**

Single mutant Exendin:
HGEGTFTSDLSKQXEEEAVRLFIEWLKNGGPSSGAPPPS **99**

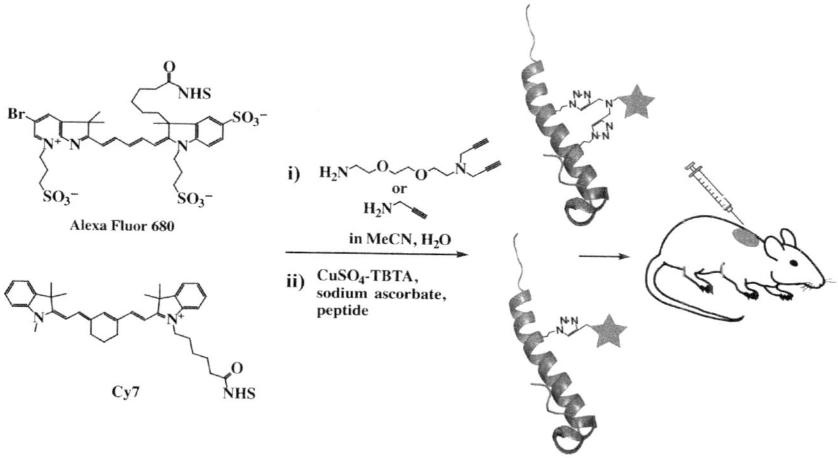

FIGURE 3.14 (i) Propargylamine or **100**, 1:1 water:acetonitrile, sodium bicarbonate, Cy7 NHS ester or Alexa Fluor 680 NHS ester (Alexa Fluor = AF), r.t., 30 min. (ii) AF680 alkyne, AF680 dialkyne, Cy7 alkyne, or Cy7 dialkyne, water:tert-BuOH, $CuSO_4$, TBTA, sodium ascorbate – single mutant or double mutant exendin, r.t., 5 h. The aminoacidic sequence of the peptides is reported (one-letter notation). X indicates the nonnatural amino acid azidohomoalanine. *TBTA*, (Tris[(1-benzyl-1*H*-1,2,3-triazol-4-yl)methyl]amine). *Reprinted with permission from Zhang, L.; Navaratna, T.; Thurber, G. M.* Bioconjug. Chem. ***2016****, 27(7), 1663–1672. Copyright 2016 American Chemical Society.*

derivative DBCO-NHS (**97**) activated as a NHS ester, which can react with free amino groups of CTX and whose triple bonds could in turn react with the azido moiety of **93** to obtain the final compound **96** (copper free reaction in aqueous media).[62]

To date, stabilized peptides address several limitations to peptide-based imaging agents and therapeutics such as poor stability and low affinity due to conformational flexibility. To have a better insight into the pharmacokinetic and bioavailability of chemically modified peptides administered by subcutaneous injection (s.c.), different fluorescent helix-stabilizing linkers with varying lipophilicity (Alexa Fluor 680, Cy7; Fig. 3.14) were linked to the glucagon-like peptide-1 (GLP-1) receptor ligand exendin (**98** and **99**) as a model system. Fluorophores are first functionalized with one or two alkynes (propargylamine or **100**) using amine-NHS chemistry, followed by either single (on **98**) or double-click reactions (on **99**) to generate the fluorescent conjugates. The mutant peptide has one or two nonnatural amino acid azidohomoalanine in their chain, indicated as X in Fig. 3.14. The click reaction is conducted in water:*tert*-butanol 1:1 followed by $CuSO_4$–TBTA (Tris[(1-benzyl-1*H*-1,2,3-triazol-4-yl)methyl]amine) and sodium ascorbate.

The stabilized peptides showed significantly higher protease resistance and increased bioavailability independent of linker hydrophilicity, and all subcutaneous (s.c.) delivered conjugates were able to successfully target the islets of Langerhans with high specificity. The lipophilic peptide variants had slower absorption and plasma clearance than their respective hydrophilic conjugates, and the absolute bioavailability was also lower likely due to the longer residence times in the skin. Their ease and efficiency make double-click helix stabilization chemistries a useful tool for increasing the bioavailability of peptide therapeutics, many of which suffer from rapid in vivo protease degradation.[63]

Several therapeutic peptides suffer from low plasma life due to metabolic inactivation, so that frequent administration is needed and poor compliance is expected. This is the case of the GLP-1 agonist somatostain and its synthetic analogs, such as octreotide or exenetide. It seems that these difficulties (i.e., rapid metabolic inactivation) can be efficiently faced. In fact, the peptide was covalently attached to Tetra-PEG hydrogel microspheres by a cleavable β-eliminative linker (a carbamate); in detail, the uniform 40 μm microspheres were fabricated using a flow-focusing microfluidic device and in situ polymerizaton within droplets; then, an azido-modified exenatide version was reacted with monofluorocyclooctyne-derivatized hydrogel microspheres containing cleavable cross-link (carbamate). In these conditions, by permeating the azid-derivatized compound into the hydrogel without the need of any additional solvent, copper-free 1,3, dipolar cycloaddition takes part, as already described by the same authors; acetonitrile:water 30:70 was used as a solvent, with a reaction time of 24 h at r.t.[64] Loading of ~2 μmol peptide/mL packed microspheres was found.

Upon s.c. injection, the exenatide is slowly released from the adduct at a rate dictated by the linker. A second β-eliminative linker with a slower cleavage rate (another carbamate) was incorporated in polymer cross-links to trigger gel degradation after drug release. Pharmacokinetic analysis showed over a 300-fold half-life extension for the peptide in the rat. Hydrogel-exenatide microspheres should support weekly or biweekly s.c. dosing in humans.[65]

Simvastatin

Simvastatin (SIM, Scheme 3.20, **101**), a widely used antilipidemic drug, has been identified as a bone anabolic agent. Its poor water solubility and the lack of distribution to the skeleton, however, have limited its application in the treatment of bone metabolic diseases. In a recent study, an amphiphilic macromolecular prodrug of SIM was designed and synthesized to overcome these limitations. The PEG-based prodrug (**103**) can spontaneously

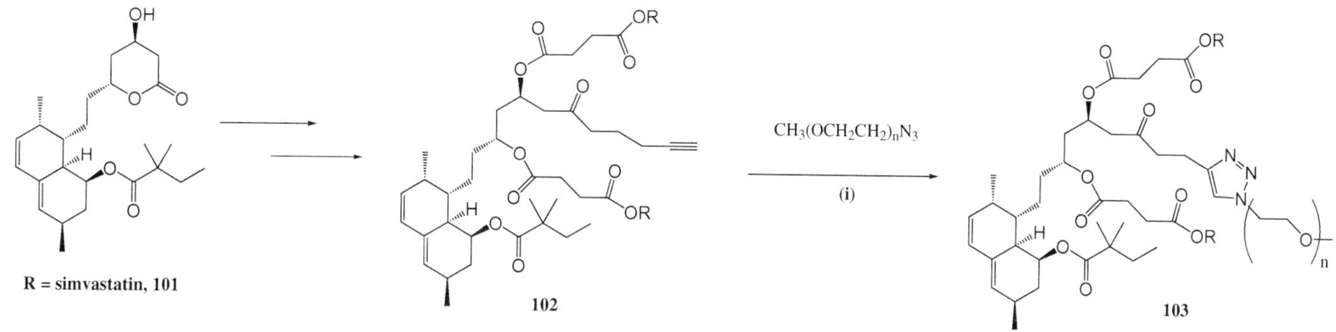

SCHEME 3.20 Synthesis of the amphiphilic macromolecular prodrug SIM-mPEG. (i) $CuSO_4$, L-ascorbic acid sodium salt, t-BuOH:H_2O, r.t., 60 h, 77%. Polyethylene glycol monomethylether (mPEG) (1.9 kDa) was used as intermediate for the synthesis of α-methoxy-ω-azido-PEG ($CH_3(OCH_2CH_2)_nN_3$).

self-assemble to form micelles. The use of SIM trimer as the prodrug's hydrophobic segment allows easy encapsulation of additional free SIM. The synthetic strategy relies on the ring opening of the lactone of SIM by 3-butyn-1-ol and following treatment with succinic anhydride. Esterification of the compound with two molecules of SIM, in which the two carboxylic acid moiety of succinic acid reacts with the alcoholic group of the 3-hydroxy-hexalactonic ring, lead to compound **102**, in which one alkyne for every three molecules of SIM is present. Compound **102** is further reacted with an azido-terminated PEG (α-methoxy-ω-azido-PEG ($CH_3(OCH_2CH_2)_nN_3$), from 1.9 kDa Polyethylene glycol monomethylether) in the presence of copper sulfate and sodium ascorbate.

The in vitro studies showed that SIM/SIM-mPEG micelles were internalized by MC3T3 cells via lysosomal trafficking and retain the biological functions of SIM. After systemic administration, optical imaging suggests that the micelles would passively target to bone fracture sites and had preferred cellular uptake by inflammatory and resident cells within the fracture callus tissue.[66]

Empty drug delivery systems

The approach of some research groups is very interesting; they have dedicated their chemical knowledge to produce optimized drug delivery systems, leaving them "empty" (i.e., not specifying the drug that can be loaded, which would be a pharmacological more than a chemical affair, in their thoughts), but giving proofs of concept and general indications for the synthesis of drug shuttles using cycloaddition reactions.

Vescicles

The synthesis of biomimicking vescicles and the study of their behavior can lead to important biological information for the subsequent development of efficient drug delivery systems.

For example, it is known that the association of polar functional molecules, among which we can count molecular recognition agents, with soft matter structures can be achieved by derivatization with alkyl chains, allowing noncovalent anchoring into amphiphilic membranes; in this regard, some authors reported the synthesis of five guanine-N9-derivative bearing C10-alkyl chains with different attachment chemistries, exploiting a synthetic pathway that allows a flexible choice of hydrophobic anchor moiety. Among the compounds synthesized and tested, a triazole derivative was prepared by a click chemistry approach.

In detail (Scheme 3.21), *N,O*-protected guanine (6-*O*-(*N,N*-diphenylcarbamoyl), 2-*N*-acetyl) **104** was alkylated at N9 in Mitsunobu conditions using 5-hexyn-1-ol and obtaining the alkyne **105**. This last compound was reacted

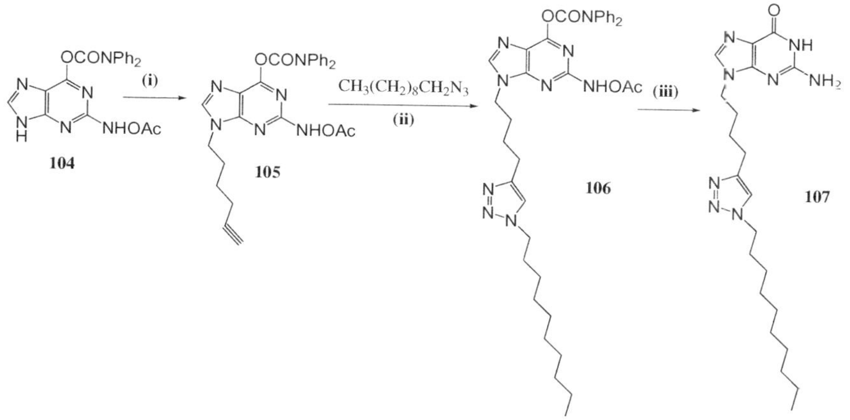

SCHEME 3.21 Synthesis of the triazole guanine-N9-derivative. (i) 5-Hexyn-1-ol, PPh_3, DIAD, THF, 70°C, 12 h, yield: 18%; (ii) 1-Azidodecane, $CuSO_4$, sodium ascorbate, THF − H_2O − t-BuOH, r.t., 12 h, yield: 41% (iii) NH_4OH (~28%, aq), MeOH, 60°C, 12 h, yield: 6%. *DIAD*, Diethyl azodicarboxylate.

with azidodecane, copper(II) sulfate and sodium ascorbate to obtain **106**, that was subsequently deprotected with ammonium hydroxide to obtain the desired compound **107**.

Incubation of this guanine conjugate with a decanoic acid vesicle suspension showed that the triazole-linked C10-anchor yielded an increased partitioning of the guanine derivative into the membranous phase compared to directly N9-linked saturated alkyl anchors. Decanoic acid vesicle membranes could be loaded with up to 5.5 mol% guanine-derivative, indicating that dugs derivatized with this guaninic compound in principle could be effectively loaded onto decanoic acid vescicles using the described strategy.[67]

Cycloaddition reactions can be used also to promote synthetic liposomes' vesicle fusion: lipid **110** (Scheme 3.22) was designed, synthesized, and proven effective for liposome preparation and derivatization. The liposomes modified with **110** were tested to assess fusion with liposomes containing azido-lipid **61**, using a click chemistry protocol (copper-free SPAAC). Larger multilamellar assemblies after liposome mixing were formed: this provides a platform for triggered fusion toward drug delivery applications. Liposome formed with PC (L-α-phosphatidylcholine) and 1,2-dioleoylsn-glycero-3-phosphoethanolamine-*N*-(7-nitro-2-1,3-benzoxadiazol-4-yl) (NBD − PE) were used; during the preparation of the liposome, the opportune lipid (**110** or **61**) is added so that is incorporated in the membrane.

The synthesis involves some of the reagents already encountered in Scheme 3.14: namely, the azo lipid **61** that is reacted according to Scheme 3.22 with Boc-protected propargylamine in presence of copper(II) sulfate and ascorbic acid to give **108**. In turn, **108** is deprotected with trifluoroacetic

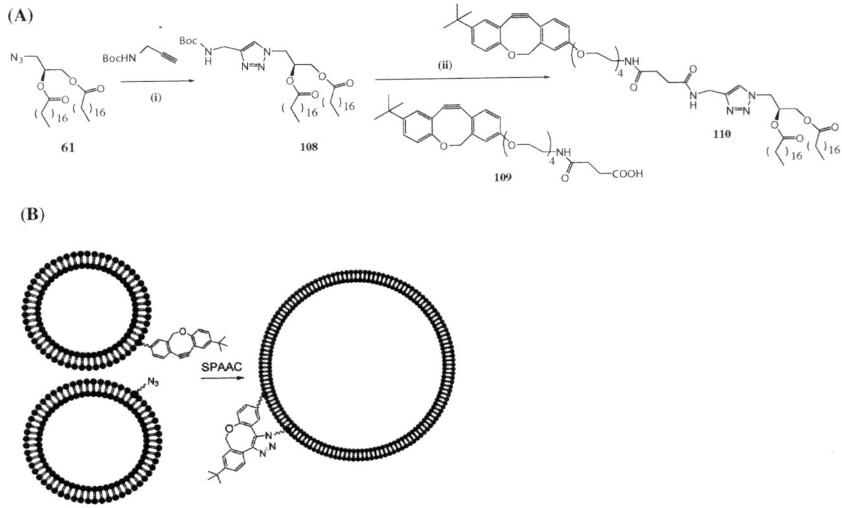

SCHEME 3.22 (A) Synthesis of the lipid **110**. (i) $CuSO_4$, sodium ascorbate, THF:tBuOH 2:1, 16 h, r.t., yield: 60%; (ii) (1) TFA:DCM 1:1, 2 h, r.t., then (2) (1-ethyl-3-(3-dimethylaminopropyl)carbodiimide hydrochloride (EDC*HCl), DMAP, DIEA, DMF, 12 h, r.t., yield (1 + 2): 55%. (B) Schematic of the fusion of azide liposomes and alkyne functionalized liposome. *DIEA*, Diisopropylethylamine; *DMAP*, 4-dimethylaminopyridine. *Reprinted with permission from Whitehead, A.; McNitt, C. D.; Mattern-Schain, S. I.; Carr, A. J.; Alam, S.; Popik, V. V.; Best, M. D.* Bioconjug. Chem. ***2017***, *28, 923–932. Copyright 2017 American Chemical Society.*

acid and reacted with 1-((2-(tert-Butyl)-11,12-didehydro-6*H*-dibenzo[*b,f*]-oxocin-8-yl)oxy)-13-oxo-3,6,9-trioxa-12-azahexadecan-16-oic acid (**109**) with coupling reagents (EDC*HCl, DMAP, DIEA).[68]

Nanoparticles

Scaling down from microstructure and from liposome to nanostructured vectors, we can find a plethora of empty drug delivery systems, some of which were prepared making use of cycloaddition reactions.

Among them, an interesting approach was based on the synthesis of an amphiphilic graft copolymer platform (Scheme 3.23) composed of hydrophilic backbone poly(2-methyl-2-oxazoline-*co*-2-pentyl-2-oxazoline) (P(MeOx-*co*-PentOx)) **113**, prepared via cationic ROP of 2-methyl-2-oxazoline **111** and 2-(5-azidopentyl)-2-oxazoline **112**. In parallel, alkynyl-terminated PLA oligomers **115** were synthesized by the transesterification of high-molecular-weight PLA **114** in the presence of propargyl alcohol and dibutyltin dilaurate. Hydrophobic poly-(D,L-lactide) (PLA) chains of **115** were grafted on the backbone of **113** via Huisgen 1,3-dipolar cycloaddition. The click copper-catalyzed cycloaddition was performed using commercially available copper NPs with nominal size from 20 to 40 nm, in combination with microwave reaction

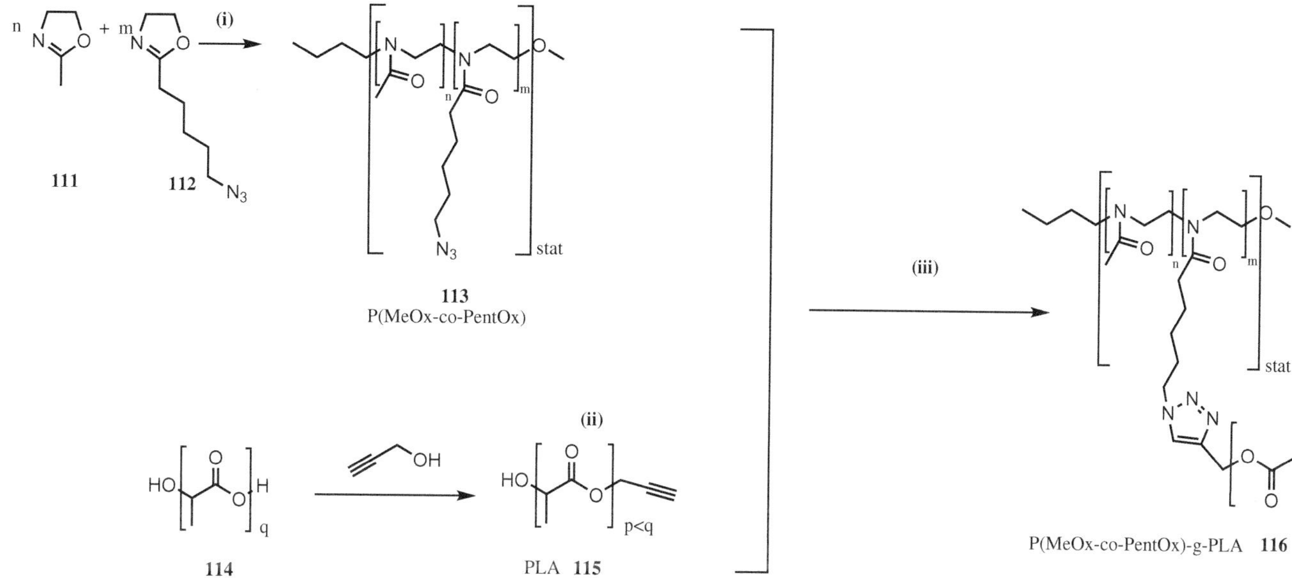

SCHEME 3.23 Synthesis of functionalized homopolymers **113** and **115** and the resultant graft copolymer [poly(2-methyl-2-oxazoline-*co*-2-pentyl-2-oxazoline)-graft-poly(D,L-lactide)] (P(MeOx-*co*-PentOx)-*g*-PLA) **116**. (i) 1-iodobutane, acetonitrile, 80°C; copolymerization was stopped by adding an excess of basic solution of methanol; (ii) dibutyltin dilaurate, chloroform, yield: 90% – 95%, (iii) copper nanoparticles, 80°C (maintained modulating microwave power), 20 min, yield: 65%–86%.

conditions. These copolymers, named (P(MeOx-*co*-PentOx)-*g*-PLA) **116**, when nano-precipitated in water (this last being the hydrophilic backbone elective solvent), obtained various structures of NPs with a plurality of morphologies, going from unilamellar, multilamellar, and large compound vesicles to core–shell NPs, depending on the PLA molecular weight. First evaluation of their stealthiness by studying the stability and the interaction of these nano-objects with a model protein revealed the role played by the P (MeOx-*co*-PentOx) chains (**113**) in these interactions, demonstrating the utility of this amphiphilic graft copolymer platform with well-defined architectures for the design of nanocarriers in drug delivery applications.[69,70]

As already described in the selected cases of the drug shuttles employable for the delivery of antineoplastic, nanocarriers with carbohydrates on the surface represent an interesting class of drug vehicles because sugars are involved in biomolecular recognition events in vivo. For example, a biocompatible miktoarm star copolymers (**119**) comprising glycopolypeptide and poly (ε-caprolactone) chains was synthesized using ROP and click chemistry of the acetyl-protected alkynyl glycopeptide **117** and azide-functionalized polycaprolactone **118.** The Huisgen cycloaddition was performed using $CuSO_4$, tris (3-hydroxypropyltriazolylmethyl)amine and sodium ascorbate in THF:methanol:water:DMSO (2:1:0,2:1). Deprotection of the acetyl esters was done by using hydrazine monohydrate (Fig. 3.15). The amphiphilic copolymers were self-assembled in water into morphologies such as nanorods, polymersomes, and micelles with carbohydrates displayed on the surface. The precise nanostructure that is obtained could be tuned by varying the chain length of the blocks and was not affected by the type of sugar residue. It is possible to sequester both hydrophobic as well as hydrophilic molecules within the nanostructures, making the structures possible candidates for drug delivery. These noncytotoxic mannosylated rods and polymersomes were efficiently taken up by MDA-MB-231 breast cancer cells, demonstrating their potential as nanocarriers for the targeted delivery of antineoplastic agents.[71]

Among the nanostructures that can be functionalized by cycloaddition reactions, semiconductor inorganic nanowires (NWs) are gaining significant importance. Efficient and controlled immobilization of biomolecules on the NW surface is crucial and can be achieved by the use of the CuAAC and its strain-promoted variant for the covalent functionalization of vertical NWs with peptides and proteins. For example, GaAs NWs modified with alkyne groups can be reacted with azido-terminated peptides in the presence of sodium ascorbate, copper(II) sulfate, and tris[(1-benzyl-1*H*-1,2,3-triazol-4-yl) methyl]amine in DMSO:water. The reaction takes place rapidly (90 min) at room temperatures, obtaining a surface uniformly covered by the biomolecule, which can be a physiologically active species.[72] Of course, a gallium arsenide nanocomposite can be used as a model for bioconjugation of active compounds, but due to its intrinsic toxicity its in vivo use is excluded.

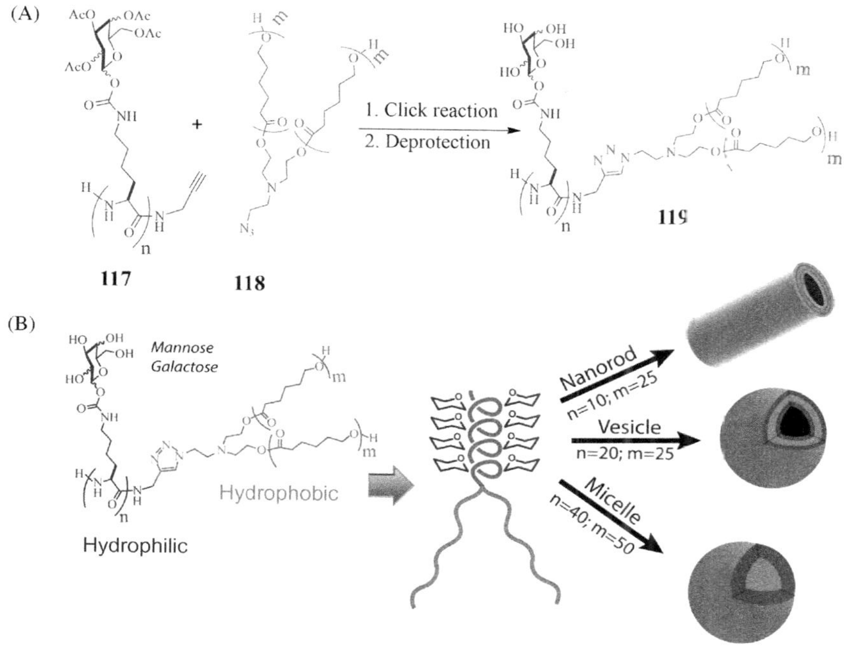

FIGURE 3.15 (A) Synthesis of miktoarm star copolymers. (B) General chemical structures of nanostructures obtained by tuning the hydrophilic and hydrophobic block length. *Reprinted with permission from Pati, D.; Das, S.; Patil, N. G.; Parekh, N.; Anjum, D. H.; Dhaware, V.; Ambade, A. V.; Sen Gupta, S.* Biomacromolecules ***2016***, 17, *466–475 Copyright 2016 American Chemical Society.*

The use of biologically derived polymers is much more interesting and has a greater chance of being used for further in vivo studies; for example, it is known that HA forms pericellular coats in many cell types that are involved in the early stages of cell adhesion by interacting with the CD44 receptor. Based on the largely recognized overexpression of the CD44 receptor in tumor tissues, a polybenzofulvene molecular brush has been enveloped into hyaluronan shells to obtain a tricomponent polymer brush (TCPB, **120**, Fig. 3.16) composed of intrinsically fluorescent backbones bearing nona(ethylene glycol) arms (poly-6-ANEGA-CMO-BF3k-GT, **122**) terminated with low molecular weight HA macromolecules (HA-FA-Pg, **121**). The crosslinking of the HA layer was obtained by click chemistry, using an alkyne bearing derivative of HA (HA-FA-Pg, **121**) and a poly(thiofulvene) derivative with an azido moiety (**122**), which reacted in the presence of $CuSO_4$ and sodium ascorbate in water:tert-BuOH as solvent.

The nanoaggregates obtained in TCPB water dispersions have a cylindrical brush architecture and showed promising prerequisites for the preparation of nanostructured drug delivery systems.[73]

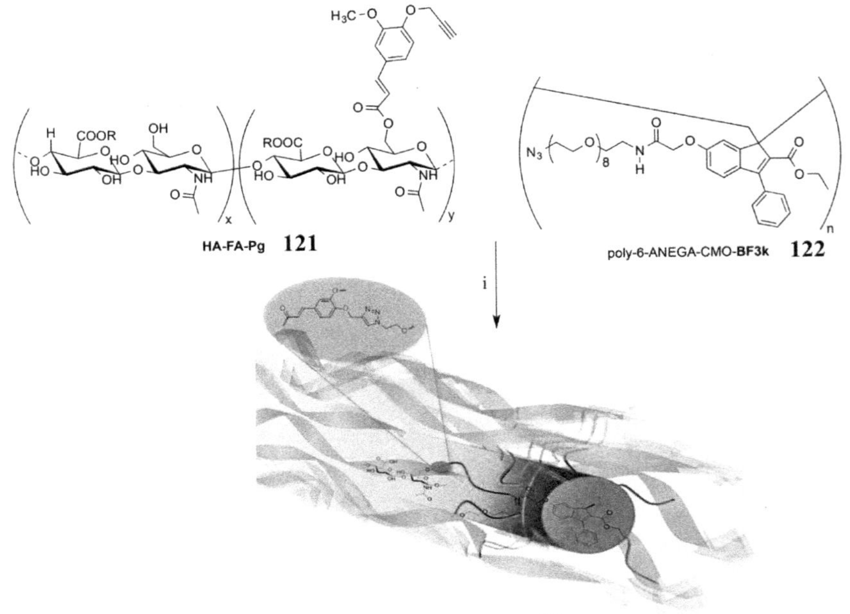

FIGURE 3.16 Click chemistry functionalization of poly-6-ANEGA-CMO-BF3k with hyaluronan synthon HA-FA-Pg. Reagents: (i) $CuSO_4$, sodium ascorbate, *tert*-BuOH: H_2O. Substituents: R = H or C_2H_5. Red (dark grey in print version): hydrophobic backbone (**122**); *stripes*: nona(ethylene glycol) branches of **122**; *bands*: hyaluronic acid residues of **121**. *Reproduced from Cappelli, A.; Paolino, M.; Grisci, G.; Razzano, V.; Giuliani, G.; Donati, A.; Bonechi, C.; Mendichi, R.; Battiato, S.; Samperi, F.; Scialabba, C.; Giammona, G.; Makovec, F.; Licciardi, M.* Polym. Chem. ***2016****, 7, 6529–6544, with permission of The Royal Society of Chemistry.*

The understanding of these glycopolymer-based nanostructures' formation and their lectin binding properties are of great interest to the development of drug delivery systems; apart from HA, other glycosylated structures were developed as nanocarriers. Some authors proposed, for example, the synthesis of glycosylated polypeptides from a clickable poly(L-lysine)-*b*-poly (benzyl-L-glutamate) copolypeptide, obtained by a sequential ROP. The clickable poly(L-lysine) chain was completely functionalized by introducing galactose and lactose moieties on the copolypeptide's hydrophilic block, aiming specific lectin recognition, oligo(ethylene glycol). Coupling was carried out using copper-catalyzed Huisgen cycloaddition, rendering the alkyne functionalized poly(L-lysine) block hydrophilic. In this way, an amphiphilic polypeptide-based copolymer with a sugar moiety as hydrophilic block was synthesized; the reaction was conducted at 30°C and was completed in 12 h using DMSO as a solvent, and $CuSO_4$ in association with sodium ascorbate as catalyst.

Spheres, pearl-necklace, and wormlike structures from poly(L-lysine)-*b*-poly(benzyl-L-glutamate) copolypeptides were obtained after nanoprecipitation, depending on the hydrophilic block functionalization and the hydrophobic block length. Specific interaction between the sugars on the micelles' surface with RCA_{120} lectin (*Ricinus communis* agglutinin) was observed with the nanostructures created with poly(lysine)-*b*-poly(benzyl-L-glutamate) copolymer having a longer hydrophobic block. However, unexpected and unspecific interactions were observed with some nanostructures, showing the importance of the chemical functionalization and nanostructure stabilization.[74]

Although not based on biopolymers, dendrones with ester moieties can be considered biocompatible, due to the degradation induced by intracellular esterases; moreover, redox-sensitive hyperbranched dendritic-linear polymers (HBDLPs; Fig. 3.17) can be prepared and stabilized individually as nanometric unimolecular micelles. The core—shell amphiphilic structures can be synthesized through a combination of self-condensing vinyl copolymerization and ATRP. Cleavable disulfide bonds are introduced, and by triggered reductive degradation, the HBDLPs showed a decrease in molecular weight, and the extent of degradation can be tuned by the amount of incorporated disulfides. A schematic diagram of the procedure to obtain the micelles and the precursor used is reported in Fig. 3.17; as reported in the figure, polymer with variable molar fraction of disulfide-containing brancher in the hyperbranched core were prepared. The HBDLP with pendant disulfide-linked functionalities in the hyperbranched core can be readily postfunctionalized with a drug. An instant release of the drug is expected as a response to a reductive environment (that, as already said, can be found, e.g., in some neoplastic cell lines). As a proof of concept, azide-functional cleavable pendant disulfide bonds (AzSSMA) can be used to prepare a micelle core with BBEMA, obtaining HBMI-SSN3. This last can be covered with an OEGMA shell, and azide residue can be linked through click chemistry to Disperse Red 13-prop-2-ynyl (**123**) in presence of copper sulfate and ascorbic acid (THF, 40°C, 20 h); release properties in presence of dithiothreitol (DTT) were studied (for the meaning of the abbreviations, see Fig. 3.17 caption).[75]

A really particular, completely degradable, and intriguing nanocarrier is a fragment of viral capside. Virus-like particles (VLPs) in fact constitute a promising platform in vaccine development and targeted drug delivery. To date, most applications use simple nonenveloped VLPs as human papillomavirus or hepatitis B vaccines, even though the envelope is known to be critical to retain the native protein folding and biological function. According to some authors, CuAAC reaction was exploited for the surface modification of cowpea chlorotic mottle virus (CCMV). The exposed carboxyl residues of the CCMV capsids (**124**, Scheme 3.24) were modified with propargyl amine to introduce alkyne motif on the exterior surface of CCMV capsids (**125**). These groups were then further modified with an azide, using a triazole

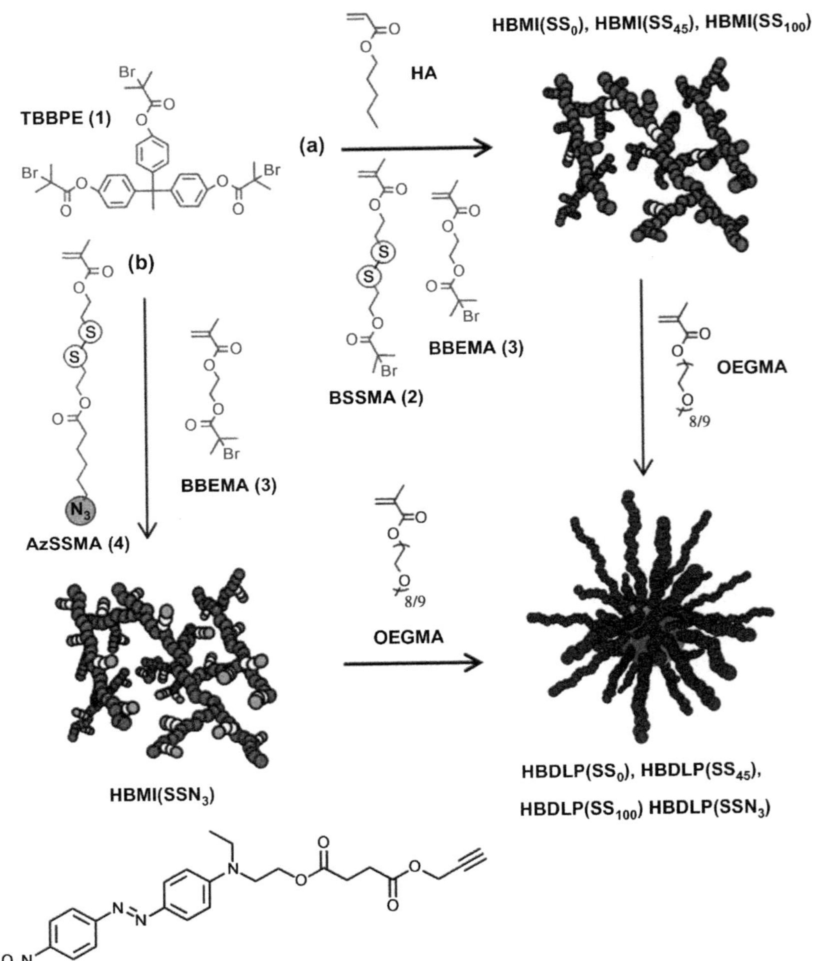

Disperse Red 13-prop-2-ynyl 123

FIGURE 3.17 Schematic illustration of the synthetic route to the unimolecular micelles with (A) backbone-cleavable disulfide bonds and (B) azide-functional cleavable pendant disulfide bonds. SSx in the polymer names denotes the molar fraction (x) of disulfide-containing brancher in the hyperbranched core. *AzSSMA*, 2-((2-(Methacryloyloxy)ethyl)disulfanyl)ethyl 6-azidohexanoate; *BBEMA*, (2-(2-bromoisobutyryloxy(ethyl methacrylate); *BSSMA*, 2-((2-((2-bromo-2-methylpropanoyl)oxy)ethyl)-disulfanyl)ethyl methacrylate; *HA*, hexyl acrylate; *HBDLPs*, redox-sensitive hyperbranched dendritic-linear polymers; *HBMIs*, hyperbranched macroinitiators; *OEGMA*, Oligo(ethylene glycol) methyl ether methacrylate; *TBBPE*, 1,1,1-tris (4′-(2″-bromoisobutyryloxy)phenyl)ethane. *Reprinted with permission from Porsch, C.; Zhang, Y.; Montanez, M. I.; Malho, J. -M.; Kostiainen, M. A.; Nystroem, A. M.; Malmstroem, E.* Biomacromolecules ***2015***, *16, 2872–2883. Copyright 2015 American Chemical Society.*

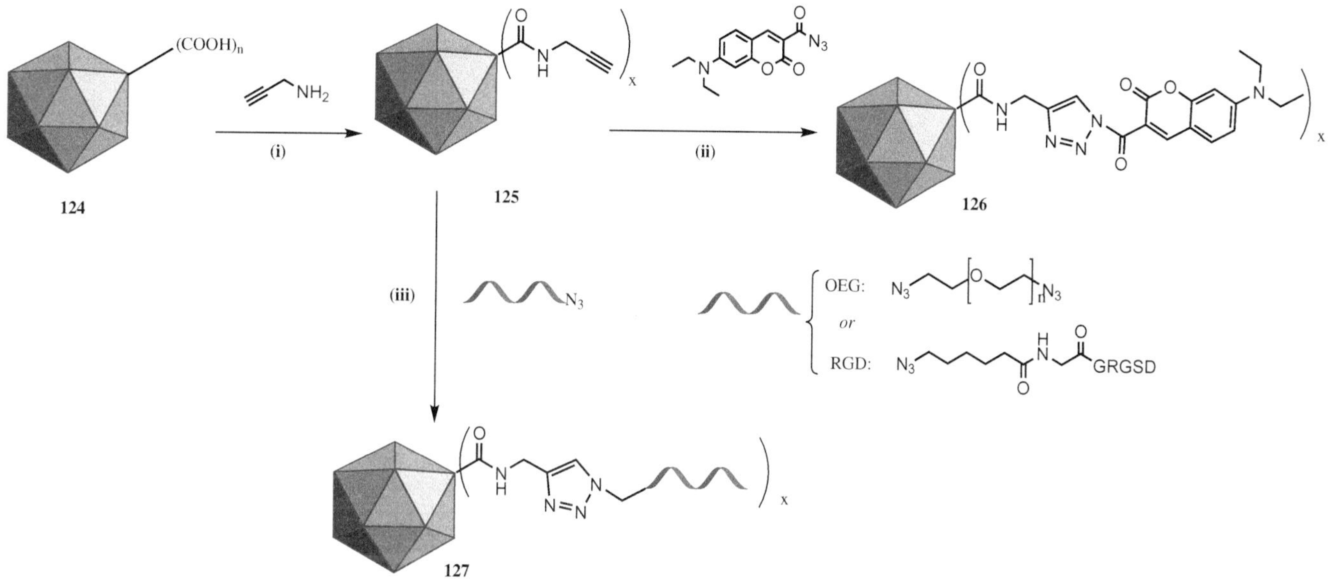

SCHEME 3.24 Schematic diagram of bioconjugation onto CCMV capsids using the CuAAC reaction. (i) propargylamine 1-ethyl-3-[3-dimethylaminopropyl] carbodiimide hydrochloride (EDC) + *N*-hydroxysulfosuccinimide sodium salt (sulfo-NHS), 24 h, 4°C. (ii) and (iii) coumarin azide, oligo-ethylene glycol (OEG) short chain or RGD-containing peptide, copper(II) sulfate, tris(2-carboxyethyl)phosphine hydrochloride, bathocuproindisulfonic acid disodium salt, 24 h, 4°C. Peptide sequence is reported using the one-letter notation.

connection in the presence of Cu(II), tris(2-carboxyethyl)phosphine hydrochloride and bathocuproin disulfonic acid disodium salt as copper ligand. Different probes were successfully grafted onto the CCMV capsids. At first, triazole connection of CCMV-alkyne with azide ligands of fluorescent coumarin azide was exploited (**126**) to follow the biological fate of the capsids when interacting with cell cultures, exploiting its fluorescence.

An OEG bis-azide short chain and an RGD peptide were also connected to the CCMV capsids via the CuAAC reaction (**127**). Identified as a pervasive cell-adhesive peptide, RGD has been extensively investigated to enhance the binding activity with mammalian cells. In contrast, OEG presents inhibition to cell binding similar to that of PEG, a water-soluble polymer. The use of the bis-azide moreover led to a cross-linked structure, which could further affect the cell-binding process.

The modification and integrity of the viral capsids were confirmed, as well as the cell-penetrating ability of the adduct with RGD, which demonstrates the potential applications in drug delivery.[76]

Similarly, tagged, enveloped VLPs (TagE-VLPs) as a valuable strategy for the production of specific-site-functionalized enveloped influenza VLPs was presented. This two-step procedure allows bioorthogonal functionalization of azide-tagged nascent influenza type A hemagglutinin proteins in the envelope of VLPs through a strain-promoted [3 + 2] alkyne-azide cycloaddition reaction; the drug (or taggant) to be loaded to the target cell was linked to a cyclooctine moiety. Labeling has no deleterious effects on VLPs biological function.[77]

In some ways, capside-like shuttles can be prepared in a simpler way using very hydrophilic polymers bearing sugars and/or proteins; these shuttles, with respect to capsids, are available in greater amounts; their structure is well defined and highly reproducible.

In a recent example, α-acetal, ω-alkyne poly(ethylene oxide) (**129**, Scheme 3.25) was synthesized by reaction of α-acetal, ω-hydroxy poly(ethylene oxide) **129** with propargyl bromide and used as building blocks of glycoconjugated poly(ε-caprolactone)-graft poly(ethylene oxide) (PCL-*g*-PEO) copolymers **134**.

The alkyne group in **129** is indeed instrumental for the PEGylation of a poly(α-azido-ε-caprolactone-*co*-ε-caprolactone) copolymer **130** by the Huisgen's 1,3 dipolar cycloaddition promoted by CuI and triethylamine, to obtain PCL-PEO **131**. Moreover, deprotection of the acetal end-group of the hydrophilic poly(ethylene oxide) grafts of **131** at acidic pH to obtain **132**, followed by reductive amination of the accordingly formed aldehyde with an aminated sugar is a valuable strategy of glycoconjugation of the graft copolymer (intermediate **133** reduced with sodium cyanoborohydride to **134**), whose nanometric micelles are then potentially a drug delivery system. As a proof of concept, a model molecule (fluoresceinamine) was considered in order to optimize the reductive amination. These conditions were then

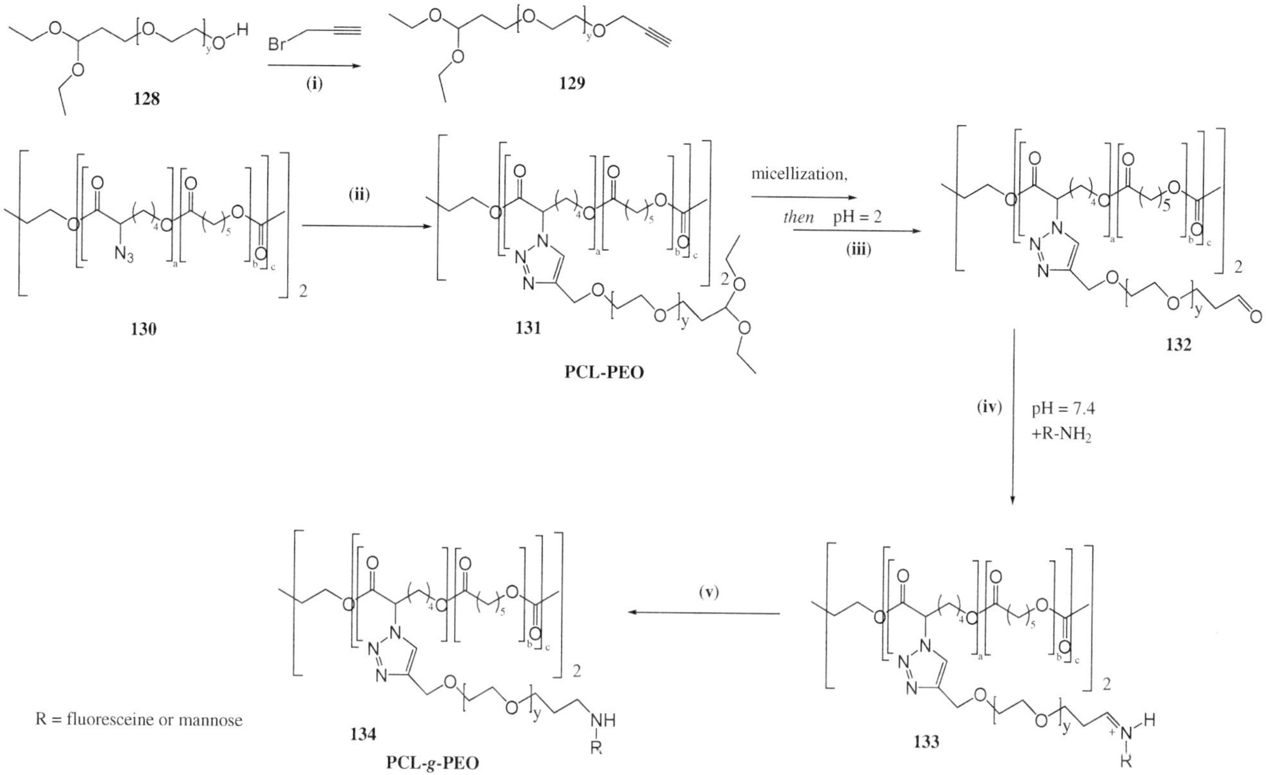

SCHEME 3.25 Strategy for the preparation of PCL-PEO copolymers and decoration of micelles by reductive amination of an amino-dye (when R = fluorescein) or amino-sugar (when R = mannose) with reactive PCL-PEO micelles. (i) potassium naphthalenide, THF, r.t., 12 h, yield: 90%; (ii) THF, CuI, trimethylamine, 35°C, 35 h, yield: 75%; (iii) water/THF, HCl to pH = 2, 2 h (not isolated); then (iv) phosphate buffer (500 mM) is added to pH, and then the amino derivative was added. After 1 h, $NaBH_3CN$ is added and after 96 h the reaction is stopped. Yield: 98% (from the polymer). R = fluoresceine or mannose.

extended to the decoration of the graft copolymer micelles with mannose by reaction with 2-aminoethyl-α-D-mannopyranoside, which is a targeting agent of dendritic cells and macrophages.[78]

Among other biological-based drug shuttles, DNA has emerged as a versatile building block for programmable self-assembly. A variety of strategies have been developed to functionalize these nanostructures. In a study, a facile click-based strategy to incorporate a metal chelating ligand and a fluorescent tag into a three-point-star DNA tile containing 2′-*O*-propargyl groups was reported.

The three-point-star motif, a three-arm planar structure that is composed of seven DNA strands where each of the three arms is a four-arm junction (Fig. 3.18), was used.

The motif is threefold symmetric, thus requiring three unique strands for assembly: a long central strand (strand L), three identical medium strands (strand M), and three identical short strands (strand S). Strands M and S were modified to contain a 2′-*O*-propargyl group, and the authors synthesized two variations each of the strands M and S containing the 2′-*O*-propargyl modifications at two different sites on the sequence, and assembled four species of the three-point-star motif using these modified strands (see Fig. 3.18). A azido-modified fluorescent tag (6-Carboxyfluorescein-azide: 6-FAM-azide, **135**) or a bipyridine azide ligand (**136**) was selected as the functional moiety to be attached onto these strands via click chemistry. This bipyridine ligand could be used as a metal chelating ligand and as a tool for chiral molecular recognition. The click reaction takes place in the presence of CuBr (water/acetonitrile, 12 h, r.t.).

Such a strategy opens up the possibility of functionalizing preassembled DNA strands to construct platforms for metal or drug delivery.[79]

Microparticles

Microparticles also have received considerable attention, as functionalized polymeric microparticles possess significant potential for controlled drug delivery and biosensing applications, but yet current fabrication techniques face challenges in simple and scalable fabrication and biofunctionalization.

Microparticles with controlled 2D and 3D shapes and dimensions can be prepared by micromolding methods combined with biopolymeric conjugation handles and bioorthogonal click reaction via SPAAC and tetrazine-trans-cyclooctene ligation reactions. The aminopolysaccharide chitosan was used as an efficient conjugation handle. The reaction takes place in absence of any catalyst, and microparticles can be purified and recovered by a repeated centrifugation/washing procedure.[80]

The cited example is a proof that metabolic sugar labeling followed by the use of reagent-free click chemistry can be considered an established technique for in vitro cell targeting. Similar to the example discussed earlier, but

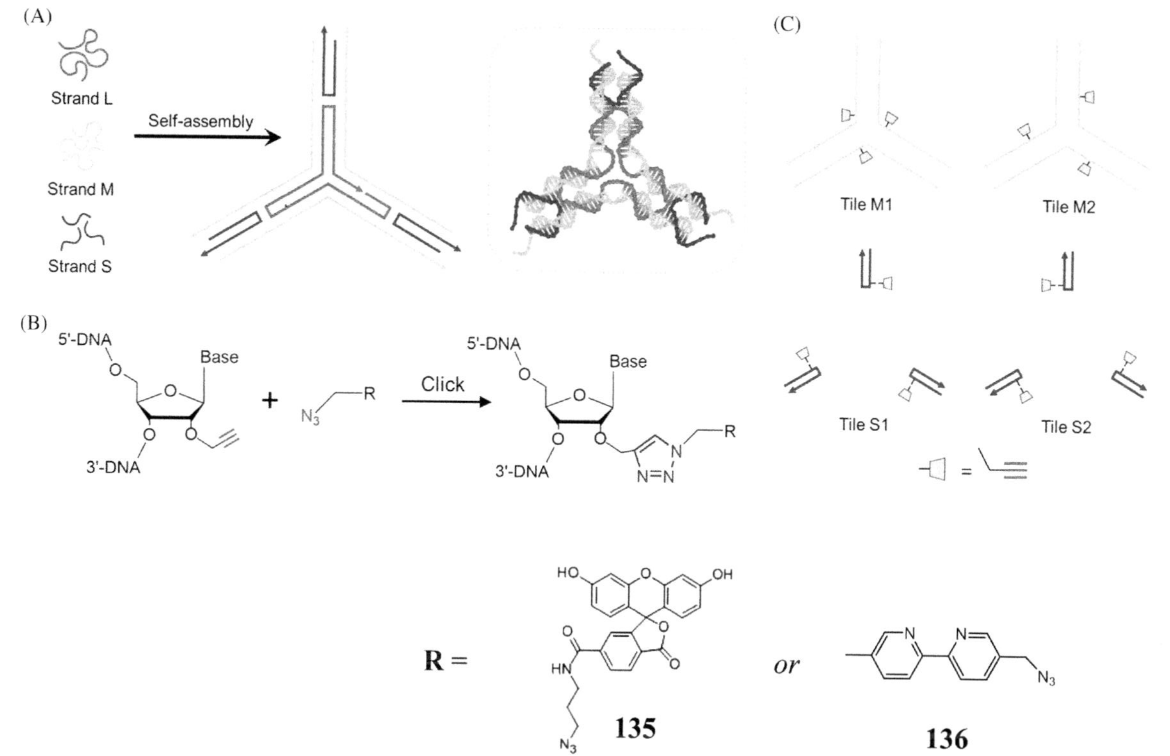

FIGURE 3.18 Click functionalization of a DNA tile. (A) Assembly of the three-point-star tile and a molecular model (inset). (B) 2′-*O*-propargyl modification and alkyne–azide click strategy. (C) The four modified tiles used. Tiles M1 and M2 have 2′-*O*-propargyl modifications at two different sites on strand M. Tiles S1 and S2 have 2′-*O*-propargyl modifications at two different sites on strand S. *Reproduced from Valsangkar, V.; Chandrasekaran, A. R.; Wang, R.; Haruehanroengra, P.; Levchenko, O.; Halvorsen, K.; Sheng, J. J.* Mater. Chem. B: Mater. Biol. Med. ***2017****, 5, 2074–2077, with permission of The Royal Society of Chemistry.*

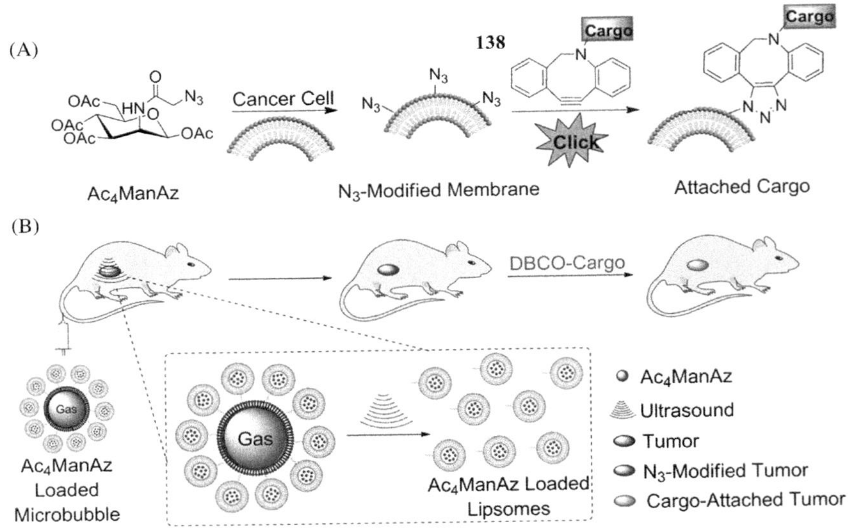

FIGURE 3.19 (A) Metabolic labeling of tetraacetyl *N*-azidoacetylmannosamine (Ac_4ManAz) and subsequent targeting of DBCO-bearing therapeutics (DBCO-cargo) using copper-free click chemistry. (B) Ultrasound-assisted accumulation and metabolic expression of Ac_4ManAz in the tumor area and subsequent tumor-targeted delivery of DBCO-cargo using click chemistry. *DBCO*, Dibenzocyclooctyne. *Reproduced from Wang, H.; Gauthier, M.; Kelly, J. R.; Miller, R. J.; Xu, M.; O'Brien, W. D, Jr; Cheng, J.* Angew. Chem. Int. Ed. ***2016***, *55, 5452–5456, with permission of John Wiley and Sons.*

with a remarkable technological leap, the use of targeted ultrasound pulses to induce the release of tetraacetyl *N*-azidoacetylmannosamine (Ac_4ManAz, **137**, Fig. 3.19) from microbubbles (MBs) was reported. These amine-functionalized, perfluorobutane-gas-filled MBs with a mean diameter of 870 nm were prepared using the one-step sonication method, and they were conjugated with azido-sugar loaded liposomes by the coupling reaction between activated carboxylates and amine groups (Fig. 3.19B).

Ac_4ManAz-loaded MBs showed great stability under physiological conditions, but rapidly collapsed in the presence of tumor-localized ultrasound pulses. A secondary effect of MB destruction is the temporary increase of capillary permeability and production of transient poration in cell membranes, using cancer-localized ultrasound pulses, the released Ac_4ManAz from MBs was able to selectively label quiet 4T1 tumor cells. The azido groups on Ac_4ManAz significantly improved the tumor accumulation of DBCO-bearing therapeutics/markers (DBCO-cargo, **138**) by subsequent SPAAC. As a proof-of-concept study, DABCO linked with a cyanine dye (DBCO-Cy5) was used, and in vivo experiments performed with mice grafted with subcutaneous 4T1 tumor, showing that preferential accumulation of the targeted compound in neoplastic tissue takes part.[81]

Macroscopic structures

Incorporation of unnatural amino acids (UAAs) bearing bioorthogonal reactive groups into proteins could be a powerful tool for developing novel protein-based biomaterials with innovative and controlled performance, as already discussed in the preceding sections, where viruses were incubated with UAAs bearing azido groups to produce azido-modified capsids. Also superior organisms can do the same things as viruses, and for example *Bombyx mori* was exploited for this purpose. *B. mori* in fact produce silk fibroin, a naturally derived protein material extensively studied for biomaterial development due to its mechanical strength and biocompatibility. It is in fact possible to incorporate 4-azidophenylalanine (AzPhe) into silk fibroin, which can be spun in varied material forms (thread, film, and porous sponge) and further modified by click chemistry. Oral administration of AzPhe mixed in a commercially available standard diet to the transgenic, mutant strain of *B. mori* H03 larvae led to efficient production of an AzPhe-incorporated silk fibroin, named AzidoSilk. AzidoSilk threads were subjected to CuAAC with carboxyrhodamine 110 alkyne (Tris–HCl pH 8, $CuSO_4$, Tris[(1-benzyl-1*H*-1,2,3-triazol-4-yl)methyl]amine: TBTA, sodium ascorbate) or SPAAC with sulforhodamine B DBCO or sulfo-DBCO-biotin conjugate (sulfo-DBCO-biotin) at room temperature (Tris–HCl, pH 8). To demonstrate the possibility of modifying AzidoSilk sponges with active proteins by click chemistry, green fluorescent protein, modified with DBCO-PEG4-NHS ester beforehand, was reacted with AzidoSilk sponge, and resulted selectively attached to the sponge, demonstrating that active proteins could be used to modify AzidoSilk to enhance its function as a biomaterial.

Using this methodology, photolithographic micropatterning of fluorescent molecules directly onto silk fibroin film was achieved and opened the field to the availability of silk-based biomaterials for drug delivery, tissue scaffolds, and implantable devices.[82]

While silk-based materials are scarcely biodegradables in the host tissue, biocompatible and degradable injectable materials prepared via bioorthogonal reactions in situ and administered in a minimally invasive way are highly promising for biomedical applications. In a work, a PEG-based injectable hydrogel was fabricated via a copper-free, SPAAC click chemistry. Azide and cyclooctyne moieties on the PEG backbones (**139** and **140** in Fig. 3.20, respectively) underwent a rapid click reaction to trigger the formation of the hydrogel within several minutes when the precursors are injected together in vivo. Resulting from the introduction of ester groups into the cross-linked network, the hydrogel presented pH-dependent hydrolysis and biological compatibility and fast degradability.

The injectable hydrogel may be useful as biomaterials such as embolic agents for interventional therapy.[83]

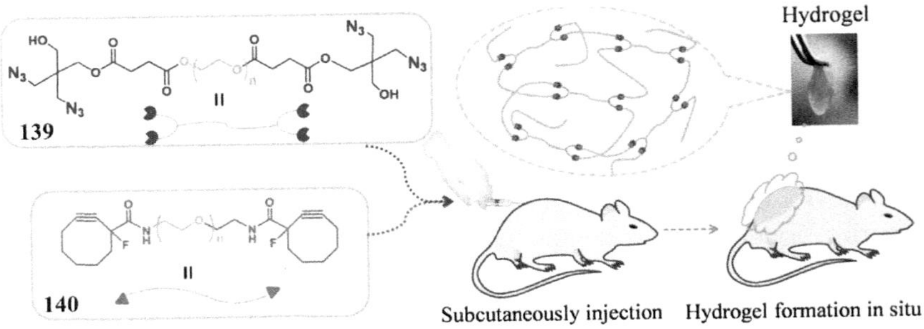

FIGURE 3.20 Precursor structures and hydrogel formation in vivo based on SPAAC click chemistry. *SPAAC*, Strain-promoted azide-alkyne cycloaddition. *Reproduced from Jiang, H.; Qin, S.; Dong, H.; Lei, Q.; Su, X.; Zhuo, R.; Zhong, Z.* Soft Matter ***2015***, 11, *6029–6036, with permission of The Royal Society of Chemistry.*

Preformed hydrogels are also interesting, as they can be loaded with the drug before their implantation in the body. For this purpose, a gel based on temperature-responsive copolymer (**150** in Scheme 3.26) composed of poly [2-(2-methoxyethoxy)ethyl methacrylate-*co*-OEGMA-*co*-*N*-hydroxymethyl acrylamide], P(MEO2MA-*co*-OEGMA-*co*-HMAM), **145**, and an alkyne derivatized poly(methacrylic acid), (PMOH/PMAA hydrogel) **149**, was synthesized by atom transfer radical polymerization and RAFT polymerization. In detail, the precursor PM-OH **144** was synthesized by atom transfer radical polymerization of *N*-hydroxymethyl acrylamide **141**, 2-(2-methoxyethoxy) ethyl methacrylate **142**, and OEGMA **143**, and finally the hydroxyl groups on **143** were transformed into azide groups (**145**).

In parallel, RAFT polymerization of methacrylic acid **146** in presence of *S*-1-dodecyl-*S*′-(α,α′-dimethyl-α″-acetic acid)-trithiocarbonate **147** chain transfer agent led to PMMA **148**, and the carboxyl groups on PMAA were transformed into alkyne groups (**149**).

The temperature and pH responsive hydrogel was fabricated by click chemistry between the azide-groups of **145** and alkyne groups of **149** in the presence of $CuSO_4$ and sodium ascorbate in aqueous solution. The rheological kinetics of gel formation demonstrated that gelation had commenced within 5 min at 25°C, and proved that the formed hydrogel had temperature and pH sensitivities. Bovine serum albumin was used as a model to evaluate the sustained release of the hydrogel; the results indicated that the hydrogel was a promising candidate for controlling protein drug delivery.[84]

Inspired by upregulated levels of fucosylated proteins on the surfaces of multiple types of cancer cells, micelles carrying β-L-fucose and β-D-glucose were prepared. A range of block copolymers were synthesized by reacting 2-azidoethyl β-L-fucopyranoside ($FucEtN_3$) and 2-azideoethyl β-D-glucopyranoside ($GlcEtN_3$) with poly(propargyl methacrylate)-block-poly(Bu acrylate) (PPMA-*b*-PBA) using CuAAC. This last was prepared

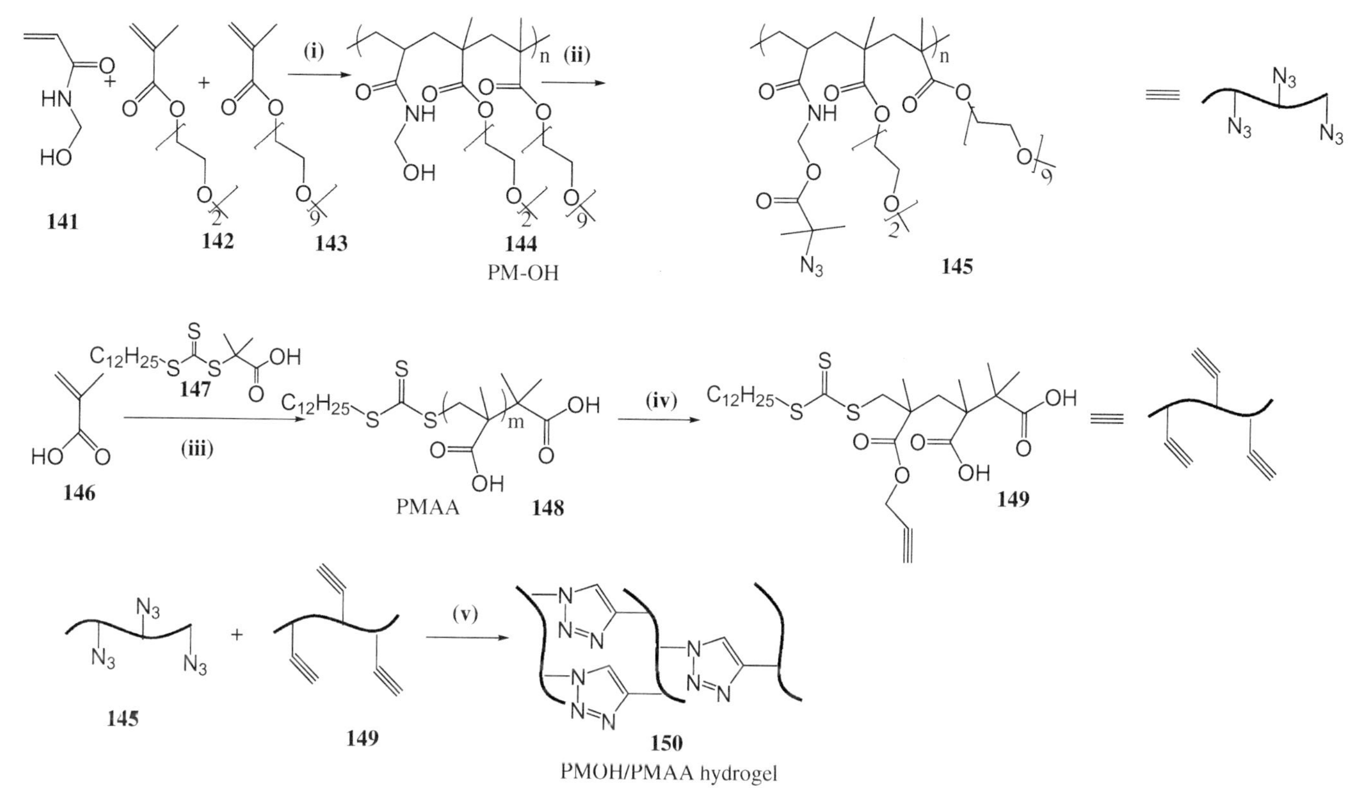

SCHEME 3.26 The synthetic routes of PMOH/PMAA hydrogel and its building blocks. (i) CuCl, 2,2′-bipyridine, 60°C, 8 h; (ii) (1) 2-Bromoisobutyryl bromide, DMF, triethylamine, 24 h, then (2) NaN_3, DMF, 24 h; (iii) ethanol, 60°C, AIBN, 12 h; (iv) DMSO; propargyl bromide, 1,8-diazabicyclo-[5.4.0]-7-undecene (DBU) 3 h, r.t. v) sodium ascorbate, $CuSO_4$, water, r.t., several minutes [sic].

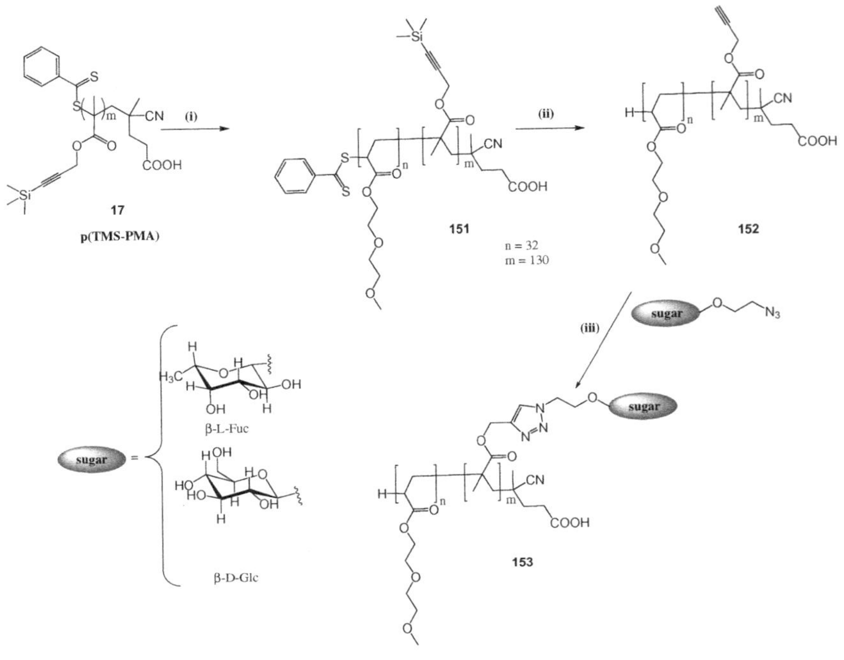

pPMA-*b*-p(OEGMA-DMAEMA)-R

SCHEME 3.27 Schematic Representation of the synthesis of amphiphilic block glycopolymers with various Glc/Fuc ratios. (i) *n*-butyl acrylate, AIBN, toluene, 70°C, 4 h; (ii) TBAF, THF, 0°C, 30 min; (iii) PMDETA, CuBr, DMF r.t., 12 h. *AIBN*, Azobis(isobutyronitrile); *PMDETA*, *N*, *N*,*N*′,*N*″,*N*″-Pentamethyldiethylenetriamine.

starting from p(TMS-PMA), (already described as **17** in Scheme 3.5), using it as macro RAFT agent and chain-extended with *n*-butyl acrylate (*n*-BA), to obtain **151** (Scheme 3.27). The polymer was deprotected with tetrabutylammonium fluoride (TBAF) in THF to obtain **152**. Lastly, **152**, the carbohydrate azide, PMDETA, CuBr, and DMF were reacted to produce **153**.

The micelles that can be prepared using the synthesized polymers are nanometric and reveal an increased uptake by pancreatic, lung, and ovarian carcinoma cell lines, whereas the uptake by the nonneoplastic cell lines (CHO) is negligible. This finding suggests that these micelles can be used for targeted drug delivery.[85]

Acknowledgment

Financial support by THE University of Pavia and MIUR (PRIN 2011, CUP: F11J12000210001) is gratefully acknowledged.

References

1. Lallana, E.; Sousa-Herves, A.; Fernandez-Trillo, F.; Riguera, R.; Fernandez-Megia, E. Click Chemistry for Drug Delivery Nanosystems. *Pharm. Res.* **2012,** *29*, 1–34.
2. Kolb, H. C.; Finn, M. G.; Sharpless, K. B. *Angew. Chem. Int. Ed.* **2001,** *40*, 2004–2021.
3. Reichart, F.; Horn, M.; Neundorf, I. *J. Pept. Sci.* **2016,** *22*, 421–426.
4. Ashley, G. W.; Henise, J.; Reid, R.; Santi, D. V. *Proc. Natl. Acad. Sci. U.S.A.* **2013,** *110*, 2318–2323.
5. Karimi, M.; Solati, N.; Amiri, M.; Mirshekari, H.; Mohamed, E.; Taheri, M.; Hashemkhani, M.; Saeidi, A.; Estiar, M. A.; Kiani, P.; Ghasemi, A.; Basri, S. M.; Aref, A. R.; Hamblin, M. R. *Expert Opin. Drug Deliv.* **2015,** *12*, 1071–1087.
6. Martinez, A.; Ortiz Mellet, C.; Garcia Fernandez, J. M. *Chem. Soc. Rev.* **2013,** *42*, 4746–4773.
7. Gallego-Yerga, L.; Lomazzi, M.; Franceschi, V.; Sansone, F.; Ortiz Mellet, C.; Donofrio, G.; Casnati, A.; Garcia Fernandez, J. M. *Org. Biomol. Chem.* **2015,** *13*, 1708–1723.
8. Mendez-Ardoy, A.; Guilloteau, N.; Di Giorgio, C.; Vierling, P.; Santoyo-Gonzalez, F.; Ortiz Mellet, C.; Garcia Fernandez, J. M. *J. Org. Chem.* **2011,** *76*, 5882–5894.
9. Guo, J.; Ogier, J. R.; Desgranges, S.; Darcy, R.; O'Driscoll, C. *Biomaterials* **2012,** *33*, 7775–7784.
10. Ma, D.; Zhang, H.-B.; Chen, Y.-Y.; Lin, J.-T.; Zhang, L.-M. *J. Colloid Interface Sci.* **2013,** *405*, 305–311.
11. Srinivasachari, S.; Fichter, K. M.; Reineke, T. M. *J. Am. Chem. Soc.* **2008,** *130*, 4618–4627.
12. Anderson, K.; Sizovs, A.; Cortez, M.; Waldron, C.; Haddleton, D. M.; Reineke, T. M. *Biomacromolecules* **2012,** *13*, 2229–2239.
13. Mai, K.; Zhang, S.; Liang, B.; Gao, C.; Du, W.; Zhang, L.-M. *Carbohydr. Polym.* **2015,** *123*, 237–245.
14. Gosangi, M.; Rapaka, H.; Ravula, V.; Patri, S. V. *Bioconj. Chem.* **2017**. Ahead of Print.
15. Yin, Q.; Yin, L.; Wang, H.; Cheng, J. *Acc. Chem. Res.* **2015,** *48*, 1777–1787.
16. Cheng, Y.; Sellers, D. L.; Tan, J.-K. Y.; Peeler, D. J.; Horner, P. J.; Pun, S. H. *Biomaterials* **2017,** *127*, 89–96.
17. Gao, Y.; Yin, Q.; Chen, L.; Zhang, Z.; Li, Y. *Bioconjug. Chem.* **2011,** *22*, 1153–1161.
18. Cortez, M. A.; Godbey, W. T.; Fang, Yunlan; Payne, M. E.; Cafferty, B. J.; Kosakowska, K. A.; Grayson, S. M. *J. Am. Chem. Soc.* **2015,** *137*, 6541–6549.
19. Wei, H.; Chu, D. S. H.; Zhao, J.; Pahang, J. A.; Pun, S. H. *ACS Macro Lett.* **2013,** *2*, 1047–1050.
20. Mauri, E.; Moroni, I.; Magagnin, L.; Masi, M.; Sacchetti, A.; Rossi, F. *React. Funct. Polym.* **2016,** *105*, 35–44.
21. Gandioso, A.; Massaguer, A.; Villegas, N.; Salvans, C.; Sanchez, D.; Brun-Heath, I.; Marchan, V.; Orozco, M.; Terrazas, M. *Chem. Comm.* **2017,** *53*, 2870–2873.
22. Sun, Q.; Kang, Z.; Xue, L.; Shang, Y.; Su, Z.; Sun, H.; Ping, Q.; Mo, R.; Zhang, C. *J. Am. Chem. Soc.* **2015,** *137*, 6000–6010.
23. Chang, Y.; Yang, K.; Wei, P.; Huang, S.; Pei, Y.; Zhao, W.; Pei, Z. *Angew. Chem. Int. Ed.* **2014,** *53*, 13126–13130.
24. Patil, S. P.; Kim, S.-H.; Jadhav, J. R.; Lee, J.-h; Jeon, E. M.; Kim, K.-T.; Kim, B. H. *Bioconjug. Chem.* **2014,** *25*, 1517–1525.
25. Iwata, R.; Nishina, K.; Yokota, T.; Wada, T. *Bioorg. Med. Chem.* **2014,** *22*, 1394–1403.
26. Sanchez-Nieves, J.; Fransen, P.; Pulido, D.; Lorente, R.; Munoz-Fernandez, M. A.; Albericio, F.; Royo, M.; Gomez, R.; de la Mata, F. J. *Eur. J. Med. Chem.* **2014,** *76*, 43–52.

27. Noureddine, A.; Gary-Bobo, M.; Lichon, L.; Garcia, M.; Zink, J. I.; Wong Chi Man, M.; Cattoen, X. *Chemistry* **2016,** *22*, 9624–9630.
28. Mauriello-Jimenez, C.; Croissant, J.; Maynadier, M.; Cattoen, X.; Wong Chi Man, M.; Vergnaud, J.; Chaleix, V.; Sol, V.; Garcia, M.; Gary-Bobo, M., et al. *J. Mater. Chem. B: Mater. Biol. Med.* **2015,** *3*, 3681–3684.
29. Yang, Y.; Aw, J.; Chen, K.; Liu, F.; Padmanabhan, P.; Hou, Y.; Cheng, Z.; Xing, B. *Chemistry* **2011,** *6*, 1381–1389.
30. Li, N.; Guo, C.; Duan, Z.; Yu, L.; Luo, K.; Lu, J.; Gu, Z. *J. Mater. Chem. B: Mater. Biol. Med.* **2016,** *4*, 3760–3769.
31. a) Wang, C. E.; Wei, H.; Tan, N.; Boydston, J.; Pun, S. H. *Biomacromolecules* **2016,** *17*, 69–75. b) Wei, H.; Wang, C. E.; Tan, N.; Boydston, A. J.; Pun, S. H. *ACS Macro Lett.* **2015,** *4*, 938–941.
32. Fu, L.; Sun, C.; Yan, L. *ACS Appl. Mater. Interfaces* **2015,** *7*, 2104–2115.
33. Oh, S. S.; Lee, B. F.; Leibfarth, F. A.; Eisenstein, M.; Robb, M. J.; Lynd, N. A.; Hawker, C. J.; Soh, H. T. *J. Am. Chem. Soc.* **2014,** *136*, 15010–15015.
34. Qin, M.; Zong, H.; Kopelman, R. *Biomacromolecules* **2014,** *15*, 3728–3734.
35. Zhang, C.; Pan, D.; Luo, K.; Li, N.; Guo, C.; Zheng, X.; Gu, Z. *Polym. Chem* **2014,** *5*, 5227–5235.
36. Yu, Y.; Chen, C.-K.; Law, W.-C.; Weinheimer, E.; Sengupta, S.; Prasad, P. N.; Cheng, C. *Biomacromolecules* **2014,** *15*, 524–532.
37. Li, N.; Cai, H.; Jiang, L.; Hu, J.; Bains, A.; Hu, J.; Gong, Q.; Luo, K.; Gu, Z. *ACS Appl. Mater. Interfaces* **2017,** *9*, 6865–6877.
38. Bianchi, A.; Arosio, D.; Perego, P.; De Cesare, M.; Carenini, N.; Zaffaroni, N.; De Matteo, M.; Manzoni, L. *Org. Biomol. Chem.* **2015,** *13*, 7530–7541.
39. Zhu, W.; Wang, Y.; Cai, X.; Zha, G.; Luo, Q.; Sun, R.; Li, X.; Shen, Z. *J. Mater. Chem. B: Mater. Biol. Med.* **2015,** *3*, 3024–3031.
40. Heckert, B.; Banerjee, T.; Sulthana, S.; Naz, S.; Alnasser, R.; Thompson, D.; Normand, G.; Grimm, J.; Perez, J. M.; Santra, S. *ACS Macro Lett.* **2017,** *6*, 235–240.
41. Zhang, F.; Zhang, S.; Pollack, S. F.; Li, R.; Gonzalez, A. M.; Fan, J.; Zou, J.; Leininger, S. E.; Pavia-Sanders, A.; Johnson, R., et al. *J. Am. Chem. Soc.* **2015,** *137*, 2056–2066.
42. Abandansari, H. S.; Nabid, M. R.; Rezaei, S. J. T.; Niknejad, H. *Polymer* **2014,** *55*, 3579–3590.
43. Chen, W.; Zhang, J. Z.; Hu, J.; Guo, Q.; Yang, D. J. *Polym. Sci., A: Polym. Chem.* **2014,** *52*, 366–374.
44. Lai, C.-H.; Chang, T.-C.; Chuang, Y.-J.; Tzou, D.-L.; Lin, C.-C. *Bioconjug. Chem.* **2013,** *24*, 1698–1709.
45. Xiao, K.; Li, Y.; Lee, J. S.; Gonik, A. M.; Dong, T.; Fung, G.; Sanchez, E.; Xing, L.; Cheng, H. R.; Luo, J.; Lam, K. S. *Cancer Res.* **2012,** *72*, 2100–2110.
46. Ren, K.; Zhang, M.; He, J.; Ni, P.; Wu, Y. *ACS Appl. Mater. Interfaces* **2015,** *7*, 11263–11271.
47. Aydin, O.; Youssef, I.; Yuksel Durmaz, Y.; Tiruchinapally, G.; ElSayed, M. E. H. *Mol. Pharmaceutics* **2016,** *13*, 1413–1429.
48. Du, X.; Sun, Y.; Zhang, M.; He, J.; Ni, P. *ACS Appl. Mater. Interfaces* **2017,** *9*, 13939–13949.
49. Zhang, Q.; He, J.; Zhang, M.; Ni, P. *J. Mater. Chem. B: Mater. Biol. Med.* **2015,** *3*, 4922–4932.
50. Wang, H.; Xu, M.; Xiong, M.; Cheng, J. *Chem. Commun.* **2015,** *51*, 4807–4810.
51. Wang, H.; Tang, L.; Tu, C.; Song, Z.; Yin, Q.; Yin, L.; Zhang, Z.; Cheng, J. *Biomacromolecules* **2013,** *14*, 3706–3712.

52. Alam, S.; Alves, D. S.; Whitehead, S. A.; Bayer, A. M.; McNitt, C. D.; Popik, V. V.; Barrera, F. N.; Best, M. D. *Bioconjug. Chem.* **2015,** *26*, 1021–1031.
53. Joshi, M.; Kumar, P.; Kumar, R.; Sharma, G.; Singh, B.; Katare, O. P; Raza, K. *Mater. Sci. Eng., C: Mater. Biol. Appl.* **2017,** *75*, 1376–1388.
54. Choi, E. J.; Shin, J.; Khaleel, Z. H.; Cha, I.; Yun, S.-H.; Cho, S.-W. Song, C. *Polym. Chem.* **2015,** *6*, 4473–4478.
55. Li, M.; Neoh, K. G.; Xu, L.; Yuan, L.; Leong, D. T.; Kang, E.-T.; Chua. K. L.; Hsu, L. Y. *Pharm. Res.* **2016,** *33*, 1161–1174.
56. Vasilyeva, S. V.; Levina, A. S.; Li-Zhulanov, N. S.; Shatskaya, N. V.; Baiborodin, S. I.; Repkova, M. N.; Zarytova, V. F.; Mazurkova, N. A.; Silnikov, V. N. *Bioorg. Med. Chem.* **2015,** *23*, 2168–2175.
57. Borreguero, A. M.; Munoz, M.; De Haro, J. C.; Carmona, M.; Rodriguez, J. F. *React. Funct. Polym.* **2016,** *101*, 1–8.
58. Castro, L. S.; Kviecinski, M. R.; Ourique, F.; Parisotto, E. B.; Grinevicius, V. M.; Correia, J. F.; Wilhelm Filho, D.; Pedrosa, R. C. *Redox Biol.* **2016,** *10*, 90–99.
59. Yhaya, F.; Binauld, S.; Kim, Y.; Stenzel, M. H. *Macromol. Rapid Commun.* **2012,** *33*, 1868–1874 S1868/1-S1868/17.
60. Cao, X. T.; Kim, Y. H.; Park, J. M.; Lim, K. T. *Europ. Polym. J.* **2016,** *78*, 264–273.
61. Oude Blenke, E.; Sleszynska, M.; Evers, M. J. W.; Storm, G.; Martin, N. I.; Mastrobattista, E. *Bioconjug. Chem.* **2017,** *28*, 574–582.
62. Ekholm, F. S.; Pynnoenen, H.; Vilkman, A.; Pitkaenen, V.; Helin, J.; Saarinen, J; Satomaa, T. *ChemMedChem* **2016,** *11*, 2501–2505.
63. Zhang, L.; Navaratna, T.; Thurber, G. M. *Bioconjug. Chem.* **2016,** *27* (7). 1663–1672.
64. Henise, J.; Hearn, B. R.; Ashley, G. W.; Santi, D. V. *Bioconjug. Chem.* **2015,** *26*, 270–278.
65. Schneider, E. L.; Henise, J.; Reid, R.; Ashley, G. W.; Santi, D. V. *Bioconjug. Chem.* **2016,** *27*, 1210–1215.
66. Jia, Z.; Zhang, Y.; Chen, Y. H.; Dusad, A.; Yuan, H.; Ren, K.; Li, F.; Fehringer, E. V; Purdue, P. E.; Goldring, S. R. *J. Contr. Rel.* **2015,** *200*, 23–34.
67. Wamberg, M. C.; Pedersen, P. L.; Loffler, P. M. G.; Albertsen, A. N.; Maurer, S. E; Nielsen, K. A.; Monnard, P.-A. *Bioconjug. Chem.* **2017**. Ahead of Print.
68. Whitehead, A.; McNitt, C. D.; Mattern-Schain, S. I.; Carr, A. J.; Alam, S.; Popik, V. V; Best, M. D. *Bioconjug. Chem.* **2017,** *28*, 923–932.
69. Le Fer, G.; Le Coeligatur, C.; Guigner, J.-M.; Amiel, C.; Volet, G. *Langmuir* **2017,** *33*, 2849–2860.
70. Lav, T.-X.; Lemechko, P.; Renard, E.; Amiel, C.; Langlois, V.; Volet, G. *React. Funct. Polym.* **2013,** *73*, 1001–1008.
71. Pati, D.; Das, S.; Patil, N. G.; Parekh, N.; Anjum, D. H.; Dhaware, V.; Ambade, A. V; Sen Gupta, S. *Biomacromolecules* **2016,** *17*, 466–475.
72. Vutti, S.; Schoffelen, S.; Bolinsson, J.; Buch-Manson, N.; Bovet, N.; Nygard. J.; Martinez, K. L.; Meldal, M. *Chem. Eur. J.* **2016,** *22*, 496–500.
73. Cappelli, A.; Paolino, M.; Grisci, G.; Razzano, V.; Giuliani, G.; Donati, A.; Bonechi, C; Mendichi, R.; Battiato, S.; Samperi, F.; Scialabba, C.; Giammona, G.; Makovec, F; Licciardi, M. *Polym. Chem.* **2016,** *7*, 6529–6544.
74. Gauche, C.; Lecommandoux, S. *Polymer* **2016,** *107*, 474–484.
75. Porsch, C.; Zhang, Y.; Montanez, M. I.; Malho, J.-M.; Kostiainen, M. A.; Nystroem, A. M.; Malmstroem, E. *Biomacromolecules* **2015,** *16*, 2872–2883.
76. Wu, Y.; Yang, H.; Jeon, Y.-J.; Lee, M.-Y.; Li, J.; Shin, H.-J. *Biotechnol. Bioprocess Eng.* **2014,** *19*, 747–753.

77. Carvalho, S. B.; Freire, I.; Joao, M.; Moleirinho, M. G.; Monteiro, F.; Gaspar, D.; Castanho, M. A. R. B.; Carrondo, M. J. T.; Alves, P. M.; Bernardes, G. J. L.; Peixoto, C. *Bioconjug. Chem.* **2016,** *27*, 2386–2399.
78. Freichels, H.; Alaimo, D.; Auzely-Velty, R.; Jerome, C. *Bioconjug. Chem.* **2012,** *23*, 1740–1752.
79. Valsangkar, V.; Chandrasekaran, A. R.; Wang, R.; Haruehanroengra, P.; Levchenko, O.; Halvorsen, K.; Sheng, J. *J. Mater. Chem. B: Mater. Biol. Med.* **2017,** *5*, 2074–2077.
80. Jung, S.; Choi, C.-H.; Lee, C.-S.; Yi, H. *Biotech. J.* **2016,** *11*, 1561–1571.
81. Wang, H.; Gauthier, M.; Kelly, J. R.; Miller, R. J.; Xu, M.; O'Brien, W. D., Jr; Cheng, J. *Angew. Chem. Int. Ed.* **2016,** *55*, 5452–5456.
82. Teramoto, H.; Nakajima, K.-I.; Kojima, K. *ACS Biomater. Sci. Eng.* **2016,** *2*, 251–258.
83. Jiang, H.; Qin, S.; Dong, H.; Lei, Q.; Su, X.; Zhuo, R.; Zhong, Z. *Soft Matter* **2015,** *11*, 6029–6036.
84. Sheng, W.; Liu, T.; Liu, S.; Wang, Q.; Li, X.; Guang, N. *Polym. Int.* **2015,** *64*, 1415–1424.
85. Babiuch, K.; Dag, A.; Zhao, J.; Lu, H.; Stenzel, M. H. *Biomacromolecules* **2015,** *16*, 1948–1957.

Chapter 4

Natural strategies in cycloaddition reactions

Misal Giuseppe Memeo
Department of Chemistry, University of Pavia, Pavia, Italy

Many known chemical reactions performed every day in organic chemistry laboratories have had their natural analogues on the stage for ages. Furthermore, developments achieved in organic and biological chemistry triggered the ability to synthesize completely new enzymes or selectively modified natural biomolecules, allowing researchers to make happen almost every sort of chemical reaction both in isolated cells and living systems. Thus, to properly define a natural cycloaddition (NC) is definitely trickier than what it would have been some years ago.[1] According to the title of this chapter we consider it appropriate not to include in NC all of those reactions that involve synthetic or functionalized enzymes on either natural and unnatural substrates. Even though we opted for this limitation, the number of reactions that we could consider as NC is still quite broad. An interesting example, representative of the ambiguous category of NC, is represented by macrophomate synthase.[2] This enzyme is produced by nature in the fungus *Macrophoma commelinae* and is involved in the conversion of 5-acetyl-4-methoxy-6-methyl-2-pyrone (**2**), to 4-acetyl-3-methoxy-5-methylbenzoic acid (**7**). While the overall mechanism of conversion from **2** to the final products was mainly established, some doubts concerning the transformation of **3** into **5** lasted for quite a long time.[3] In Scheme 4.1, two possible mechanisms are proposed: (A) a straightforward conversion of **3** into **5** through a Diels–Alder reaction and (B) a stepwise Michael cyclization to get to the same adduct. Then a common decarboxylation step leads to the formation of **6**.

Even though overall both pathways involve a cyclization, the two proposed mechanisms are deeply different, especially when observed from an enzymatic point of view. In path **A** the enzyme is catalyzing a reaction that does not involve any intermediate and the catalytic effect has to be played on lowering the energy of a single Transition State (TS), while in

Modern Applications of Cycloaddition Chemistry. DOI: https://doi.org/10.1016/B978-0-12-815273-7.00004-6

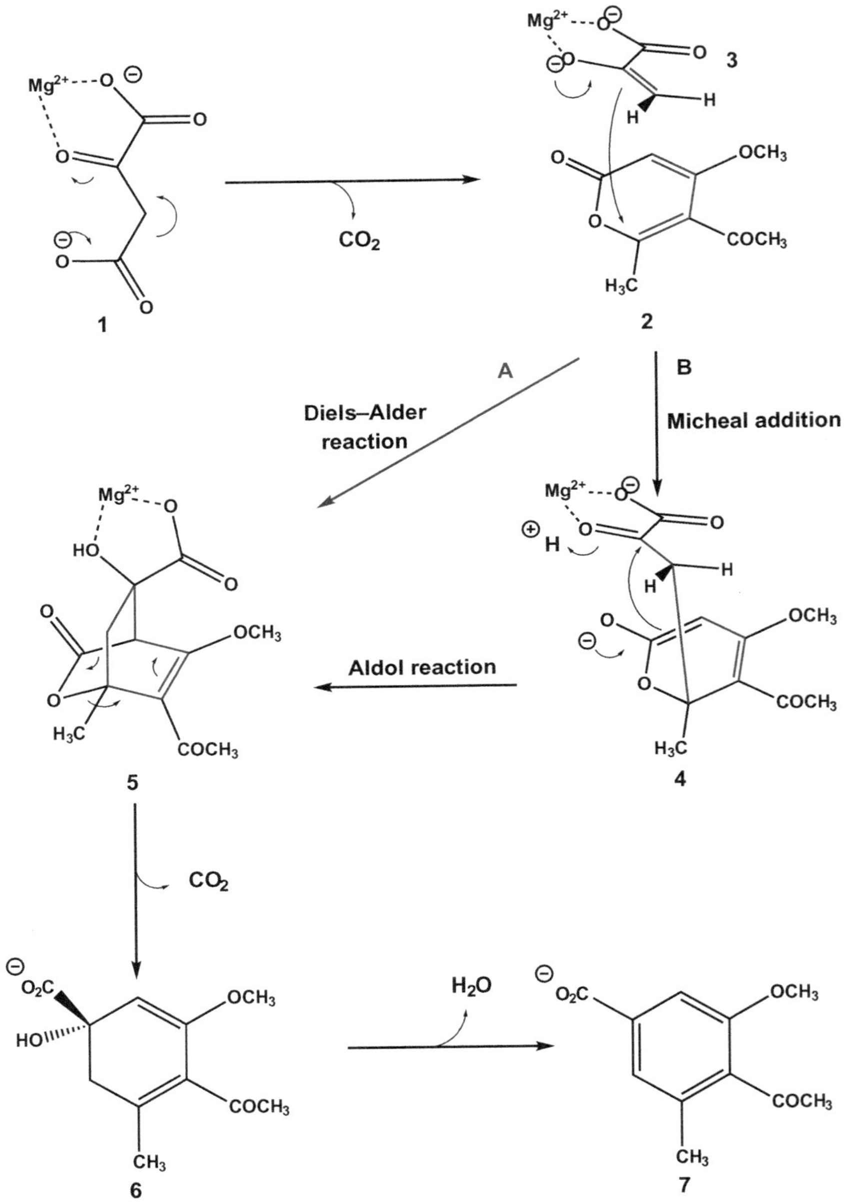

SCHEME 4.1 Conversion of 5-acetyl-4-methoxy-6-methyl-2-pyrone (2) to 4-acetyl-3-methoxy-5-methylbenzoic acid (7) through a Diels–Alder mechanism (A) and a Michael addition (B).

case **B** there is a stepwise mechanism that involves one intermediate and two TSs. Oikawa et al., supported by trapping and derivatization experiments, proposed the existence of **5** suggesting that its formation could have come from an unprecedented natural Diels–Alder cycloaddition.[3c] Years later, Jorgensen and coworkers found out that from a computational point of view the theoretical Diels–Alder was unlikely, whereas the stepwise Michael cyclization hypothesis looked more realistic.[4a] Furthermore Hilvert et al. showed that machrophomate synthase can serve as aldolase promoting the formation of β-hydroxy phenylpropanoate from oxaloacetate and benzaldehyde with decarboxylation of oxaloacetate being rate-limiting under saturating conditions.[3d] Scheme 4.2 reports an example of a transformation promoted by Macrophomate synthase (MPS) acting as aldolase on a synthetic substrate.

A second and still debated example is represented by the first reported putative Diels-Alderase (DAase) by Ichihara,[5a] who was working on a class of phytotoxins produced by two phytopathogenic fungi called *Alternaria solani* and *Aschochyta rabiei*.[5–7] These toxins, called solanapyrones, are polyketides characterized by the presence of a *trans*-decalin core whose presence suggests the possible involvement of a DAase in their biosynthesis. The natural synthesis of solanapyrons begins with a sequence of reactions that transform acetate into desmethylprosolanapyrone I **11** catalyzed by the type I polyketide synthase (PKS). Then, *O*-methyltransferase Sol2 transforms **11** into **12**, which is monooxygenated by the cytochrome P450 enzyme Sol6 to yield the corresponding alcohol prosolanapyrone II **13** that bears the two addends of the putative intramolecular Diels–Alder reaction. Because of the transconfiguration of the diene moiety this molecule is relatively stable in water, having a half-life of 6 days at 30°C. Compound **13** is then converted through oxidation of the hydroxymethylene functionality to the corresponding aldehyde prosolanapyrone III **14** by oxygen and the flavin adenine dinucleotide–dependent oxidase Sol5. At this point, the intramolecular cycloaddition takes place, with formation of the corresponding Diels–Alder adducts Solanapyrone A **15** and Solanapyrone D **16**, which, through a reduction mediated by dehydrogenase Sol3, are finally converted in the final

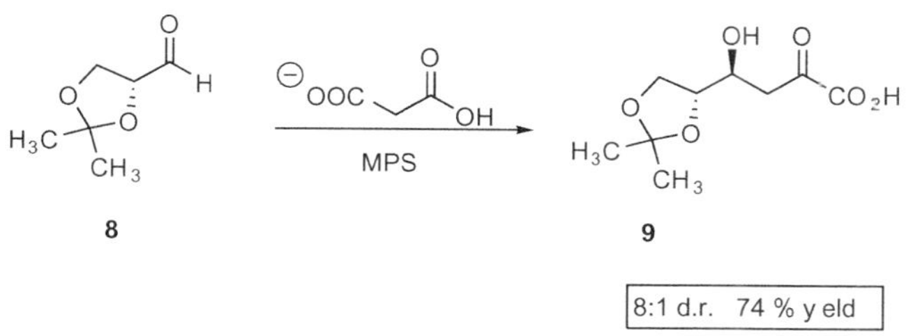

SCHEME 4.2 MPS acting as aldolase on a synthetic substrate.

products **17** and **18**, respectively. In this relatively simple biosynthesis, the role played by Sol5, beside the oxidation of the hydroxyl group, remains unclear. The nonenzymatic intramolecular Diels–Alder reaction of prosolanapyrone III **14** in aqueous solution is slightly slower than the enzymatic one and the stereochemical outcome of the thermal reaction is in favor of the *endo* product while the enzymatic transformation is in favor of the *exo* product. Oikawa et al. performed studies on both cellular extracts and isolated solanapyrone synthase, showing that the conversion rate at which **14** undergoes the cyclization is in direct proportion to the concentration of the cellular extracts, suggesting a possible active role of Sol5 in the catalysis of the cycloaddition. However, the modest contribution to the kinetic rate of the cyclization accompanied by a strong stereochemical effect on the outcome of the intramolecular Diels–Alder suggests that Sol5 might play the role of a stereochemical template rather than increasing the reactivity of the diene-dienophile couple (Scheme 4.3).[7,8]

These examples show how difficult it is to determine the chemical mechanisms underlying enzymatic transformations and at same time making the substantial difference between the activity of cyclase enzymes, and what we should eventually name as pericyclase enzymes, emerge. In organic

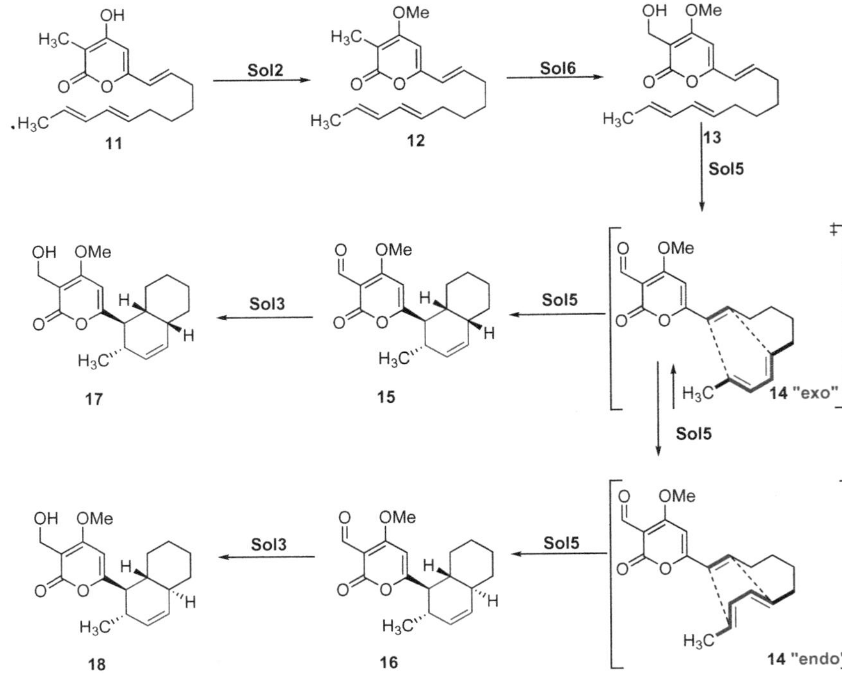

SCHEME 4.3 Solanapyrones biosynthesis.

chemistry, there is a general class of reactions called cycloadditions where the formation of a cyclic moiety happens with a net reduction of the bond multiplicity; within this category, there are reactions that can occur in a concerted manner through a single pericyclic transition state. Analogously, we think it is relevant to differentiate the general family of cyclases from the more peculiar family of pericyclases that perform cycloadditions. In this chapter we will discuss several examples in the search of NCs that occur through a pericyclic mechanism.

Biosynthesis of spinosyn A

Spinosyns are a large family of molecules that show potent activity against species that cause extensive damage to crops and some external parasites of livestock, companion animals, and humans.[9] This class of molecules is produced from the fermentation of two species of *Saccharopolyspora*, and their structure is represented by a polyketide-derived tetracyclic macrolide core with two appended saccharides (Fig. 4.1).

The main fermentation product of *S. spinose* is spinosyn A **19**, whose structure is a tetracyclic polyketide aglycone bound to a neutral saccharide substituent on the C-9 hydroxyl group and an aminosugar moiety on the C-17 hydroxyl group. The spinosyn biosynthesis is unusual because of the involvement of a processive type I PKS enzyme, which usually does not lead to the formation of polyketides containing carbocyles, while iterative PKS does. Four genes named *spnF*, *spnJ*, *spnL*, and *spnM* in the spinosyn A biosynthetic gene cluster of *S. spinose* were proposed to convert the PKS product **20** into the tetracyclic aglycone **23**. The flavin-dependent dehydrogenases enzyme produced by *spnJ* has been found responsible for the catalytic oxidation of the 15-OH of **20** to form the corresponding keto-intermediate **21** (Scheme 4.4).

In order to investigate the functions of SpnF, SpnL, and SpnM, Kim et al.[9b] overexpressed the corresponding genes in *Escherichia coli* BL21 and their products purified as N-terminal His_6-tagged proteins with a good level of purity. Isolated enzymes then were tested for activity with **21**. While SpnF and SpnL were found unreactive, SpnM completely converted **21** in

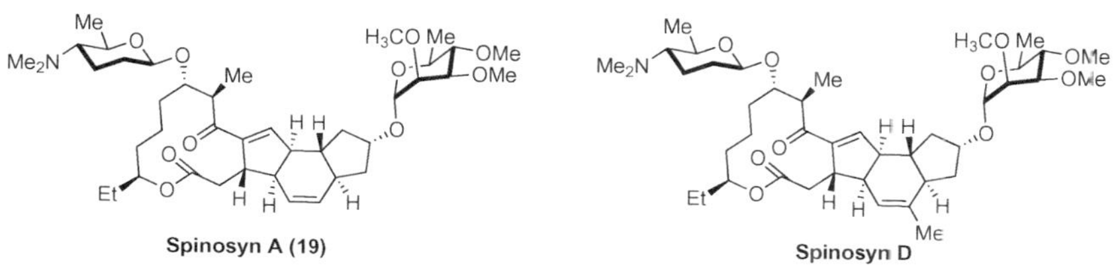

FIGURE 4.1 Structures of Spinosyn A and D.

compound **23** where a new cyclohexene moiety is fused to a cyclopentane. The authors followed this transformation by High-performance liquid chromatography (HPLC) analyses and noticed that a transient intermediate was appearing before complete transformation of the substrate into **23**. This intermediate was fully characterized as the monocyclic macrolactone **22** that comes from the dehydration of **21** with formation of a second conjugate diene. The following cyclization of **22** to **23** was then further investigated to understand the possible role played either by SpnM or one of the enzymes aforementioned.

The authors found that **22** undergoes a spontaneous [4 + 2] cyclization but that in the presence of SpnF is around 500 times faster. Finally, the remaining enzymes contribute to the production of the tetracyclic nucleus of Spinosyn A. After this seminal paper, which unequivocally showed the

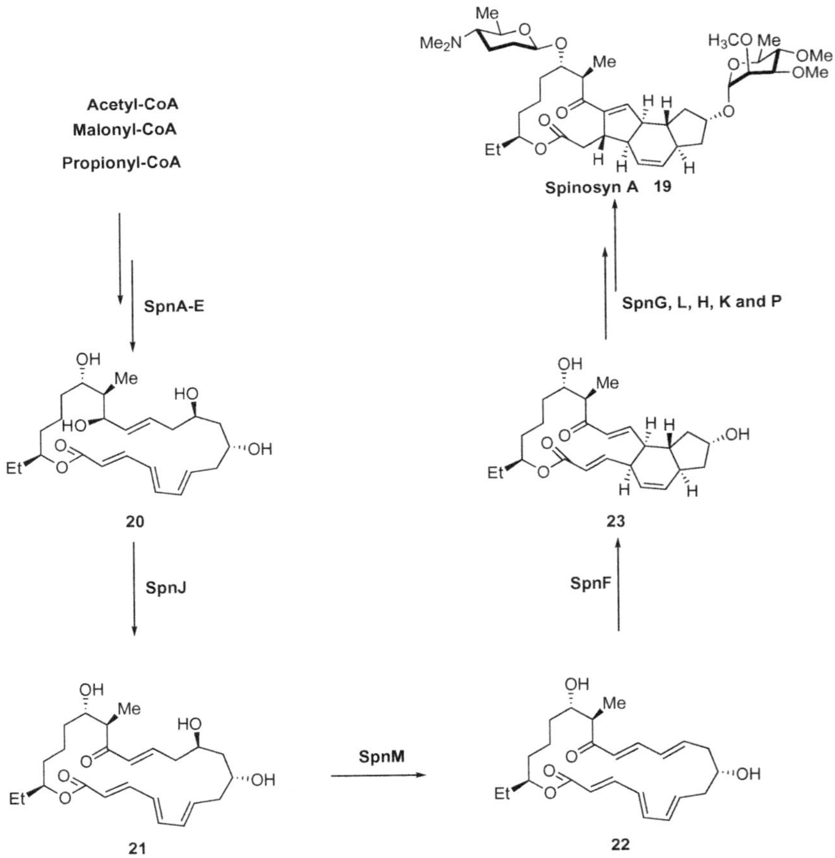

SCHEME 4.4 Biosynthesis of spinosyn A.

exclusive role played by SpnF in the cyclization of **22** into **23**, a lot of speculation was made on the reaction mechanism underlying this transformation. In the literature, two main options have been reported: (1) Diels–Alder reaction and (2) a [6 + 4] cycloaddition followed by a Cope rearrangement. While the intramolecular Diels–Alder reaction is the more intuitive pathway and the one that was initially suggested on the base of computational models of nonenzymatic reactions, Houk et al. proposed, at the end of a thorough quantum mechanics/molecular mechanics (QM/MD) analysis, a different and fascinating possible pathway (Fig. 4.2).[10a]

According to these authors, this cycloaddition cannot be described as either a concerted or stepwise process, and dynamical effects strongly influence the identity and timing of bond formation. The identified transition state was defined as ambimodal, since it can lead directly to the observed product or to an unobserved [6 + 4] cycloadduct that through a rapid cope rearrangement is converted into the final [4 + 2] adduct. However, calculated barriers were much higher than the experimental ones, and above all, these calculations were not taking into consideration the possible role played by the enzyme. In 2015, the X-ray structure of SpnF was finally resolved, making it possible for Chen et al. to study the role of the protein by means of biosimulation.[10e] These authors, by QM(DFT)/MM/MD simulations involving the

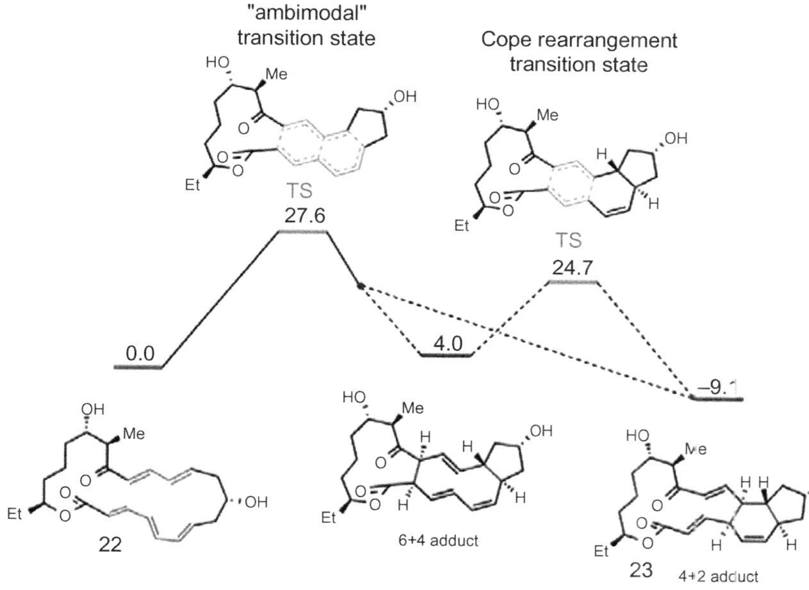

FIGURE 4.2 Calculated barriers for the mechanism suggested by Houk et al. *Patel, A.; Chen, Z.; Yang, Z.; Gutierrez, O.; Liu, H. W.; Houk, K. N.; Singleton, D. A.* J. Am. Chem. Soc. ***2016***, 138, *3631–3634.*

full enzyme model, showed the pivotal role of three residues that cooperate with other residues to control the direction of the cycloaddition. These findings show that the Diels–Alder reaction is more likely to happen than the [6 + 4] cycloaddition, with a concerted mechanism and a barrier of about 20 kcal/mol that agrees with experimental data. The catalytic function is performed by three residues, Tyr23, Thr196, and Trp256. Tyr23 and Trp256 delocalize the electrons around the C2-C3 double bond, while Thr196 reduces the double-bond character of the C13-C14 bond (Fig. 4.3). The authors evaluated also the [6 + 4] pathway, concluding that the Diels–Alder mechanism is dominant in the SpnF catalytic reaction. Remarkably, this enzyme not only exploits hydrogen bonds to catalyze the cycloaddition, but also uses an aromatic interaction to control the reaction pathway.

Pyrroindomycins biosynthesis

The pyrroindomycins (PYRs) are a class of natural compounds isolated from *Spetromyces rugosporus* bacteria and were found active against drug-resistant bacterial pathogens such as *Staphylococcus aureus* or *Enterococcus faecium*.[11] These molecules belong to the family of spirotetramates and are characterized by the presence of an aglycone moiety that contains two cyclohexene units present in both the dialkyldecalin and the tetramate spiro-conjugate portions (Fig. 4.4).

Liu et al.[11e] revealed that a modular PKS-nonribosomal peptide synthase hybrid system promotes the biosynthesis of the tetramate containing two

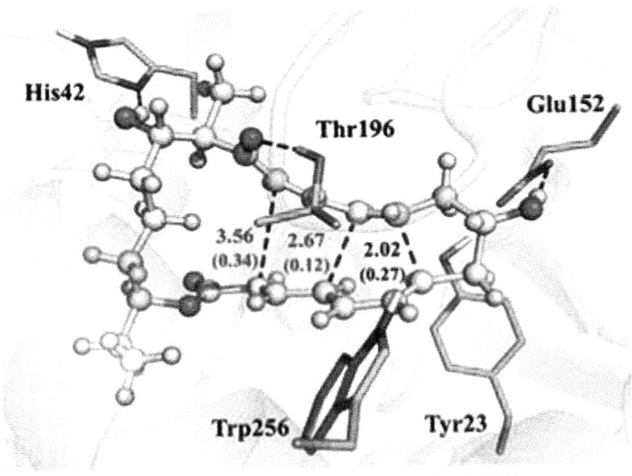

FIGURE 4.3 The simulated [4 + 2] TS in the active site of the enzyme. *Chen, N.; Zhang, F.; Wu, R.; Andess Hess, Jr., B.* ACS Catal. ***2018***, 8, *2353–2358.*

pairs of 1,3-diene and alkenes groups that are then combined through a [4 + 2] cycloaddition in the corresponding cyclohexane rings The enzymes involved in the cyclization steps work in tandem, since treatment of **24** with purified PyrE3 and PyrI4 rapidly yields **26** without the presence of any cofactor. These authors performed a detailed study, published in *Nature Chemical Biology* in 2015,[11f] where they revealed the individual contribution of PyrE3 and PyrI4 in the construction of the pentacyclic core of PYRs. Once isolated, compound **24** was treated with both enzymes separately. In the presence of PyrI4 no reaction occurred, while when incubated with PyrE3 in the absence of any added cofactor, **24** was successfully converted into **25**. Analogously, conversion of **25** into **26** could be achieved only by treatment with PyrI4, thus validating the idea that this enzyme acts as a spiro-conjugate synthase (Scheme 4.5).

Remarkably, intermediate **24** and **25** are relatively stable in water at room temperature for at least 24 h, indicating that both cycloadditions depend on the enzymatic activity. Thanks to the resolution of the crystal structure of PyrI4, it has been possible to hypothesis the mechanism by

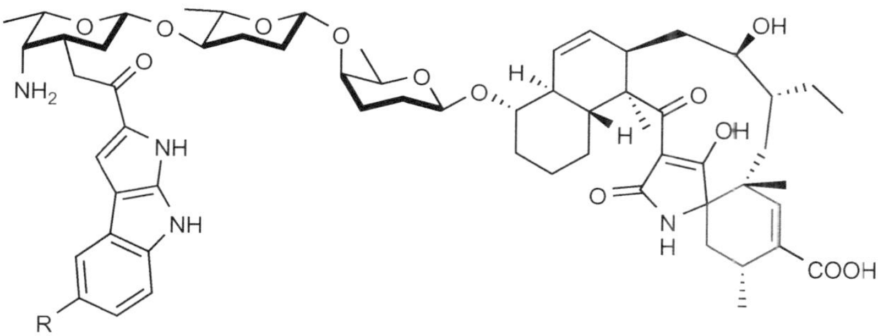

R = H: Pyrroindomicin A

R = Cl: Pyrroindomicin B

FIGURE 4.4 Structures Pyrroindomicin A and B.

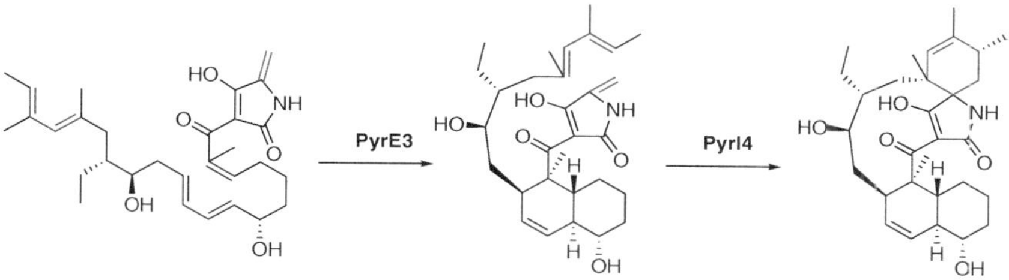

24 **25** **26**

SCHEME 4.5 Construction of the pentacyclic core of PYRs by PyrE3 and PyrI4.

which this enzyme catalyzes the formation of the spirocyclic moiety.[11g] A comparative analysis of the structure of PyrI4 in its free and bound state suggests that this enzyme appears to not have critical catalytic amino acids. The catalytic activity can be explained through a trapping mechanism whereby the lid-like action of the N-terminal tail imposes conformational constraints on the β-barrel catalytic core, which enhances the proximity and polarization effects of reactive groups to drive cyclization in a regio- and stereospecific manner (Fig. 4.5).

PyrE3 instead selectively accommodates its substrate in a highly positively charged pocket, accelerating the formation of the *trans*-decalin moiety, likely through an *endo*-selective [4 + 2] transition state (Fig. 4.6).[11h]

Biosynthesis of versipelostatin

Versipelostatin (VST) is a 17-membered macrocyclic polyketide product that contains a spirotetronate skeleton and is produced by *Spetromyces versipellis* 4083-SVS6 (Fig. 4.7).[12]

Kuzuyama et al.,[12d] from the University of Tokyo, managed to identify, through heterologous expression by mean of a bacterial artificial chromosome, the entire VST biosynthetic gene cluster. As it can be seen from the

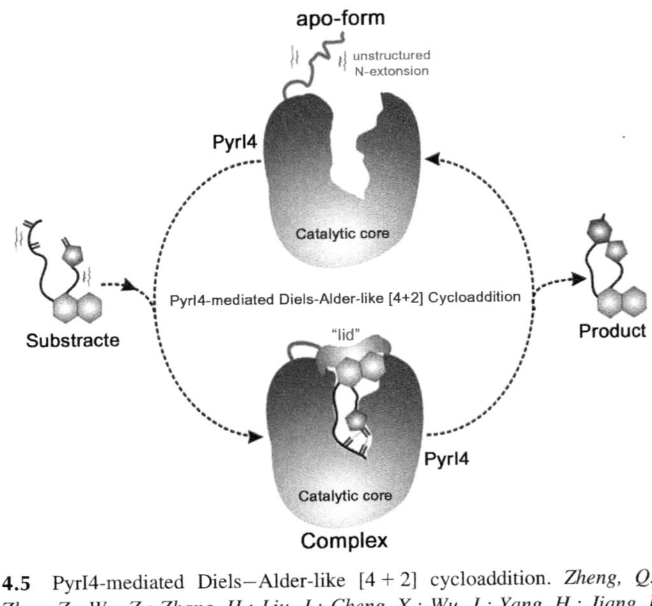

FIGURE 4.5 PyrI4-mediated Diels–Alder-like [4 + 2] cycloaddition. *Zheng, Q.; Guo, Y.; Yang, L.; Zhao, Z.; Wu, Z.; Zhang, H.; Liu, J.; Cheng, X.; Wu, J.; Yang, H.; Jiang, H.; Pan, L.; Liu, W.* Cell Chem. Biol. ***2016***, *23, 352–360.*

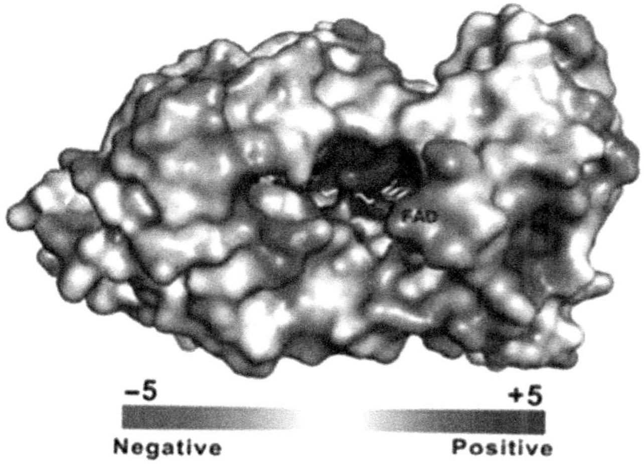

FIGURE 4.6 Highly positively charged pocket of PyrE3. *Zheng, Q.; Gong, Y.; Guo, Y.; Zhao, Z.; Wu, Z.; Zhou, Z.; Chen, D.; Pan, L.; Liu, W.* Cell Chem. Biol. ***2018***, 25, 718–727.

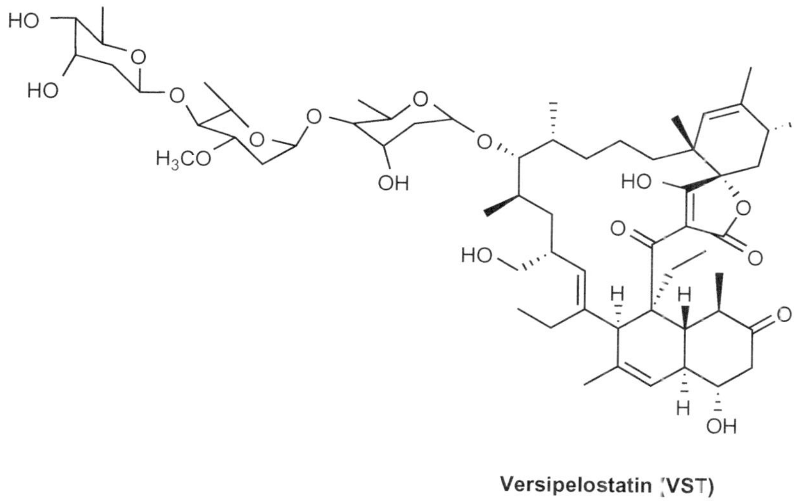

FIGURE 4.7 Versipelostatin structure.

molecular structure of VST, a spirocyclic moiety that can be expected to be synthesize through a [4 + 2]-cycloaddition analogously to what we have already seen in the previous example is present.

The same authors managed to identify the gene and the corresponding protein, named *vstJ* and VstJ, respectively, essential in the biosynthesis of

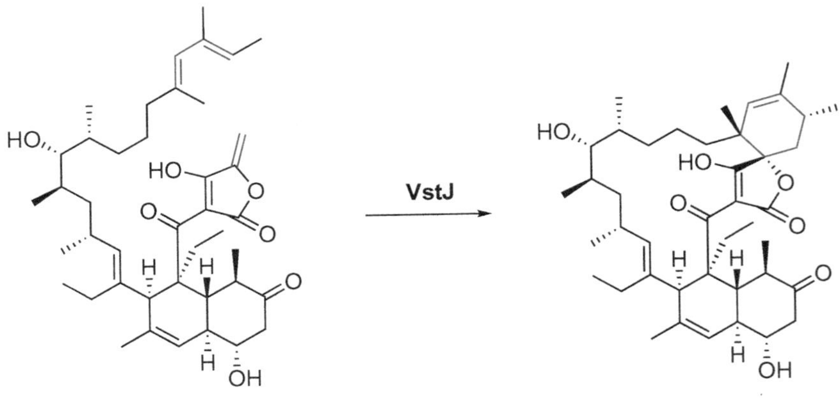

SCHEME 4.6 Conversion of **27** in to **28** by Vstj.

VST, along with compound **27** as possible precursor of VST. In order to test this hypothesis, VstJ was overexpressed in *E. coli* and then isolated and purified. Compound **27** was then incubated with the isolated enzyme resulting in the formation of the corresponding 37-deoxy-VST aglycone **28**, a product that shares the same stereochemistry of VST and is not thermally produced by treatment of **27** with temperature at 30°C. That catalytic activity of VstJ can be possibly explained by stabilization of the *s-cis* conformation along with favoring the spatial vicinity of the two cycloaddends. Even though this is a plausible explanation, further studies concerning the mechanism of catalysis are needed (Scheme 4.6).

Biosynthesis of abyssomicin C

The spirotetronate abyssomicin C **33** was first isolated from the marine actinomycete *Verrucosispora maris* AB-18-032 and was found active against different bacterial pathogens. The proposed biosynthesis of this antibiotic proceeds through the crucial formation of a spiro-bicyclic system that can occur via an enzyme-catalyzed Diels–Alder reaction, a hypothesis further supported by analogy with other spirotetronates and spirotetramates biosynthesis (Scheme 4.7).[13]

In order to verify this hypothesis, Byrne et al.[13d] from the United Kingdom, purified a N-terminally hexa-histidine variant of AbyU considered responsible for the catalysis of the cyclization, and tested in the conversation of **34** and **36** that art analogues of the proposed AbyU natural substrate **31**. These experiments showed unequivocally the ability of AbyU to perform the catalysis of intramolecular [4 + 2]-cycloadditions, speeding up the kinetics of those that would occur even in the absence of enzyme (**34**), and

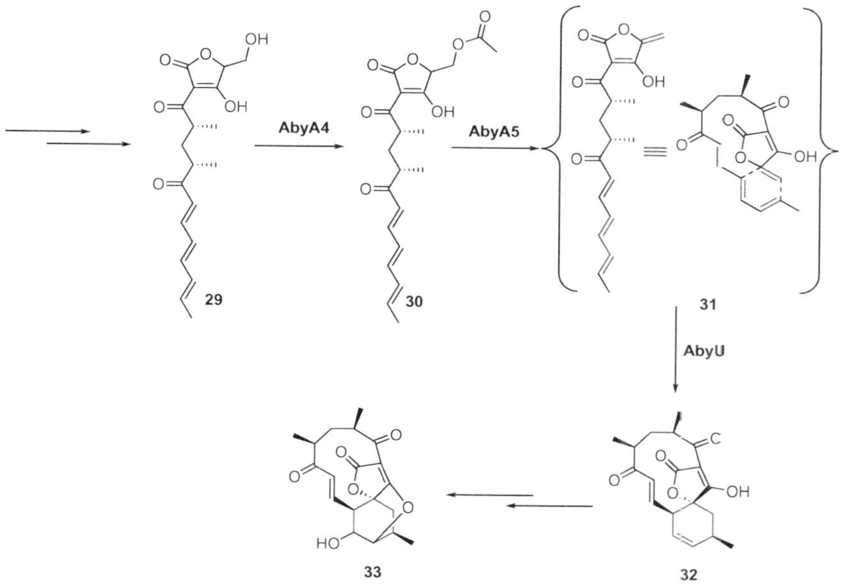

SCHEME 4.7 Biosynthesis of abyssomicin C.

undergoing the conversion of substrates that thermally do not readily cyclize (**36**) (Scheme 4.8).

The same authors, with the support of computational tools, managed to localize the binding site and demonstrated through QM/MM calculations that the substrate can react from the proposed binding pose through a concerted, asynchronous Diels–Alder mechanism (Fig. 4.8).

Myceliothermophins biosynthesis

From *Myceliophthora thermophile* is possible to isolate a class of cytotoxic compounds named myceliothermophins, that have shown a remarkable activity against a variety of cancer cell lines (Fig. 4.9).[14]

Within this class of natural compounds, Myceliothermophin E (**38**) and A (**39**) are suspected to be formed through a biosynthetic pathway that include a natural Diels–Alder reaction responsible for the formation of the *trans*-fused decalin ring system. In order to verify this hypothesis, Tang and Houk groups[14k] performed a detailed study intending to investigate the biosynthesis of **38** and **39** with particular focus on the formation of the bicyclic system. First of all, these authors studied the genome of *M. thermophile* ATCC 42462, identifying the gene cluster responsible for the production of these compounds. Deletion of *mycA* and *mycB* genes through gene replacement techniques of the wild-type strain of *M. thermophila* resulted in the missing

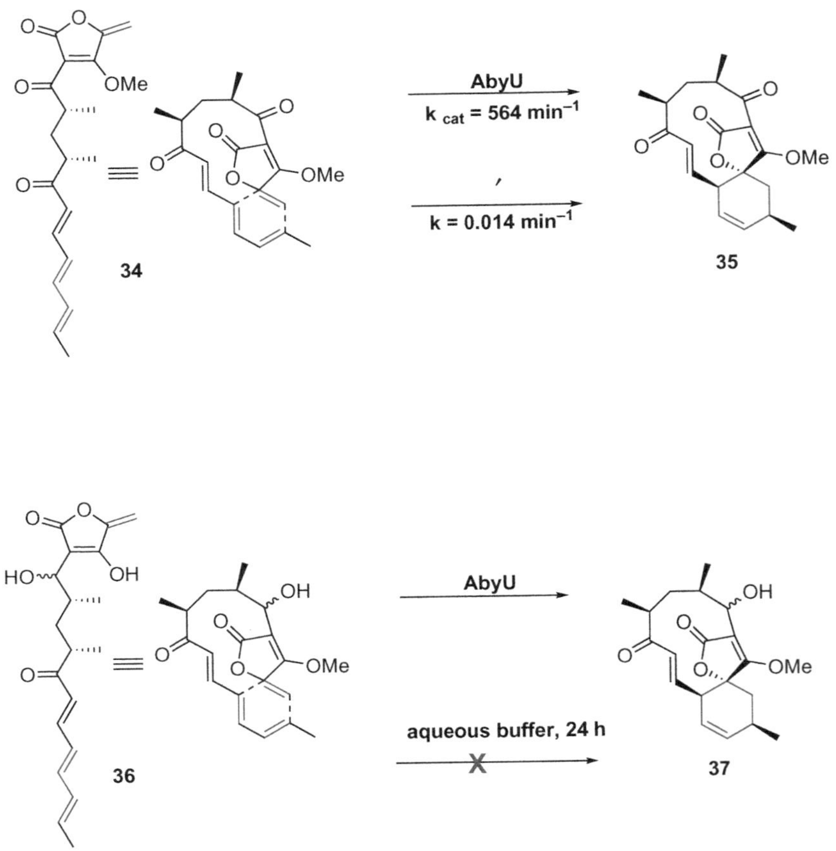

SCHEME 4.8 AbyU activity in the conversion of **34** and **36**.

formation of compounds **38** and **39**. Trace LC–MS analyses of the wild-type products revealed the presence of two compounds that once isolated and purified could be fully characterized as **42** and **43**. Compound **42** is an acyclic polyolefine conjugated to a 4-pyrrolin-2-one moiety, and it is likely to be the enolized form of the PKS-NRPS/ER product. To further investigate the formation of this compound, *mycA* and *mycC* were heterologously expressed in *Aspergillus nidulans* A1145, resulting in the formation of **42**, confirming that the *myc* PKS-NRPS and its ER partner are responsible for the formation of the heterocyclic moiety. The cyclized product **43**, that is the cyclized form of **42**, is not recovered in the *ΔmycB* strain of *M. thermophila* and is instead found in the *mycABC* strain of *A. nidulans* extracts. By deletion of the *mycB* gene, both in *M. thermophile* and *A. nidulans* the acyclic form of **38** was accumulated (**44**).

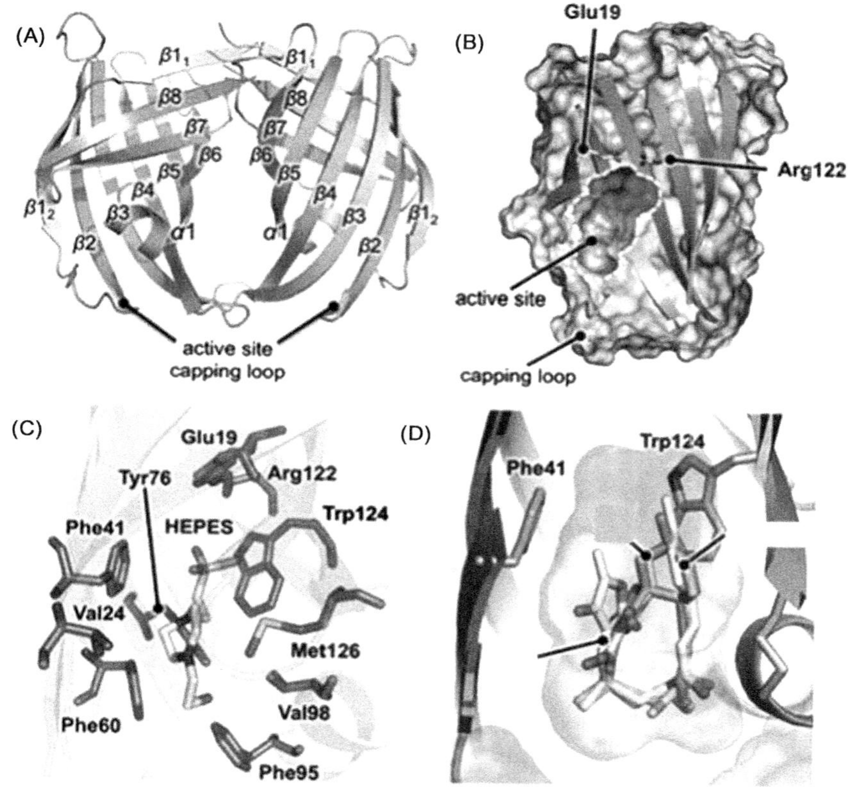

FIGURE 4.8 Crystal structure of AbyU and substrate binding mode. (A) Overall fold of the AbyU dimer. (B) Cut-away view of the active site of AbyU. (C) Detailed view of the AbyU active site bound to its substrate. (D) Superposition of the computationally predicted binding modes of **31**, **32**, and the atropisomer of **32** within the active site of AbyU. *Byrne, M. J.; Lees, N. R.; Li-Chen, H.; van der Kamp, M. W.; Mulholland, A. J.; Stach, J. E. M.; Willis, C. L.; Race, P. R.* J. Am. Chem. Soc. ***2016***, 138, *6095–6098.*

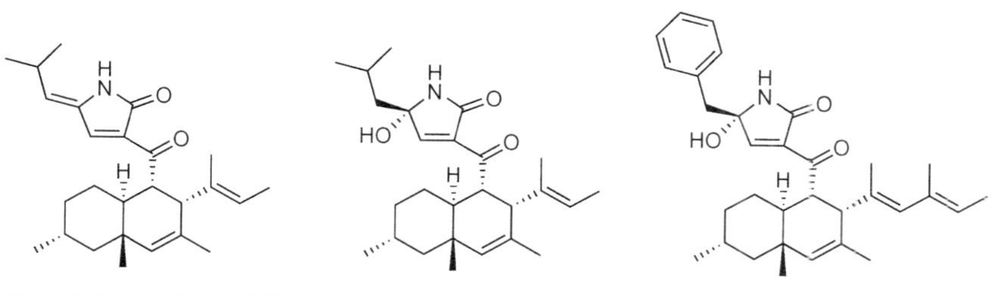

FIGURE 4.9 Structures of Myceliothermophin E (**38**) and A (**39**) and Oteromycin.

In order to support the evidence that clearly supports *mycB* being the gene responsible for the catalytic activity in the formation of the decalin moiety, the authors prepared the *N*-FLAG-tagged recombinant enzyme MycB through heterologous expression from *Saccharomyces cerevisiae* BJ5464-NpgA. With the isolated enzyme in their hands, they could observe that MycB does not catalyze the conversion of **42** into **43**, while instead it catalyzes the cyclization of **44** into **38**. This different behavior can be likely addressed to a differential reactivity of the keton and the enol substrates toward the cycloaddition reaction. Taking into consideration these experimental data, along with a computational analysis of the single transformations, the authors proposed the possible biosynthetic route of **38** as reported in Schemes 4.8 and 4.9.

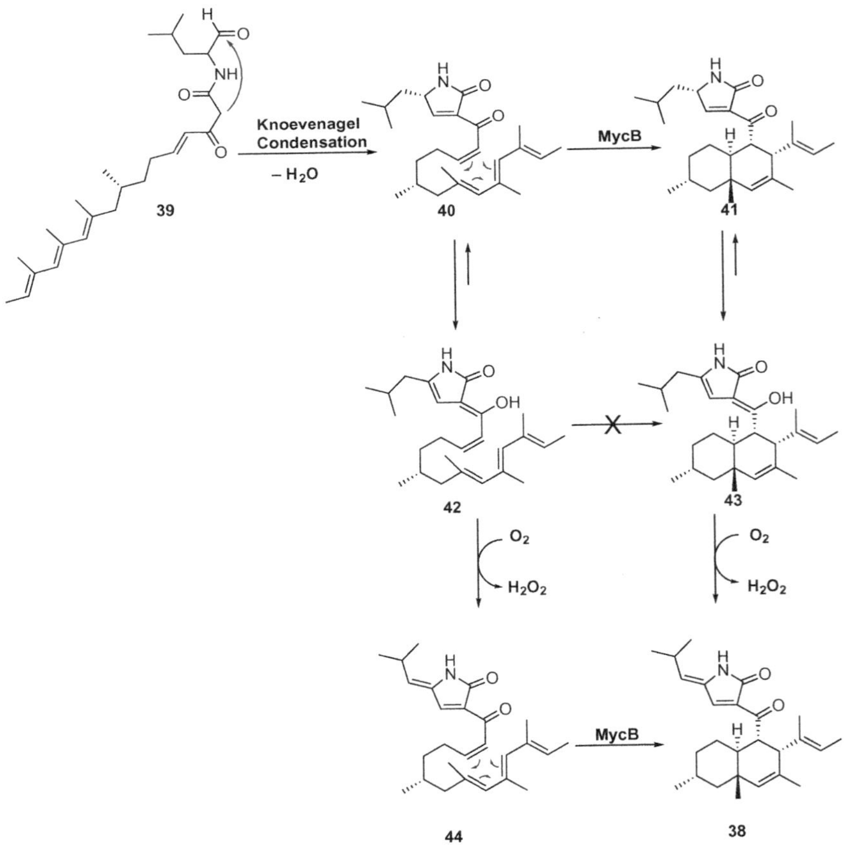

SCHEME 4.9 Biosynthesis of **39**.

Biosynthesis of equisetin

Another example of the formation of a decaling-containing product catalyzed by a DAase is represented by the equisetin (**45**) biosynthesis, a potent HIV-1 integrase inhibitor, produced in *Fusarium* sp. FN080326.[15] This compound is produced by PKS/NRPS hybrid enzymes where a linear polyene intermediate, formed by the action of a highly reducing, iterative type I PKS module in collaboration with a *trans*-acting enoyl reductase, serves as a substrate for a putative Diels–Alder reaction that is responsible for the formation of the decalin system.

Kato et al.[15e,f] conducted a genetic analysis of *Fusarium* sp. FN080326 that allowed the identification of the *fsa* cluster responsible for the biosynthesis of **45** (Fig. 4.10).

Within this gene cluster, throughout knockout experiments, several genes were located downstream of *fsa1* and their activity in the biosynthetic pathway elucidated. Thanks to the study of deletion mutants, the authors managed to define the role of five *fsa* genes, among which *fsa2* was found involved in the formation of the decalin ring.

The *Δfsa2* mutant revealed that in the absence of the corresponding enzyme, **46** is formed in a reduced amount along with a second compound that once characterized was revealed as the diasteroisomer **47** with opposite configuration at the C-3 and C-6 positions that are formed during the cycloaddition (Scheme 4.10).

This stereoisomer is not found otherwise, indicating that *fsa2* gene is apparently responsible for the stereocontrol of the reaction addressing the *endo*-selectivity of the Diels–Alder reaction that leads to the formation of **45**.

In order to further asses this hypothesis the same group searched for an *fsa2* homologue involved in the biosynthesis of an enantiomerically opposite analogue of **45** to perform experiments of gene replacement aimed at the evaluation of the stereochemical outcome of the corresponding products. The authors found a natural compound named phomasetin **49**, produced by *Pyrenochaetopsis* sp. RK10-F058, whose stereochemistry of the decalin moiety is the opposite of equisetin.

Analogously to what was done for the elucidation of the biosynthesis of equisetin, the gene cluster responsible for the biosynthesis of phomasetin

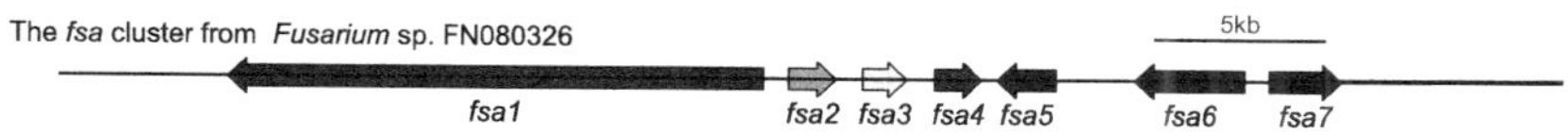

FIGURE 4.10 *fsa* cluster responsible for the biosynthesis of **45**. *Kato, N.; Nogawa, T.; Hirota, H.; Jang, J.-H.; Takahashi, S.; Ahn, J. S.; Osada, H.* Biochem. Biophysical Res. Comm. ***2015***, 460, *210–215.*

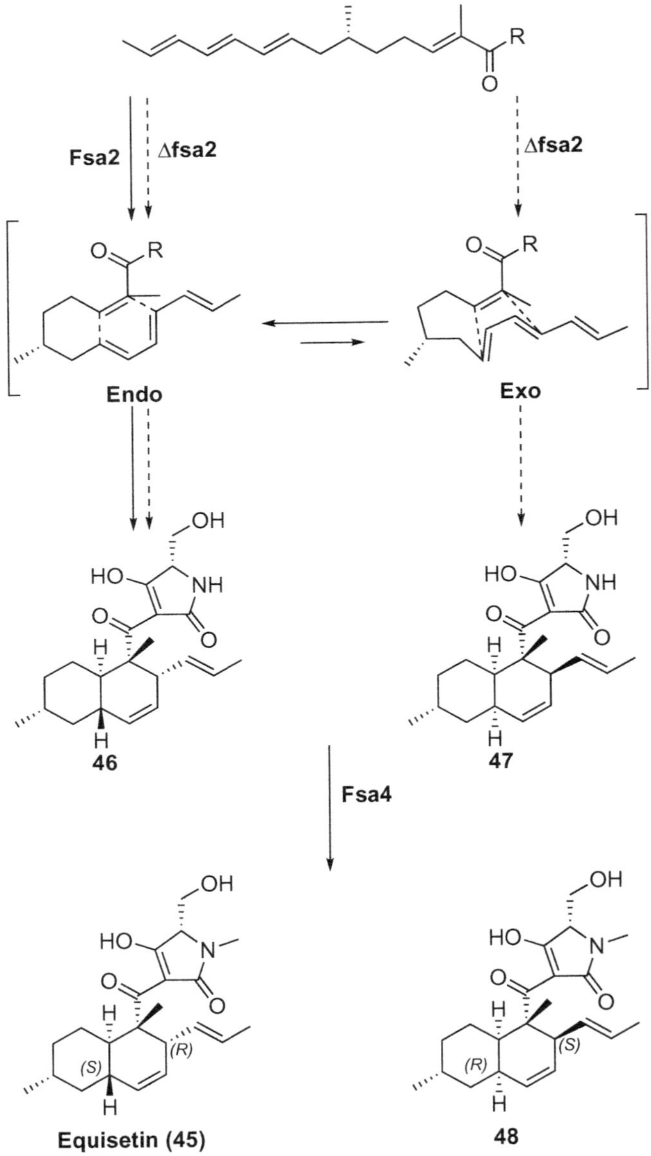

SCHEME 4.10 Formation of Equisetin **45** and its diastereoisomer **48**.

was identified and named *phm*. The gene *phm7* encodes for Phm7 that shares overall sequence identity of 16–50% to Fsa2. In order to get more insight on the biosynthesis of phomasetin, different gene knockout experiments were performed (Scheme 4.11).

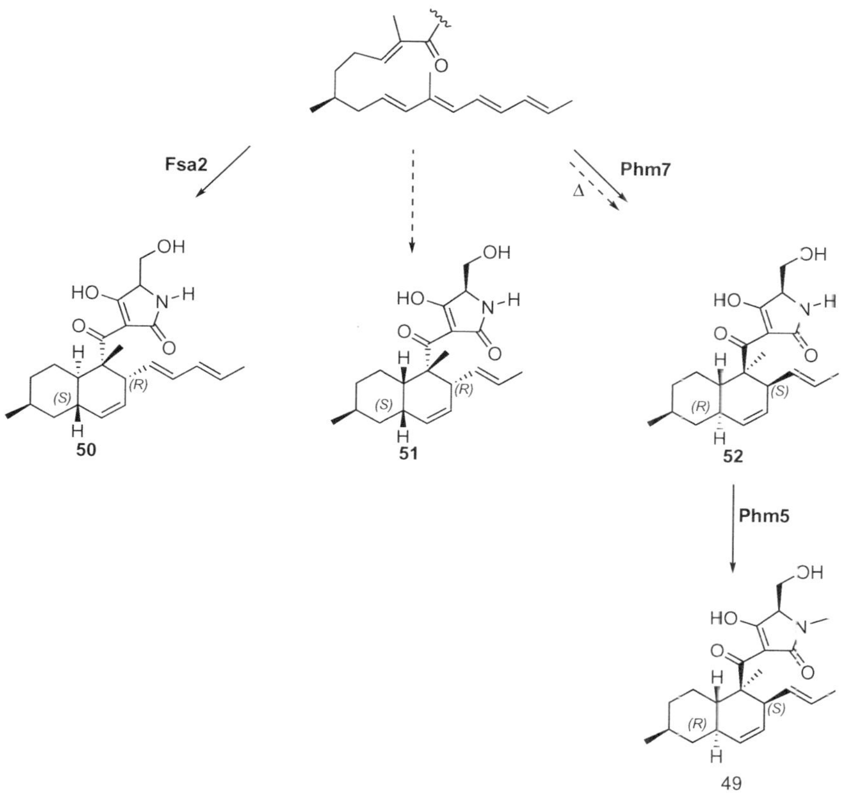

SCHEME 4.11 Comparison of the activity of Fsa2 and Phm7.

Deletion of *phm1* and *phm4* led to no production of **49**, while $\Delta phm5$ allowed the isolation and characterization of **52**, indicating that *phm5* acts as *N*-methyltransferase in the last step of the biosynthetic pathway. Analogously to what happened in the deletion of *fsa2*, in the $\Delta phm7$ culture, **49** was found in a reduced amount and a new compound subsequently characterized as **51** showed opposite configurations at the C3 and C6 stereocenters with respect to **49**, thus confirming the role played by Fsa2 and Phm7 in the stereochemistry of the Diels–Alder reaction.

In order to evaluate the role played by these two enzymes in the control of the stereochemistry of the corresponding cycloadducts, gene-swapping experiments were performed. By replacement of *phm7* with *fsa2* in strain RK10-F058 was possible to isolate compound **50**. Characterization of the four chiral centers of **50** revealed, as expected, that the configuration of C2, C3, C6, and C11 are opposite to those of cycloadduct **52** as expected.

These results clearly highlight the synthetic potentiality of the selective manipulation of natural enzymes that catalyze Diels–Alder reactions.

Decarboxylation of phenylacrylic acids catalyzed by ferulic acid decarboxylase

Generally speaking, enzyme-catalyzed decarboxylation reactions on unactivated substrates are an important class of reactions that endow the preparation of relevant molecules otherwise difficult to obtain. Unfortunately, these reactions are difficult to achieve because of the incipient negative charge during the formation of the transition state that increases the activation energy. To overcome this problem, enzymes evolved, developing a strategy where prosthetic groups promote the stabilization of the negative charge reducing the energetic barrier needed to undergo the transformation.[16]

Recently,[17] a modified form of flavin mononucleotide (FMN) denominated prenylated FMN (prFMN, **53**) was demonstrated to act as a decarboxylation cofactor of various aromatic carboxylic acids in different microbes along with their attendant prFMN synthases. This cofactor differs from the standard FMN molecule because of the addition of an isopentyl group between the C6 and N5 positions of the isoalloxazine flavin nucleus that results in the formation of a six-membered ring. prFMN are typically synthesized starting from the reduced form of FMN and dimethylallyl phosphate by dedicated enzymes called prenyl transferase. Belonging to this class, ferulic acid decarboxylase (FDC) plays a pivotal role in the decaboxylation of ring-substituted phenylacrylic acids yielding the corresponding substituted styrene derivatives that have a remarkable range of applications.

On the basis of a seminal paper published in *Nature* in 2015,[17d] it was proposed that prFMN catalytic activity is carried out by an initial step that involves a 1,3-dipolar cycloaddition of the cofactor on the double bond of the phenylacrylic acid to form the corresponding pentacyclic adduct. The following decarboxylation step occurs by elimination of carbon dioxide through a Grob-type elimination mechanism. This intermediate is subsequently protonated by a glutamate residue with formation of the styrene adduct that by cyclo-elimination conclude the catalytic cycle yielding the styrene derivative and prFMN (Fig. 4.11).

Since no 1,3-dipolar enzymatic reactions were never reported before, this proposed mechanism attracted a lot of interest because even if this mechanism is perfectly consistent with the known reactivity of 1,3-dipoles and conjugated double bonds, to exclude *a priori* that the reaction does not follow a Michael addition mechanism is questionable.

Marsh et al. from the University of Michigan managed to trap the pentacyclic cycloadduct by means of an analogue of the natural substrate, corroborating the suggested enzymatic mechanism of prFMN and so confirming the first reported case of a 1,3-dipolar cycloaddition in nature.[17g]

The first evidence came from the measurement of the secondary deuterium kinetic effect showing that CO_2 release occurs after protonation of the reaction product while a linear free energy analysis of the reaction through

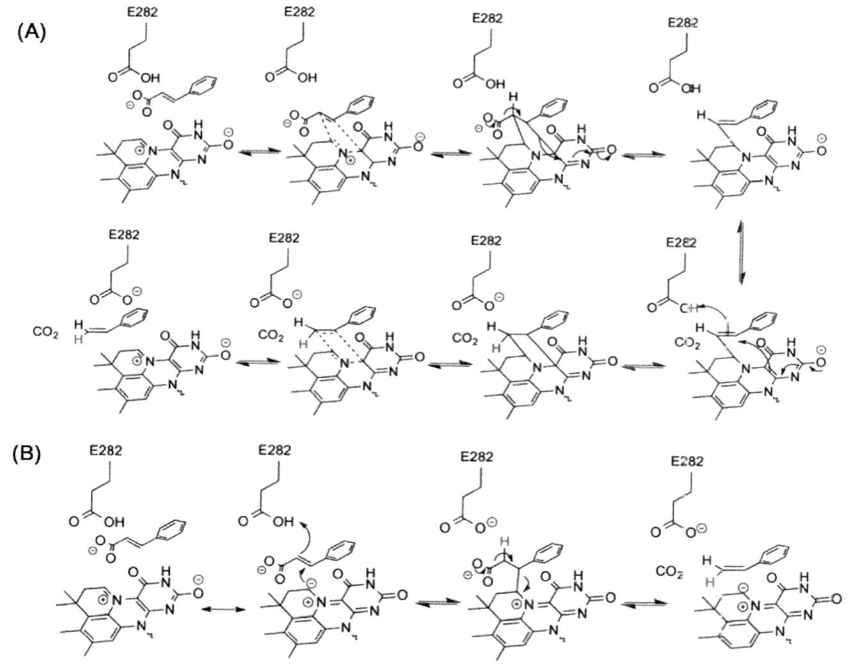

FIGURE 4.11 Proposed mechanisms for the FDC-catalyzed decarboxylation of phenylacrylic acid proceeding through mechanisms (A) and (B). *FDC*, Ferulic acid decarboxylase. *Ferguson, K. L.; Arunrattanamook, N.; Marsh, E. N.* Biochemistry ***2016***, *55*, *2857–2863*.

the use of different *para*- and *meta*-substituted phenylacrylic acids demonstrated that the rate-determining step is the cyclo-elimination of the styrene molecule. These results allowed the identification of a suitable probe that could mimic the reactivity of the natural substrate while inhibiting the following proposed biosynthetic steps.

The choice fell on (*Z*)-2-fluoro-2-nitro-2-vinylbenzene (FNVB, **54**), a well-known dipolarophile with a lower energy lowest unoccupied molecular orbital (LUMO) with respect to the phenylacrylic acid with consequent increased reactivity with the highest occupied molecular orbital (HOMO) of the nitrogen ylide, and bearing a nitro group and a fluorine atom that are respectively good isosteres of the carboxylate group and the hydrogen of the natural substrate (Scheme 4.12).

Thus a solution of FDC was reacted with FNVB in a citrate buffer at 25°C and after 10 min, assay of the enzyme with phenylacrylic acid revealed FDC to be essentially inactive.

The reaction between the enzyme and FNVB was monitored by measurements of the UV-absorption spectrum of the dipolarophile, revealing a spectra with an isosbestic point indicative of a single chemical reaction with a

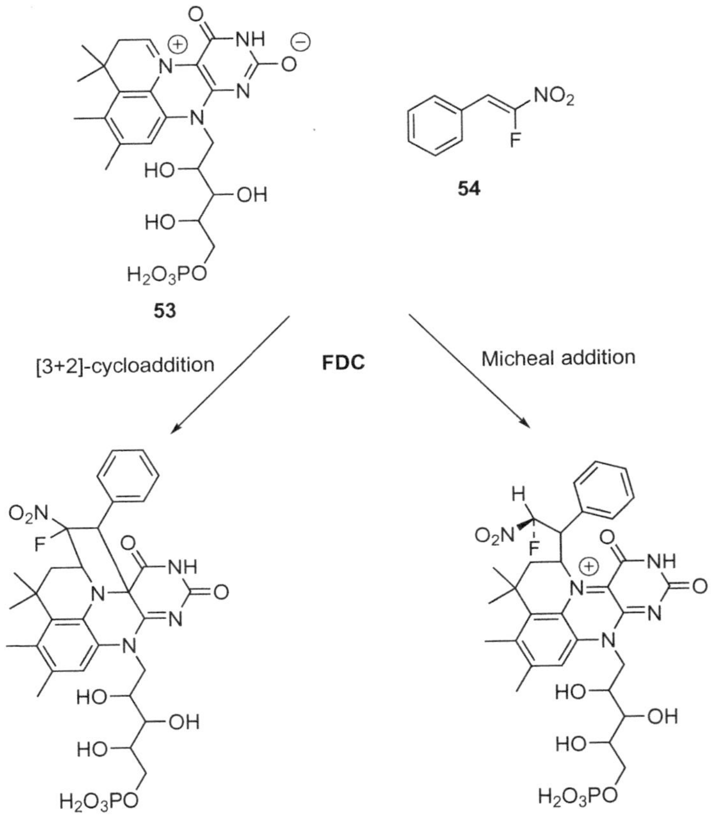

SCHEME 4.12 Formation of the cycloaddition product **55** and of the Michael addition **56**.

first-order kinetic model. Further, the presence of a broad absorbance band at 425 nm is consistent with the formation of the expected cycloadduct **55** rather than the Michael addition adduct **56**, since the last one is expected to show two absorption bands centered at 350 and 550 nm.

Additional evidence that the observed reaction is a [3 + 2]-cycloaddition came from native mass spectrometry. In this experiment, a solution of FDC in citrate buffer was reacted with FNVB and then rapidly desalted to get the inactivate holoenzyme. Once introduced into a mass spectrometer by electrospray ionization, the cofactor dissociates from the holoenzyme revealing a peak at $m/z = 730.167$, indicative of the presence of the cycloaddition product of prFMN and FNVB **55**.

All these experiments ultimately confirm the mechanism by which prFMN catalyzes along with the attendant enzyme the decarboxylation of phenylacrylic acid, revealing an unprecedented 1,3-dipolar cycloaddition.

In the future, because of the wide range of applications that these reactions can have, a great effort should be placed in studying other enzymatic decarboxylation to evaluate whether this pericyclic mechanism is a unique example or common among this class of natural transformations in order to master new biomimetic synthetic processes.

Debated cases

As of this writing, many researchers are pursuing active and exciting experiments in the search of new NCs mediated by enzymes that have been shaped by the evolution to solely catalyze pericyclic reactions. In this section of the chapter we will discuss open questions concerning putative cycloaddition pericyclases currently under investigation.

Biosynthesis of lovastatin

Lovastatin (**57**) belongs to the family of statins, a class of compounds that are important cholesterol lowering drugs (Fig. 4.12).[18] This molecule has been found in several fungal species such as *Monascus ruber*, *Aspergillus terreus*, and *Pleurotus ostreatus*. Lovastatin is a polyketide characterized by

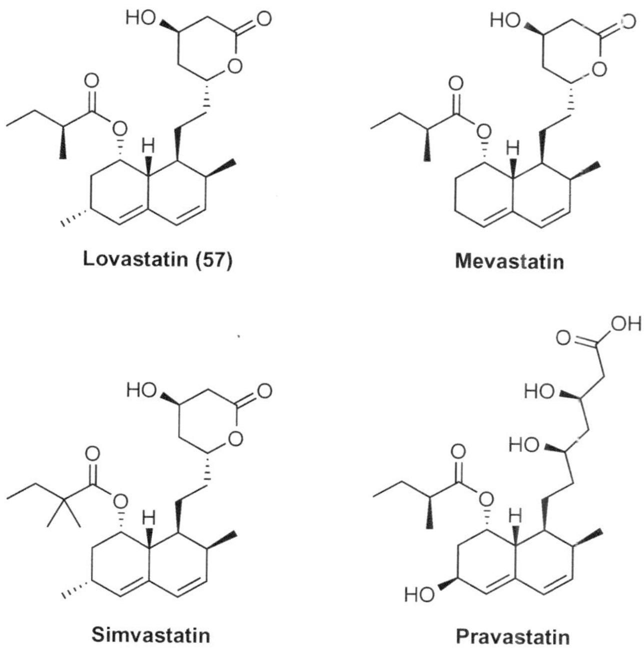

FIGURE 4.12 Examples of molecules belonging to the statins family.

the presence of a hexahydronaphthalene group that has been hypothesized to result from a biological Diels–Alder.

The biosynthetic pathway of Lovastatin identified in the *A. terreus* involves several enzymatic transformations.[19] Type I PKS LovB and type II PKS LovC work together, forming an enzymatic complex that seems responsible for the formation of the *trans*-decalin ring system through a Diels–Alder reaction. This assumption is supported by the formation of dihydromonacolin L **58** by a heterologous host containing the LovB/LovC complex, while the sole expression of LovB in the same host leads to pyrones **59** and **60** (Scheme 4.13).

More details came from the same article by Vederas et al. from the University of Alberta who studied the intermolecular cyclization of the hexaketide triene *N*-acetylcisteamine NAC thioester to ascertain the role played by LovB.[19i] Nuclear Magnetic Resonance (NMR) studies showed that compound **61** cyclizes spontaneously in aqueous media in the absence of LovB to give a 1:1 mixture of *exo* and *endo* adducts, and the same reaction performed in CDCl3 is remarkably slower. The stereoisomer corresponding to the natural product **58** was not found. Instead, in the presence of LovB, compound **65** was obtained in a **63**:**64**:**65** ratio of 15:15:1. The same reaction, conducted in the presence of denaturated LovB, yields to the formation of **63** and **64**, demonstrating that the active state of LovB is needed for the formation of **65** (Scheme 4.14).

Even though the mechanism by which LovB catalyzes the intermolecular Diels–Alder reaction in the synthesis of Lovastatin has not be fully

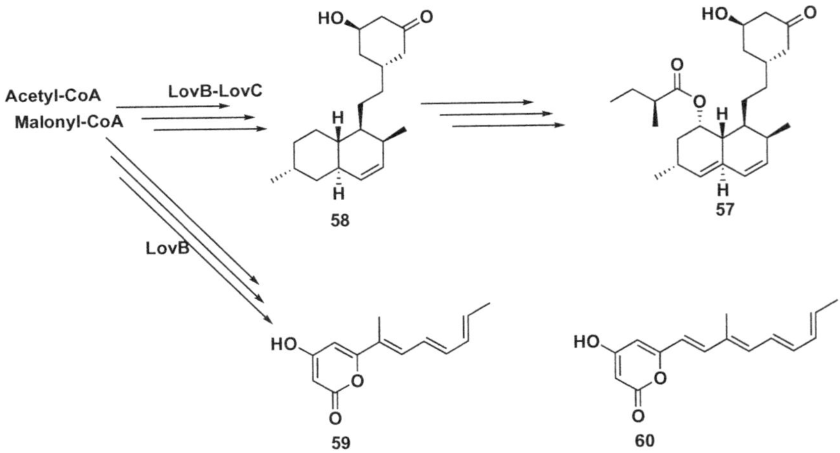

SCHEME 4.13 Formation of dihydromonacolin L **58** by a heterologous host containing the LovB/LovC complex and the sole expression of LovB leading to pyrones **59** and **60**.

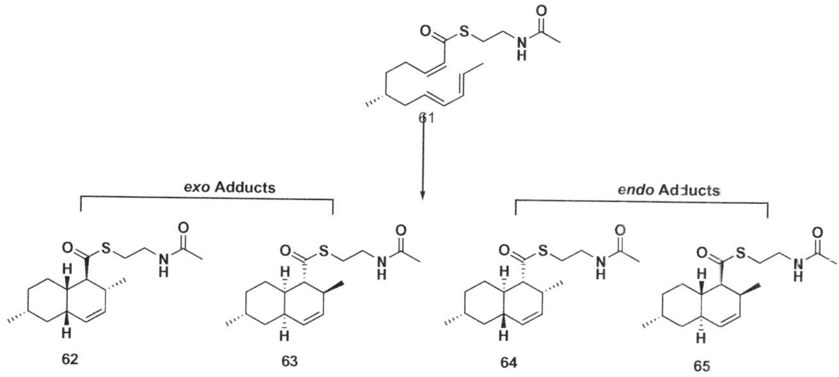

	62		63		64		65
	/	:	1	:	1	:	/
LovB	/	:	15	:	15	:	1

SCHEME 4.14 Intermolecular cyclization of the hexaketide triene *N*-acetylcisteamine NAC thioester **61** in the presence and absence of LovB.

explained yet, these results suggest that this enzyme plays an active role in directing the stereochemical outcome of the reaction and increases the cyclization rate.

Biosynthesis of leporin C

Whether or not the synthesis of leporin C **66** is clearly mediated by the intervention of LepI, an enzyme that mediates different pericyclic reactions as demonstrated in the fascinating article published in *Nature* in 2017, the attribution of LepI to the category of pericyclases is ambiguous.[20]

The hypothesis that a highly selective hetero-DAase could be involved in the biosynthesis of leporin C emerged when the first total synthesis of leporins was published in 1996, and the authors showed that the precursor of leporin C undergoes a mixture of intramolecular Diels–Alder and hetero-Diels–Alder reactions, yielding five different products, including the desired natural product (Scheme 4.15).

The exclusive production of **66** in the biosynthesis that happens in the *Aspergillus* sp. suggests that the reaction sequence from the PKS-derived precursor **67** via **68** and **69** to leporin C has to be catalyzed by a highly selective hetero-DAase.

The enzymes involved in this transformation were identified by heterologous reconstitution of the leporin pathway in *A. nidulans*.[20c] Enzymes LepA, LepG, and LepH led to the formation of **67**. Addition of LepF to ketone **67** resulted in the formation of **70–72** as well as **66** and **73**. This can be

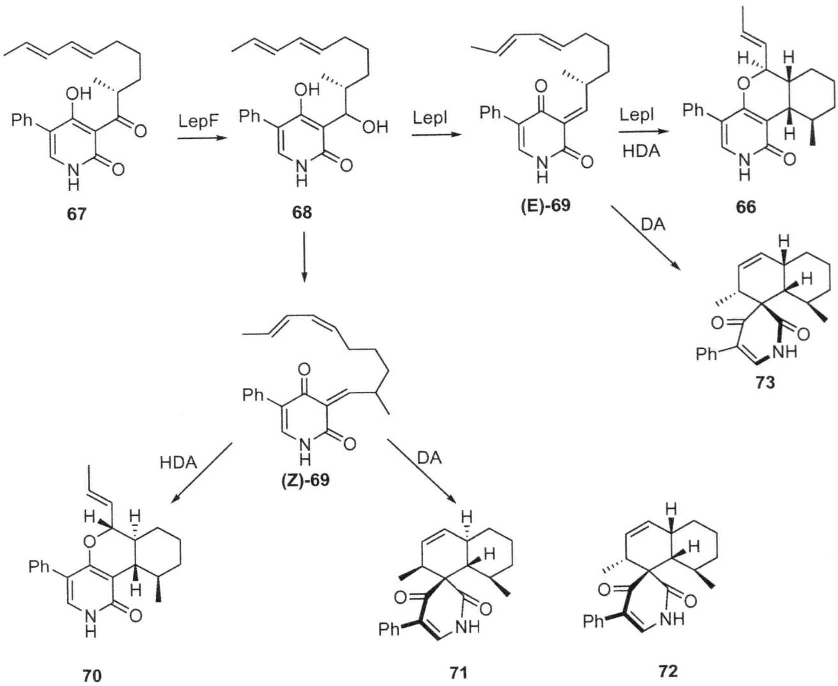

SCHEME 4.15 Suggested biosynthesis of leporin C.

explained considering that the reduction of **67** into **68** triggers the already observed thermal transformations in the observed products. Inclusion of the *lepI* gene in the host conducted to the exclusive formation of leporin C. This means that LepI is both catalyzing the stereoselective dehydration of **68** into **69** and the hetero Diels–Alder reaction of **69** into **66**. Monitoring the reaction in vitro of **69** with LepI revealed the expected formation of the hetero-Diels–Alder (HDA) product **66**, and surprisingly the transient formation of **73**, that by prolonged treatment was finally converted in **66** too. A separate experiment could unequivocally demonstrate that by incubation of **73** with LepI full conversion to **66** was obtained, suggesting that LepI can catalyze a retro-Claisen rearrangement (Scheme 4.16).

Further experiments showed also that the cofactor S-Adenosyl methionine (SAM) is an essential component of LepI catalytic activities.

The LepI-catalyzed reaction cascade was also investigated from a theoretical point of view by density functional calculations that revealed the ambimodal nature of the single transition state for both the *endo*-Diels–Alder and the HDA reactions and the role played by the sulfonium moiety of the cofactor that significantly reduce the barrier of the TSs that connect **73** to **66** (Fig. 4.13).

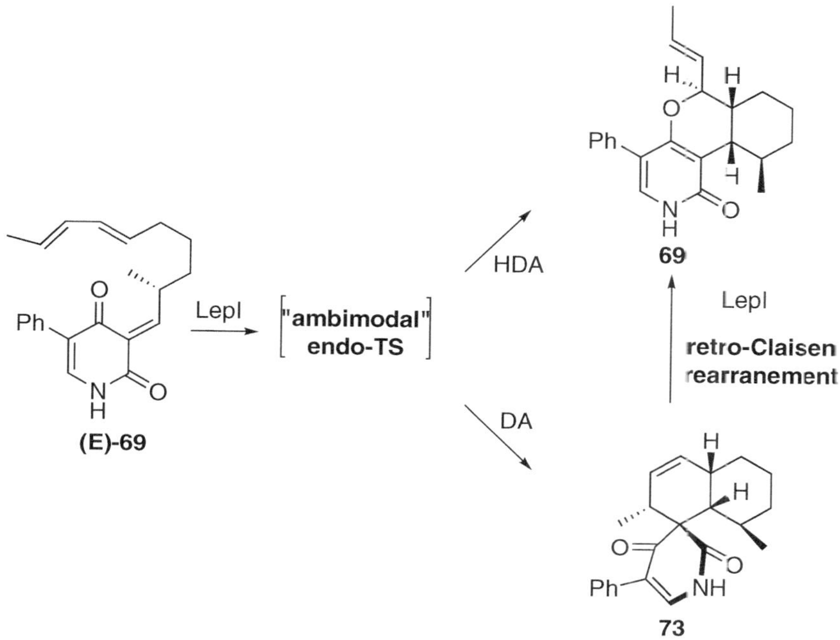

SCHEME 4.16 Reaction in vitro of **69** with LepI.

Stepping back to the first consideration of this section, even though the role of this enzyme in addressing the synthesis of leporin C has been deeply investigated, the fact that LepI both catalyzes a stereoselective dehydration and the following reaction cascade, let us question the classification of these enzymes as DAase.

Thiazolyl peptide biosynthesis

Thiazolyl peptides are a class of peptide-derived natural products with a significant activity versus antibiotic-resistant bacteria.[21] Their macrocyclic structures are particularly complex and highly functionalized, featuring multiple azoline heterocycles alternating with modified peptide side chains and cyclized on a trisubstituted pyridine or piperidine core. Because of the great synthetic difficulties needed to prepare these compounds, the standard approach to explore new libraries of thiazolyl peptides are semisynthetic preparations.[22] These thiopeptides are produced by many gram-positive bacteria and biosynthesized through a series of posttranslational modifications of a precursor peptide. This prepeptide, containing an N-terminal leader peptide followed by a C-terminus rich in serine and cysteine residues, undergoes several transformations that lead to the formation of

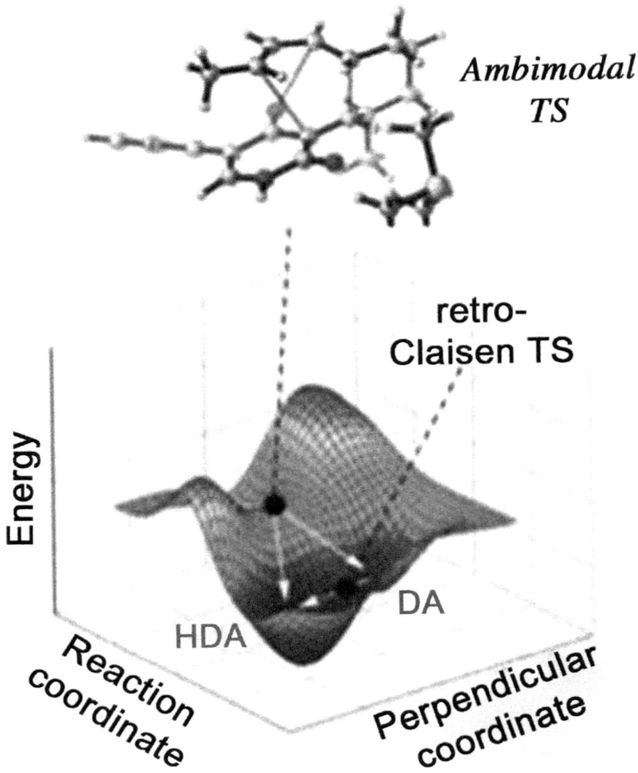

FIGURE 4.13 Potential energy surface of the transformation of the ambimodal TS in the corresponding products. *Glockle, A.; Gulder, T. A. M.* Angew. Chem. Int. Ed. ***2018***, 57, *2754–2756.*

dehydroalanine residues and thiazole rings.[23] Then by formation of a nitrogen-containing six-membered ring the linear precursor is transformed in the final macrocycle with elimination of the leader peptide. Different authors suggested that the macrocyclization driven by the formation of a piridine or piperidine ring were consistent with a formal Diels–Alder reaction. Knockout studies in *Bacillus cereus* ATCC 14579 allowed the identification of a gene cluster responsible for the production of thiocillin **74** and the enzyme responsible for the cycloaddition between two dehydroalanine residues to form the pyridine ring. Knockout experiments of the *tclM* gene led to accumulation of the acyclic precursor of thiocillin **76**, suggesting that TclM could be responsible for the catalysis of the putative hetero Diels–Alder reaction. Strong evidence arrived when isolated TclM, incubated with acyclic precursors of thiocillin analogue, led to the formation of the expected macrocycle (Fig. 4.14).

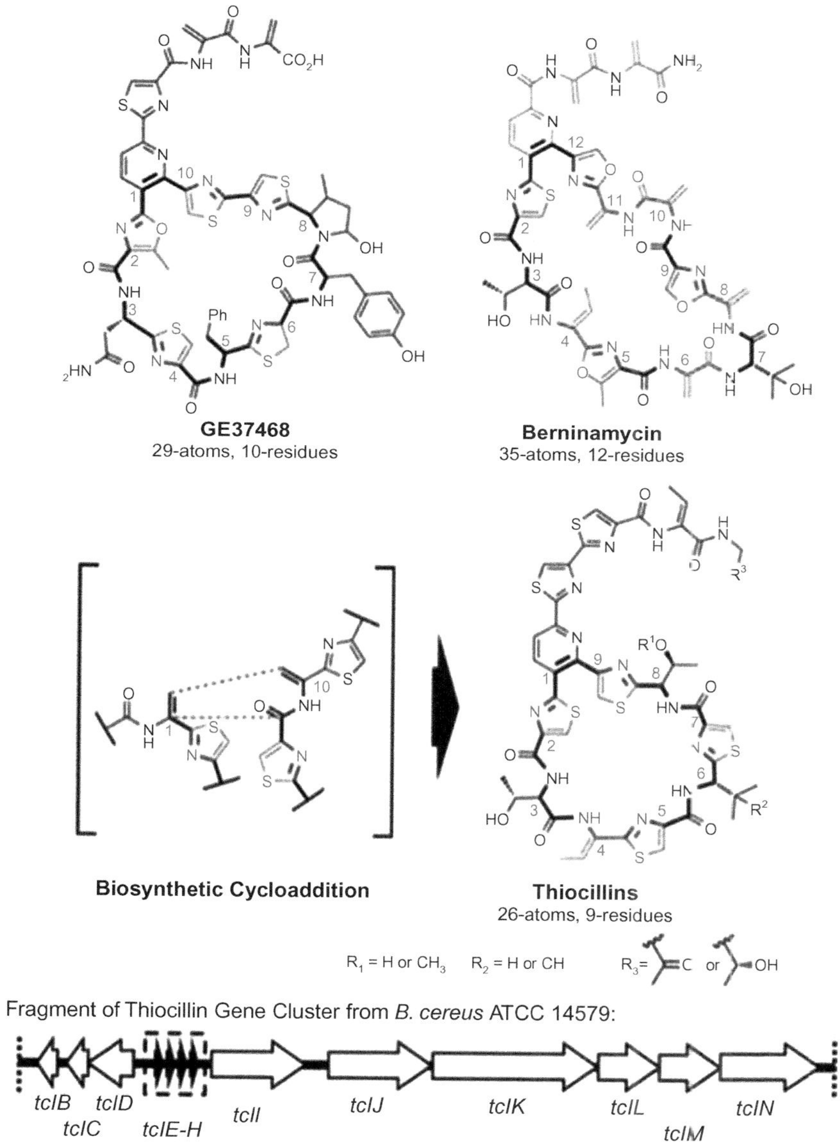

FIGURE 4.14 Structures of naturally occurring thiazolyl peptide ring size variants and fragment of thiocillin gene cluster from *Bacillus cereus* ATCC 14579. *Bowers, A. A.; Acker, M. G.; Young, T. S.; Walsh, C. T. J. Am. Chem. Soc. 2012, 134, 10313–10316.*

Later, two other homologues of TclM were identified and successfully proved as catalyst for the formation of the pyridine ring of their corresponding thipeptide, namely TbtD and PbtD.[23]

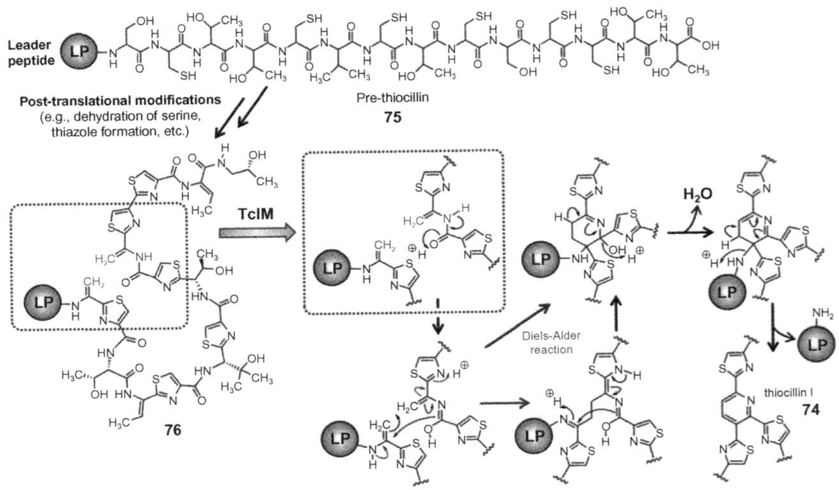

FIGURE 4.15 Current mechanistic hypotheses for the reactions catalyzed by Tclm in the biosynthesis of heterocycles in macrocyclic thiopetide antibiotics. *Jeon, B.-s.; Wang, S.-A.; Ruszczycky, M. W.; Liu, H.-w.* Chem. Rev. ***2017***, 117, *5367 – 5388.*

Even though there are few doubts left concerning the catalytic role of these enzymes in catalyzing the formation of the pyridine ring, more details concerning the pericyclic nature of the enzymatic mechanism of action are needed in order to finally label them as DAase enzymes (Fig. 4.15).[23c]

Biosynthesis of Ucs1025A

The Tang group proposed the biosynthetic pathway of UCS1025A **77**,[24] a fungal polyketide featured by a tricyclic furopyrrolizidine connected to a *trans*-decalin fragment, that has been found active in the inhibition of telomerase. This natural product and its analogues are isolated from *Acremonium* sp. KY4917 and since they all share strong antibacterial and antitumor properties many researchers dealt with their total synthesis and in understanding their biosynthesis.[24,25] By sequencing of the producing strain it was possible to identify the *ucs* gene cluster likely responsible for the biosynthesis of UCS1025A.[24c] Because of the genetic intractability of *Acremonium* sp. KY4917 and low titer of **77**, the authors had to activate the *ucs* gene cluster in a heterologous host in order to perform knockout gene experiments for the evaluation of the biosynthetic pathway. Throughout this approach **77** was confirmed to be derived from a PKS-NRPS assembly line, since by deletion of *ΔucsA* the biosynthesis was fully inhibited. Generation of the *ΔucsH* strain led to the accumulation of a pair of diastereoismers containing the pyrrolizidine scaffold but lacking in the *trans*-decalin system, indicating that analogously to the biosynthesis of myceliothermophin, the

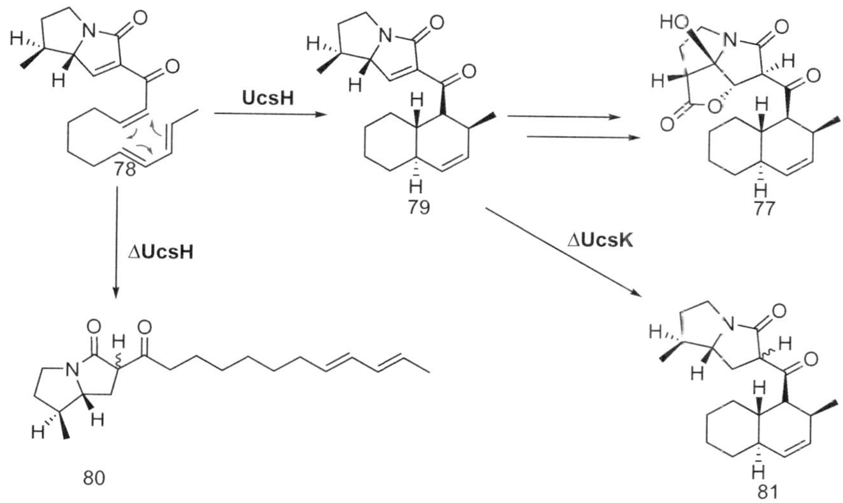

SCHEME 4.17 Biosynthesis of the *trans*-decalin system and deletion experiments of genes ucsH and ucsK.

formation of decalin system is subsequent to the Knoevenagel condensation as reported in Scheme 4.17.

By deletion of *ucsK*, a mixture of diasteroisomer **81** was detected, indicating that the putative intramolecular Diels–Alder occurred while the downstream formation of 1 was inhibited. Following experiments revealed that by heterologous reconstitution of UcsA, L, H, F, and H in *A. nidulans* **81** was obtained, further confirming the proposed biosynthesis and bringing more evidences in support of the hypothesis that UcsH is a DAase, even if additional confirmation should be obtained possibly through isolation of the protein and performing experiments aimed to elucidate the enzymatic mechanism of action (Fig. 4.16).

Biosynthesis of Sch 210972

The tetramic acid-containing metabolite Sch 210972 **82** extracted from *Chaetmium globosun* has shown potent inhibitory activity against a cell surface receptor that allows the entry of HIV-1 into cells.[26a] Structural features of this compound suggest analogies with already discussed natural compounds in which biosynthesis has been here reported. In particular, the tetramic acid and the *trans*-decalin moieties of this compound quite resemble the functionalities found in equisetin, suggesting that a DAase can be involved in its biosynthesis. In order to verify this hypothesis, Sato et al. identified the *C. globosum* gene cluster that code for the enzymes responsible for the production of **82**.[26c] The proposed biosynthetic pathway starts with the

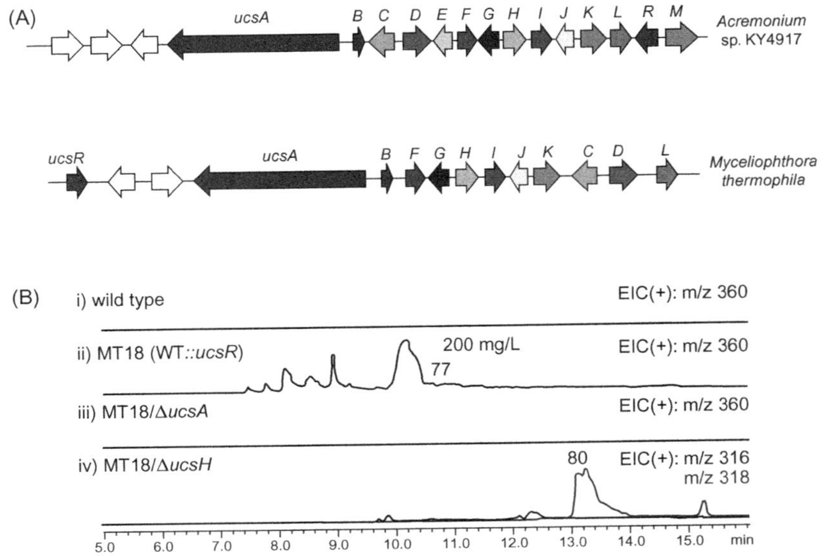

FIGURE 4.16 (A) ucs cluster responsible for the biosynthesis of **77**. The cluster is found in *Acremonium sp.* KY4917 and *Myceliophthora thermophila.* (B) Genetic analysis of the cluster found in *Myceliophthora thermophila.* Chromatograms of extracts from (i) wild type strain; (ii) MT18: the ucsR overexpression strain; (iii) MT18/ΔucsA; (iv) MT18/ΔucsH. *Li, L.; Tang, M.-C.; Tang, S.; Gao, S.; Soliman, S.; Hang, L.; Xu, W.; Ye, T.; Watanabe, K.; Tang, Y.* J. Am. Chem. Soc. ***2018***, 140, *2067–2071.*

formation of an unusual γ-hydroxymethyl-L-glutamic acid attributed to CghB and following incorporation in the linear precursor of the final product by CghG (Scheme 4.18).[26b]

Through homology studies with other proteins, the authors identified CghA as the possible candidate enzyme that catalyzes the intramolecular Diels–Alder that leads to the final product. Surprisingly, by deletion of the corresponding gene in an engineered strain of *C. globosum* named CCGKW14, **82** was still detected even if in a reduced amount, along with a new compound. Thanks to a detailed structural characterization, it was eventually possible to determine that **84** corresponds to the *exo* adduct of the proposed Diels–Alder reaction, while 1 corresponds to the *endo* adduct; this mixture is originated by the spontaneous thermal cycloaddition of the acyclic precursor of **82**. In order to probe the possible role played by CghA in the cycloaddition, the corresponding gene was reintroduced to the *ΔcghA*/CGKW14 strain resulting in the exclusive formation of **82**, unveiling that CghA is directly responsible for ensuring the stereoselective formation of Sch 210972. Further confirmation of these findings came from heterologous expression of *cghA*, *cghB*, *cghC*, and *cghG* to *A. nidulans* A1145 that

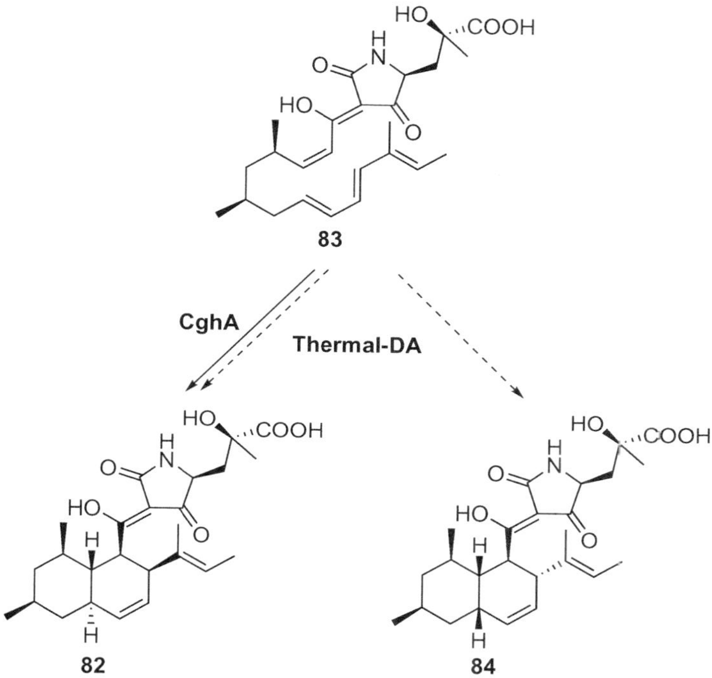

SCHEME 4.18 Exclusive formation of **82** by treatment with CghA.

resulted in the formation of **82**, whereas deletion of *cghA* led to the formation of the thermal mixture.

At the moment the question concerning the mechanism by which this enzyme catalyzes the formal [4 + 2]-cycloaddition remains open even if the authors, relying on the sequence analysis of CghA homologues and computational calculations on the cycloaddition itself, suggested that the stereoselectivity can be addressed at the ability of correctly orienting the acyclic substrate, along with an activation of the enone carbonyl of the substrate by hydrogen bonding in the active site of the protein.

Conclusion

In this chapter we have reported and discussed several examples of natural enzymes that somehow promote the formation of hexa- and pentacyclic structures throughout reactions that can be formally considered as [4 + 2]-cycloadditions. In order to identify those proteins that might or might not be labeled pericyclases, we have discussed possible mechanisms involved in the catalytic activity of such enzymes as a key discerning element. From a general perspective, it seems that the catalytic activity is usually performed through activation of the substrate by hydrogen bonding or imposing conformations that resemble the geometry of the corresponding TSs. Furthermore, we would like to highlight the fact that this hypothetical new class of enzymes is quite heterogeneous, since we met mono- and bifunctional enzymes that do not share a common motif and are phylogenetically distinct from each other.

Even though this need of speculation on the mechanistic aspects can sound trivial, we think instead that a deep understanding of these aspects will help in the design of tailor-made enzymes able to catalyze the formation of targeted complex structures otherwise difficult to obtain or to explore new portions of the chemical space.

References

1. (a) Laschat, S. *Angew. Chem. Int. Ed. Engl.* **1996,** *35*, 289–291. (b) Minami, A.; Oikawa, H. *J. Antibiot.* **2016,** *69*, 500–506. (c) Jeon, B.-s; Wang, S.-A.; Ruszczycky, M. W.; Liu, H.-w *Chem. Rev.* **2017,** *117*, 5367–5388.
2. (a) Sakurai, I.; Suzuki, H.; Shimizu, S.; Yamamoto, Y. *Chem. Pharm. Bull.* **1985,** *33*, 5141–5143. (b) Oikawa, H.; Yagi, K.; Watanabe, K.; Honma, M.; Ichihara, A. *Chem. Commun.* **1997,** 97–98. (c) Oikawa, H.; Watanabe, K.; Yagi, K.; Ohashi, S.; Mie, T.; Ichihara, A.; Honma, M. *Tetrahedron Lett.* **1999,** *40*, 6983–6986.
3. (a) Watanabe, K.; Oikawa, H.; Yagi, K.; Ohashi, S.; Mie, T.; Ichihara, A.; Honma, M. *J. Biochem.* **2000,** *127*, 467–473. (b) Ose, T.; Watanabe, K.; Mie, T.; Honma, M.; Watanabe, H.; Yao, M.; Oikawa, H.; Tanaka, I. *Nature* **2003,** *422*, 185–189. (c) Watanabe, K.; Mie, T.; Ichihara, A.; Oikawa, H.; Honma, M. *J. Biol. Chem.* **2000,** *275*, 38393–38401. (d) Serafimov, J. M.; Gillingham, D.; Kuster, S.; Hilvert, D. *J. Am. Chem. Soc.* **2008,** *130*, 7798–7799. (e) Serafimov, J. M.; Westfeld, T.; Meier, B. H.; Hilvert, D. *J. Am. Chem. Soc.* **2007,** *129*, 9580–9581.
4. (a) Guimarães, C. R. W.; Udier-Blagovic, M.; Jorgensen, W. L. *J. Am. Chem. Soc.* **2005,** *127*, 3577–3588. (b) Izard, T.; Blackwell, N. C. *EMBO J.* **2000,** *19*, 3849–3856.
5. (a) Ichihara, A.; Tazaki, H.; Sakamura, S. *Tetrahedron Lett.* **1983,** *24*, 5373–5376. (b) Ichihara, A.; Miki, M.; Tazaki, H.; Sakamura, S. *Tetrahedron Lett.* **1987,** *28*, 1175–1178. (c) Oikawa, H.; Suzuki, Y.; Naya, A.; Katayama, K.; Ichihara, A. *J. Am. Chem. Soc.* **1994,** *116*, 3605–3606. (d) Oikawa, H.; Yokota, T.; Ichihara, A.; Sakamura, S. *J. Chem. Soc., Chem. Commun.* **1989,** 1284–1285. (e) Alam, S. S.; Bilton, J. N.; Slawin, A. M. Z.; Williams, D. J.; Sheppard, R. N.; Strange, R. N. *Phytochemistry* **1989,** *28*, 2627–2630.
6. (a) Oikawa, H.; Kobayashi, T.; Katayama, K.; Suzuki, Y.; Ichihara, A. *J. Org. Chem.* **1998,** *63*, 8748–8756. (b) Lygo, B.; Bhatia, M.; Cooke, J. W. B.; Hirst, D. J. *Tetrahedron Lett.* **2003,** *44*, 2529–2532. (c) Wilson, R. M.; Jen, W. S.; MacMillan, D. W. C. *J. Am. Chem. Soc.* **2005,** *127*, 11616–11617.
7. (a) Oikawa, H.; Yokota, T.; Abe, T.; Ichihara, A.; Sakamura, S.; Yoshizawa, Y.; Vederas, J. C. *J. Chem. Soc., Chem. Commun.* **1989,** 1282–1284. (b) Kasahara, K.; Miyamoto, T.; Fujimoto, T.; Oguri, H.; Tokiwano, T.; Oikawa, H.; Ebizuka, Y.; Fujii, I. *ChemBioChem* **2010,** *11*, 1245–1252. (c) Kim, W.; Park, C.-M.; Park, J.-J.; Akamatsu, H. O.; Peever, T. L.; Xian, M.; Gang, D. R.; Vandemark, G.; Chen, W. *Mol. Plant-Microbe Interact.* **2015,** *28*, 482–496.

8. (a) Oikawa, H.; Katayama, K.; Suzuki, Y.; Ichihara, A. *J. Chem. Soc., Chem. Commun.* **1995,** 1321–1322. (b) Koetter, J. W. A.; Schulz, G. E. *J. Mol. Biol.* **2005,** *352*, 418–428. (c) Katayama, K.; Kobayashi, T.; Chijimatsu, M.; Ichihara, A.; Oikawa, H. *Biosci., Biotechnol., Biochem.* **2008,** *72*, 604–607. (d) Katayama, K.; Kobayashi, T.; Oikawa, H.; Honma, M.; Ichihara, A. *Biochim. Biophys. Acta, Protein Struct. Mol. Enzymol.* **1998,** *1384*, 387–395.
9. (a) Kirst, H. A. *J. Antibiot.* **2010,** *63*, 101–111. (b) Kim, H. J.; Ruszczycky, M. W.; Choi, S.-h; Liu, Y.-N.; Liu, H.-W. *Nature* **2011,** *473*, 109–112. (c) Fage, C. D.; Isiorho, E. A.; Liu, Y.; Wagner, D. T.; Liu, H. W.; Keatinge-Clay, A. T. *Nat. Chem. Biol.* **2015,** *11*, 256–258.
10. (a) Patel, A.; Chen, Z.; Yang, Z.; Gutierrez, O.; Liu, H. W.; Houk, K. N.; Singleton, D. A. *J. Am. Chem. Soc.* **2016,** *138*, 3631–3634. (b) Medvedev, M. G.; Zeifman, A. A.; Novikov, F. N.; Bushmarinov, I. S.; Stroganov, O. V.; Titov, I. Y.; Chilov, G. G.; Svitanko, I. V. *J. Am. Chem. Soc.* **2017,** *139*, 3942–3945. (c) Zheng, Y. Y.; Thiel, W. *J. Org. Chem.* **2017,** *82*, 13563–13571. (d) Chen, N.; Wang, S.; Smentek, L.; Hess, B. A., Jr.; Wu, R. *Angew. Chem., Int. Ed.* **2015,** *54*, 8693–8696. (e) Chen, N.; Zhang, F.; Wu, R.; Andess Hess, B., Jr. *ACS Catal.* **2018,** *8*, 2353–2358.
11. (a) Ding, W.; Williams, D. R.; Northcote, P.; Siegel, M. M.; Tsao, R.; Ashcroft, J.; Morton, G. O.; Alluri, M.; Abbanat, D.; Maiese, W. M.; Ellestad, G. A. *J. Antibiot.* **1994,** *47*, 1250–1257. (b) Singh, M. P.; Petersen, P. J.; Jacobus, N. V.; Mroczenski- Wildey, M. J.; Maiese, W. M.; Greenstein, M.; Steinberg, D. A. *J. Antibiot.* **1994,** *47*, 1258–1265. (c) Abbanat, D.; Maiese, W.; Greenstein, M. *J. Antibiot.* **1999,** *52*, 117–126. (d) Wu, Q.; Wu, Z.; Qu, X.; Liu, W. *J. Am. Chem. Soc.* **2012,** *134*, 17342–17345. (e) Jia, X.-Y.; Tian, Z.-H.; Shao, L.; Qu, X.-D.; Zhao, Q.-F.; Tang, J.; Tang, G.-L.; Liu, W *Chem. Biol.* **2006,** *13*, 575–585. (f) Tian, Z.; Sun, P.; Yan, Y.; Wu, Z.; Zheng, Q.; Zhou, S.; Zhang, H.; Yu, F.; Jia, X.; Chen, D.; Mándi, A.; Kurtán, T.; Liu, W. *Nat. Chem. Biol.* **2015,** *11*, 259–265. (g) Zheng, Q.; Guo, Y.; Yang, L.; Zhao, Z.; Wu, Z.; Zhang, H.; Liu, J.; Cheng, X.; Wu, J.; Yang, H.; Jiang, H.; Pan, L.; Liu, W. *Cell Chem. Biol.* **2016,** *23*, 352–360. (h) Zheng, Q.; Gong, Y.; Guo, Y.; Zhao, Z.; Wu, Z.; Zhou, Z.; Chen, D.; Pan, L.; Liu, W. *Cell Chem. Biol.* **2018,** *25*, 718–727.
12. (a) Park, H. R.; Chijiwa, S.; Furihata, K.; Hayakawa, Y.; Shin-Ya, K. *Org. Lett.* **2007,** *9*, 1457. (b) Ueda, J.-y; Chijiwa, S.; Takagi, M.; Shin-ya, K. *J. Antibiot.* **2008,** *61*, 752. (c) Zhao, P.; Ueda, J.-y; Kozone, I.; Chijiwa, S.; Takagi, M.; Kudo, F.; Nishiyama, M.; Shin-ya, K.; Kuzuyama, T. *Org. Biomol. Chem.* **2009,** *7*, 1454–1460. (d) Hashimoto, T.; Hashimoto, J.; Teruya, K.; Hirano, T.; Shin-ya, K.; Ikeda, H.; Liu, H.-w; Nishiyama, M.; Kuzuyama, T. *J. Am. Chem. Soc.* **2015,** *137*, 572–575.
13. (a) Riedlinger, J.; Reicke, A.; Zähner, H.; Krismer, B.; Bull, A. T.; Maldonado, L. A.; Ward, A. C.; Goodfellow, M.; Bister, B.; Bischoff, D.; Süssmuth, R. D.; Fiedler, H.-P. *J. Antibiot.* **2004,** *57*, 271. (b) Bister, B.; Bischoff, D.; Ströbele, M.; Riedlinger, J.; Reicke, A.; Wolter, F.; Bull, A. T.; Zähner, H.; Fiedler, H. P.; Süssmuth, R. D. *Angew. Chem., Int. Ed.* **2004,** *43*, 2574. (c) Gottardi, E. M.; Krawczyk, J. M.; Von Suchodoletz, H.; Schadt, S.; Mühlenweg, A.; Uguru, G. C.; Pelzer, S.; Fiedler, H. P.; Bibb, M. J.; Stach, J. E. M.; Süssmuth, R. D. *ChemBioChem* **2011,** *12*, 1401–1410. (d) Byrne, M. J.; Lees, N. R.; Li-Chen, H.; van der Kamp, M. W.; Mulholland, A. J.; Stach, J. E. M.; Willis, C. L.; Race, P. R. *J. Am. Chem. Soc.* **2016,** *138*, 6095–6098.
14. (a) Yang, Y. L.; Lu, C. P.; Chen, M. Y.; Chen, K. Y.; Wu, Y. C.; Wu, S. H. *Chem. Eur. J.* **2007,** *13*, 6985–6991. (b) Nicolaou, K.; Shi, L.; Lu, M.; Pattanayak, M. R.; Shah, A. A.; Ioannidou, H. A.; Lamani, M. *Angew. Chem., Int. Ed.* **2014,** *53*, 10970–10974. (c)

Shionozaki, N.; Yamaguchi, T.; Kitano, H.; Tomizawa, M.; Makino, K.; Uchiro, H. *Tetrahedron Lett.* **2012,** *53*, 5167–5170. (d) Kontnik, R.; Clardy, J. *Org. Lett.* **2008,** *10*, 4149–4151. (e) Suzuki, S.; Hosoe, T.; Nozawa, K.; Kawai, K.-i; Yaguchi, T.; Udagawa, S.-i *J. Nat. Prod.* **2000,** *63*, 768–772. (f) Osterhage, C.; Kaminsky, R.; König, G. M.; Wright, A. D. *J. Org. Chem.* **2000,** *65*, 6412–6417. (g) West, R. R.; Van Ness, J.; Varming, A. M.; Rassing, B.; Biggs, S.; Gasper, S.; McKernan, P. A.; Piggott, J. *J. Antibiot.* **1996,** *49*, 967–973. (h) Singh, S. B.; Goetz, M. A.; Jones, E. T.; Bills, G. F.; Giacobbe, R. A.; Herranz, L.; Stevens-Miles, S.; Williams, D. L., Jr *J. Org. Chem.* **1995,** *60*, 7040–7042. (i) Liu, X.; Walsh, C. T. *Biochemistry* **2009,** *48*, 8746–8757. (j) Sims, J. W.; Schmidt, E. W. *J. Am. Chem. Soc.* **2008,** *130*, 11149–11155. (k) Li, L.; Peiyuan, Y.; Tang, M.-C.; Zou, Y.; Gao, S.-S.; Hung, Y.-S.; Zhao, M.; Watanabe, K.; Houk, K. N.; Tang, Y. *J. Am. Chem. Soc.* **2016,** *138*, 15837–15840.

15. (a) Schobert, R.; Schlenk, A. *Bioorg. Med. Chem.* **2008,** *16*, 4203–4221. (b) Li, G.; Kusari, S.; Spiteller, M. *Nat. Prod. Rep.* **2014,** *31*, 1175–1201. (c) Burmeister, H. R.; Bennett, G. A.; Vesonder, R. F.; Hessel-Tine, C. W. *Antimicrob. Agents Chemother.* **1974,** *5*, 634–639. (d) Song, Z.; Cox, R. J.; Lazarus, C. M.; Simpson, T. T. *ChemBioChem* **2004,** *5*, 1196–1203. (e) Kato, N.; Nogawa, T.; Hirota, H.; Jang, J.-H.; Takahashi, S.; Ahn, J. S.; Osada, H. *Biochem. Biophysical Res. Comm.* **2015,** *460*, 210–215. (f) Kato, N.; Nogawa, T.; Takita, R.; Kinugasa, K.; Kanai, M.; Uchiyama, M.; Osada, H.; Takahashi, S. *Angew. Chem. Int. Ed.* **2018,** *57*, 9754–9758.
16. (a) Jordan, F.; Patel, H. *ACS Catal.* **2013,** *3*, 1601–1617. (b) Kourist, R.; Guterl, J.-K.; Miyamoto, K.; Sieber, V. *ChemCatChem* **2014,** *6*, 689–701. (c) Claypool, J. T.; Raman, D. R.; Jarboe, L. R.; Nielsen, D. R. *J. Ind. Microbiol. Biotechnol* **2014,** *41*, 1211–1216. (d) McKenna, R.; Thompson, B.; Pugh, S.; Nielsen, D. R. *Microb. Cell Fact.* **2014,** *13*, 123.
17. (a) Huang, Z. X.; Dostal, L.; Rosazza, J. P. N. *J. Bacteriol.* **1994,** *176*, 5912–5918. (b) Mukai, N.; Masaki, K.; Fujii, T.; Kawamukai, M.; Iefuji, H. *J. Biosci. Bioengin.* **2010,** *109*, 564–569. (c) Lin, F.; Ferguson, K. L.; Boyer, D. R.; Lin, X. N.; Marsh, E. N. G. *ACS Chem. Biol.* **2015,** *10*, 1137–1144. (d) Payne, K. A. P.; White, M. D.; Fisher, K.; Khara, B.; Bailey, S. S.; Parker, D.; Rattray, N. J. W.; Trivedi, D. K.; Goodacre, R.; Beveridge, R.; Barran, P.; Rigby, S. E. J.; Scrutton, N. S.; Hay, S.; Leys, D. *Nature* **2015,** *522*, 497–501. (e) White, M. D.; Payne, K. A. P.; Fisher, K.; Marshall, S. A.; Parker, D.; Rattray, N. J. W.; Trivedi, D. K.; Goodacre, R.; Rigby, S. E. J.; Scrutton, N. S.; Hay, S.; Leys, D. *Nature* **2015,** *522*, 502–506. (f) Waugh, M. W.; Marsh, E. N. G. *Biochemistry* **2014,** *53*, 5537–5543. (g) Ferguson, K. L.; Arunrattanamook, N.; Marsh, E. N. *Biochemistry* **2016,** *55*, 2857–2863. (h) Baunach, M.; Herweck, C. *Angew. Chem. Int. Ed.* **2015,** *54*, 12550–12552.
18. (a) Tobert, J. A. *Nat. Rev. Drug Discov.* **2003,** *2*, 517–526. (b) Endo, A.; Kuroda, M.; Tsujita, Y. *J. Antibiot.* **1976,** *29*, 1346–1348. (c) Endo, A. *J. Antibiot.* **1979,** *32*, 852–854. (d) Moore, R. N.; Bigam, G.; Chan, J. K.; Hogg, A. M.; Nakashima, T. T.; Vederas, J. C. *J. Am. Chem. Soc.* **1985,** *107*, 3694–3701. (e) Yoshizawa, Y.; Witter, D. J.; Liu, Y.; Vederas, J. C. *J. Am. Chem. Soc.* **1994,** *116*, 2693–2694. (f) Endo, A. *J. Antibiot.* **1980,** *33*, 334–336. (g) Alberts, A. W. *Proc. Natl. Acad. Sci. U.S.A.* **1980,** *77*, 3957–3961.
19. (a) Hendrickson, L.; Davis, C. R.; Roach, C.; Nguyen, D. K.; Aldrich, T.; McAda, P. C.; Reeves, C. D. *Chem. Biol.* **1999,** *6*, 429–439. (b) Kennedy, J.; Auclair, K.; Kendrew, S. G.; Park, C.; Vederas, J. C.; Hutchinson, C. R. *Science* **1999,** *284*, 1368–1372. (c) Ma, S. M.; Li, J. W.-H.; Choi, J. W.; Zhou, H.; Lee, K. K. M.; Moorthie, V. A.; Xie, X.; Kealey, J. T.; Da Silva, N. A.; Vederas, J. C.; Tang, Y. *Science* **2009,** *326*, 589–592.

(d) Auclair, K.; Kennedy, J.; Hutchinson, C. R.; Vederas, J. C. *Bioorg. Med. Chem. Lett.* **2001,** *11*, 1527–1531. (e) Ames, B. D.; Nguyen, C.; Bruegger, J.; Smith, P.; Xu, W.; Ma, S.; Wong, E.; Wong, S.; Xie, X.; Li, J. W.-H.; Vederas, J. C.; Tang, Y. Tsai, S.-C. *Proc. Natl. Acad. Sci. U.S.A.* **2012,** *109*, 11144–11149. (f) Xu, W.; Chooi, Y.-H.; Choi, J. W.; Li, S.; Vederas, J. C.; Da Silva, N. A.; Tang, Y. *Angew. Chem., Int. Ed.* **2013,** *52*, 6472–6475. (g) Xie, X.; Meehan, M. J.; Xu, W.; Dorrestein, P. C.; Tang, Y. *J. Am. Chem. Soc.* **2009,** *131*, 8388–8389. (h) Barriuso, J.; Nguyen, D. T.; Li, J. W.-H.; Roberts, J. N.; MacNevin, G.; Chaytor, J. L.; Marcus, S. L.; Vederas, J. C.; Ro, D.-K. *J. Am. Chem. Soc.* **2011,** *133*, 8078–8081. (i) Auclair, K.; Sutherland, A.; Kennedy, J.; Witter, D. J.; Van den Heever, J. P.; Hutchinson, C. J.; Vederas, J. C. *J. Am. Chem. Soc.* **2000,** *122*, 11519–11520.

20. (a) Cary, J. W.; Uka, V.; Han, Z.; Buyst, D.; Harris-Coward, P. Y.; Ehrlich, K. C.; Wei, Q.; Bhatnagar, D.; Dowd, P. F.; Martens, S. L.; Calvo, A. M.; Martins J. C.; Vanhaecke, L.; Coenye, T.; De Saeger, S.; Di Mavungu, J. D. *Fungal Genet. Biol.* **2015,** *81*, 88–97. (b) Snider, B. B.; Lu, Q. *J. Org. Chem.* **1996,** *61*, 2839–2844. (c) Ohashi, M.; Liu, F.; Hai, Y.; Chen, M.; Tang, M.-C.; Yang, Z.; Sato, M.; Watanabe, K.; Houk, K. N.; Tang, Y. *Nature* **2017,** *549*, 502–506. (d) Glockle, A.; Gulder, T. A. M. *Angew. Chem. Int. Ed.* **2018,** *57*, 2754–2756.

21. (a) Li, C.; Kelly, W. L. *Nat. Prod. Rep.* **2010,** *27*, 153–164. (b) Walsh, C. T.; Acker, M. G.; Bowers, A. A. *J. Biol. Chem.* **2010,** *285*, 27525–275531. (c) Bagley, M. C.; Dale, J. W.; Merritt, E. A.; Xiong, X. *Chem. Rev.* **2005,** *105*, 685–714. (d) Arndt, H.-D.; Schoof, S.; Lu, J.-Y. *Angew. Chem., Int. Ed.* **2009,** *48*, 6770–6773. (e) Hughes, R. A.; Moody, C. J. *Angew. Chem., Int. Ed.* **2007,** *46*, 7930–7954. (f) Wieland Brown, L. C.; Acker, M. G.; Clardy, J.; Walsh, C. T.; Fischbach, M. A. *Proc. Natl. Acad. Sci. U.S.A.* **2009,** *106*, 2549–2553.

22. (a) Houck, D. R.; Chen, L.-C.; Keller, P. J.; Beale, J. M.; Floss, H. G. *J. Am. Chem. Soc.* **1987,** *109*, 1250–1252. (b) Houck, D. R.; Chen, L.-C.; Keller, P. J.; Beale, J. M.; Floss, H. G. *J. Am. Chem. Soc.* **1988,** *110*, 5800–5806. (c) Mocek, U.; Knaggs, A. R.; Tsuchiya, R.; Nguyen, T.; Beale, J. M.; Floss, H. G. *J. Am. Chem. Soc.* **1993,** *115*, 7557–7568. (d) Mocek, U.; Zeng, Z.; O'Hagan, D.; Zhou, P.; Fan, L.-D. G.; Beale, J. M.; Floss, H. G. *J. Am. Chem. Soc.* **1993,** *115*, 7992–8001. (e) Bycroft, B. W.; Gowland, M. S. *Chem. Soc., Chem. Commun.* **1978,** 256–258.

23. (a) Bowers, A. A.; Walsh, C. T.; Acker, M. G. *J. Am. Chem. Soc.* **2010,** *132*, 12182–12184. (b) Onaka, H.; Tabata, H.; Igarashi, Y.; Sato, Y.; Furumai, T.; Goadsporin. *J. Antibiot.* **2001,** *54*, 1036–1044. (c) Wever, W. J.; Bogart, J. W.; Baccile, J. A.; Chan, A. N.; Schroeder, F. C.; Bowers, A. A. *J. Am. Chem. Soc.* **2015,** *137*, 3494–3497. (d) Hudson, G. A.; Zhang, Z.; Tietz, J. I.; Mitchell, D. A.; van der Donk, W. A. *J. Am. Chem. Soc.* **2015,** *137*, 16012–16015. (e) Morris, R. P.; Leeds, J. A.; Naegeli, H. U.; Oberer, L.; Memmert, K.; Weber, E.; LaMarche, M. J.; Parker, C. N.; Burrer, N.; Esterow, S.; Hein, A. E.; Schmitt, E. K.; Krastel, P. *J. Am. Chem. Soc.* **2009,** *131*, 5946–5955. (f) Zhang, Z.; Hudson, G. A.; Mahanta, N.; Tietz, J. I.; van der Donk, W. A.; Mitchell, D. A. *J. Am. Chem. Soc.* **2016,** *138*, 15511–15514. (g) Cogan, D. P.; Hudson, G. A.; Zhang, Z.; Pogorelov, T. V.; van der Donk, W. A.; Mitchell, D. A.; Nair, S. K. *Proc. Natl. Acad. Sci. U.S.A.* **2017,** *114*, 12928–12933.

24. (a) Robertson, J.; Stevens, K. *Nat. Prod. Rep.* **2014,** *31*, 1721–1788. (b) Robertson, J.; Stevens, K. *Nat. Prod. Rep.* **2017,** *34*, 62–89. (b) Neuman, M. G.; Cohen, L.; Opris, M.; Nanau, R. M.; Jeong, H. *J. Pharm. Pharm. Sci.* **2015,** *18*, 825–843. (c) Nakai, R.; Ogawa, H.; Asai, A.; Ando, K.; Agaisuma, T.; Maisumiya, S.; Akinaga, S.; Yamashita, Y.;

Mizukami, T. *J. Antibiot.* **2000,** *53*, 294–296. (d) Agatsuma, T.; Akama, T.; Nara, S.; Matsumiya, S.; Nakai, R.; Ogawa, H.; Otaki, S.; Ikeda, S.; Saitoh, Y.; Kanda, Y. *Org. Lett.* **2002,** *4*, 4387–4390. (e) Li, L.; Tang, M.-C.; Tang, S.; Gao, S.; Soliman, S.; Hang, L.; Xu, W.; Ye, T.; Watanabe, K.; Tang, Y. *J. Am. Chem. Soc.* **2018,** *140*, 2067–2071.

25. (a) Nicolaou, K. C.; Pulukuri, K. K.; Rigol, S.; Buchman, M.; Shah, A. A.; Cen, N.; McCurry, M. D.; Beabout, K.; Shamoo, Y. *J. Am. Chem. Soc.* **2017,** *139*, 15868–15877. (b) Nicolaou, K. C.; Shah, A. A.; Korman, H.; Khan, T.; Shi, L.; Worawalai, W.; Theodorakis, E. A. *Angew. Chem., Int. Ed.* **2015,** *54*, 9203–9208. (c) Uchida, K.; Ogawa, T.; Yasuda, Y.; Mimura, H.; Fujimoto, T.; Fukuyama, T.; Wakimoto, T.; Asakawa, T.; Hamashima, Y.; Kan, T. *Angew. Chem., Int. Ed.* **2012,** *51*, 12850–12853. (d) de Figueiredo, R. M.; Fröhlich, R.; Christmann, M. *Angew. Chem., Int. Ed.* **2007,** *46*, 2883–2886. (e) Hoye, T. R.; Dvornikovs, V. *J. Am. Chem. Soc.* **2006,** *128*, 2550–2551. (f) Lambert, T. H.; Danishefsky, S. J. *J. Am. Chem. Soc.* **2006,** *128*, 426–427.

26. (a) Yang, S. W.; Mierzwa, R.; Terracciano, J.; Patel, M.; Gullo, V.; Wagner, N.; Baroudy, B.; Puar, M.; Chan, T. M.; McPhail, A. T.; Chu, M. *J. Nat. Prod.* **2006,** *69*, 1025–1028. (b) Ishiuchi, K.; Nakazawa, T.; Yagishita, F.; Mino, T.; Noguchi, H.; Hotta, K.; Watanabe, K. *J. Am. Chem. Soc.* **2013,** *135*, 7371–7377. (c) Sato, M.; Yagishita, F.; Mino, T.; Uchiyama, N.; Patel, A.; Chooi, Y.-H.; Goda, Y.; Xu, W.; Noguchi, H.; Yamamoto, T.; Hotta, K.; Houk, K. N.; Tang, Y.; Watanabe, K. *ChemBioChem* **2015,** *16*, 2294–2298.

Chapter 5

Cycloaddition reactions in material science

Chiara Milanese
Department of Chemistry, University of Pavia, Pavia, Italy

Cycloaddition reactions for materials synthesis

The development of nanotechnology and the growing needs to address research in health and energy issues has enormously increased the research for advanced functional soft materials possessing targeted engineered properties and functional groups. Soft material systems with precise control over architecture, domain size, functionality, polarity, solubility, and reactivity are needed for an increasing number of industrial applications. Simple, cheap, and high-yielding reactions are, obviously, strongly required for the preparation of these innovative systems.[1]

Over the two past decades, a wide range of functional soft materials such as polymers and C-based composites has been prepared or suitably modified by "click chemistry" and related procedures.[1–3] The term "click chemistry"[4] refers to a set of "powerful, highly reliable, and selective reactions for the rapid synthesis of useful new compounds and combinatorial libraries. These reactions must be modular, stereospecific, give very high yields, and generate only inoffensive by-products. The required process characteristics include simple reaction conditions, readily available starting materials and reagents, the use of no solvent or a solvent that is benign or easily removed, and simple product isolation."[4,5] All these characteristics are strongly appreciated in material chemistry, a branch of chemistry characterized by reactions generally taking place under energy- and time-consuming conditions. Among the reactions included in the click chemistry category, we report many families related to the topic of this book, such as Diels–Alder cycloadditions,[6,7] azide–nitrile cycloadditions,[8,9] and copper-catalyzed azide–alkyne cycloadditions (CuAAC reactions),[10–12] dominating the recent body of literature on click chemistry.[2]

Modern Applications of Cycloaddition Chemistry. DOI: https://doi.org/10.1016/B978-0-12-815273-7.00005-8

In this chapter, a summary of the rapidly expanding area of applications of cycloaddition reactions in materials chemistry will be given. Great attention will be devoted to the use of these same reactions for the functionalization of polymers and carbon-based materials with possible biomedical, energetic, electronic, and optical applications and for the preparation of novel complex architectures and structures.

Polymers

Click chemistry is particularly appealing and applied in polymer science for the preparation of chains containing functional groups not easily incorporated by other strategies. Most of the polymers are prepared by controlled polymerization techniques and subsequently modified by click chemistry. Controlled radical polymerization (CRP) allows for the preparation of polymers with a predetermined molecular weight, chain-end functionality, and tightly controlled architecture and composition.[13,14] The most common CRP techniques are atom transfer radical polymerization (ATRP)[15,16] and reversible addition-fragmentation transfer (RAFT) polymerization.[17,18] The mechanism of such reactions is based on a fast dynamic equilibrium keeping low the concentration of radicals and suppressing chain-terminating reactions.[19] Click chemistry combines particularly well with these reactions because the species used to initiate polymerization or to mediate the dynamic equilibrium can provide functional groups (either inherently present or intentionally added) that are easily converted to ‘‘clickable’’ functionality, allowing the preparation of the desired polymeric chain.[2] One of the most common examples of these reactions is the quantitative transformation of the terminal bromine with azide groups on polymers prepared by ATRP. As an example, Lutz et al.[20] converted low-molecular weight polystyrene (PS) synthesized by ATRP to PS$-N_3$ by reacting the bromine-terminated polymer with 1.1 equivalents of NaN_3 in dimethylformamide (DMF) at room temperature and subsequently carboxylic acid, hydroxyl, and methyl vinyl moieties were attached to the ω-terminus of each polymer chain by CuAAC. This strategy will be deeply discussed later in this chapter.

The polymerization of a functional monomer allows the incorporation of clickable groups in a polymer chain. For example, 3-azidopropyl methacrylate was prepared by Sumerlin et al.[21] via ATRP and subsequently functionalized (with quantitative yield) with several model alkynes, such as propargyl alcohol, propargyl triphenylphosphonium bromide, 4-pentynoic acid, and propargyl 2-bromoisobutyrate, by CuAAC (Scheme 5.1). The reactions, probably due to the solubilization of the copper catalyst in the forming triazoles, were significantly faster than those with the azidopropyl methacrylate monomer.[22]

Block copolymers comprise two or more homopolymer subunits linked by covalent bonds. The combination of polymer science and click chemistry

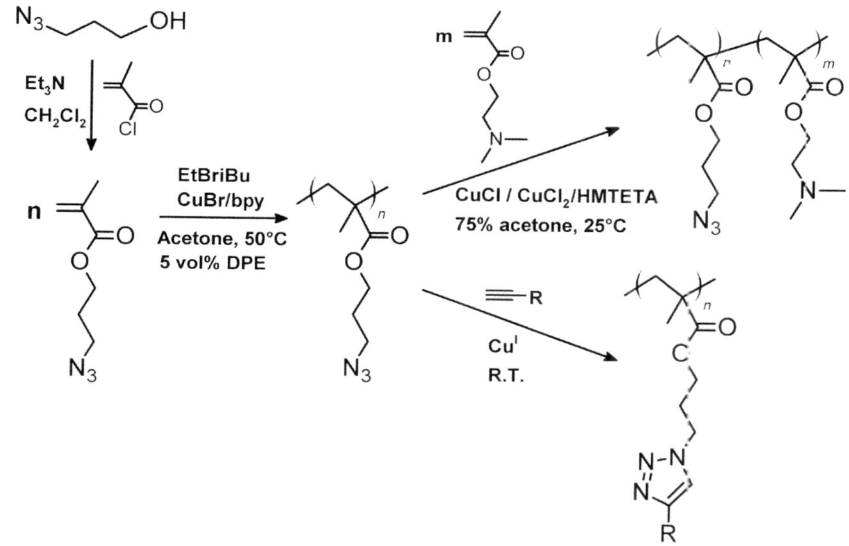

SCHEME 5.1 Synthesis of AzPMA and subsequent ATRP and postpolymerization modification with different functional alkynes by CuAAC.[21] *ATRP*, Atom transfer radical polymerization; *AzPMA*, 3-azidopropyl methacrylate; *CuAAC*, copper-catalyzed azide–alkyne cycloadditions. *Reprinted with permission from Sumerlin, B. S.; Tsarevsky, N. V.; Louche, G.; Lee, R. Y.; Matyjaszewski, K.* Macromolecules ***2005***, 38, *7540–7545. Copyright 2005 American Chemical Society.*

appears to be particularly successful for the obtainment of these molecules, which are unavailable by the conventional routes of polymerization followed by chain extension. In 2005 Opsteen and van Hest first succeeded in obtaining di- and triblock copolymers with quantitative efficiency and no residual unreacted homopolymers by conjugating an azide with alkyne-terminated polymers.[23] ATRP was used to obtain the alkyne termination by using an alkyne-modified ATRP initiator, while the azido function was incorporated by nucleophilic substitution of terminal bromine groups. This same strategy was subsequently followed for the synthesis of triphilic (hydrophilic, oleophilic, and fluorophilic) pentablock copolymers.[24]

Benicewicz et al.[25] prepared the functional monomer 2-azidoethyl methacrylate (AzMA) by dicyclohexylcarbodiimide (DCC)/DMAP-mediated coupling of methacrylic acid and 2-azidoethanol, obtaining the product in good yield under mild reaction conditions (Scheme 5.2).

Subsequently a RAFT polymerization was performed, obtaining poly (AzMA) with excellent control over the molecular weight distribution. These chains reacted with a slight excess of phenyl acetylene at room temperature by a copper-catalyzed Huisgen 1,3-dipolar cycloaddition in DMSO solvent with 0.1 equivalent of copper iodide as catalyst. The resulting functional

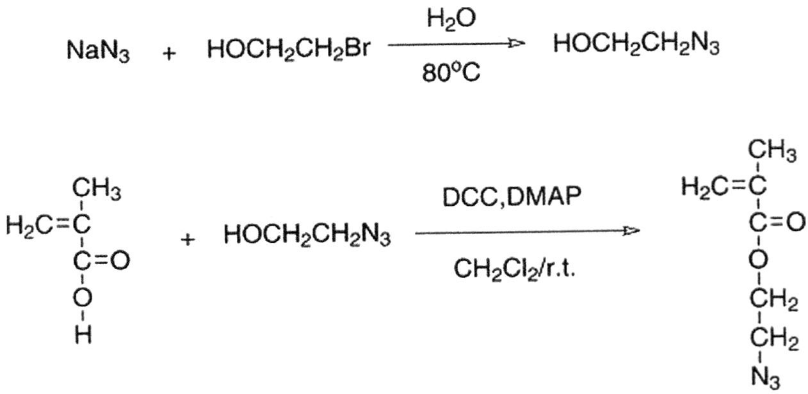

SCHEME 5.2 Synthesis of the functional monomer AzMA.[25] *AzMA*, 2-Azidoethyl methacrylate. *Reprinted with permission from Li, Y.; Yang, J.; Benicewicz, B. C.* J. Polym. Sci., Part A: Polym. Chem. ***2007****, 45, 4300–4308. Copyright 2007 John Wiley and Sons.*

polymer showed identical ^{1}H-nuclear magnetic resonance (NMR) and infrared (IR) spectra with respect to a polymer of the same molecular structure prepared by a prefunctionalization approach (AzMA was first reacted with phenyl acetylene by click coupling reaction to prepare TriazoleMA, which was then polymerized under RAFT conditions to give polyTriazoleMA), confirming the retention of the azide side chains during the RAFT polymerization of AzMA (Scheme 5.3).

The ready availability of azide salts and the high reactivity of N_3^- in nucleophilic displacement reactions have also prompted a growing effort in preparing azide-functionalized polymers through postpolymerization strategies.[1] For example, Tunca et al.[26] developed a facile synthesis of copolymers of styrene and *p*-chloromethylstyrene that can easily be converted into random heterograft copolymers, demonstrating the orthogonality and the efficiency of the click reactions (Cu-catalyzed Huisgen and Diels–Alder reactions) in preparing advanced macromolecular architectures from simple building blocks. First of all, they prepared random copolymers of styrene and the functionalizable moomer *p*-chloromethylstyrene via nitroxide mediated radical polymerization. Subsequently, the anthracene functionality was attached to the preformed copolymer by the *o*-etherification procedure and the remaining $-CH_2Cl$ was converted into the azide functionality (Scheme 5.4).

Finally, by using double-click reactions in one-pot technique, maleimide-end-functionalized poly(methyl methacrylate) (PMMA–MI) and alkyne end-functionalized poly(ethylene-glycol) (PEG-alkyne) were introduced onto the copolymer-bearing pendant anthryl and azide moieties P(S-Anth-Azide) (Scheme 5.5) giving the PS-*g*-(PMMA–PEG) heterograft copolymer.

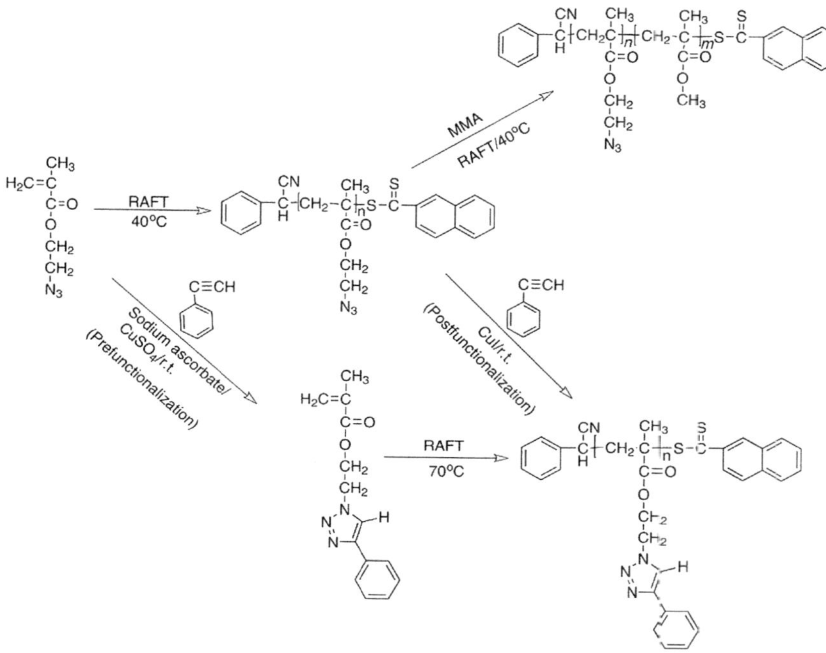

SCHEME 5.3 Preparation of poly(AzMA) and poly(AzMA) derivatives by combining RAFT polymerization and click chemistry.[25] *AzMA*, 2-Azidoethyl methacrylate; *RAFT*, reversible addition-fragmentation transfer. *Reprinted with permission from Li, Y.; Yang, J.; Benicewicz, B. C.* J. Polym. Sci., Part A: Polym. Chem. ***2007****, 45, 4300–4308. Copyright 2007 John Wiley and Sons.*

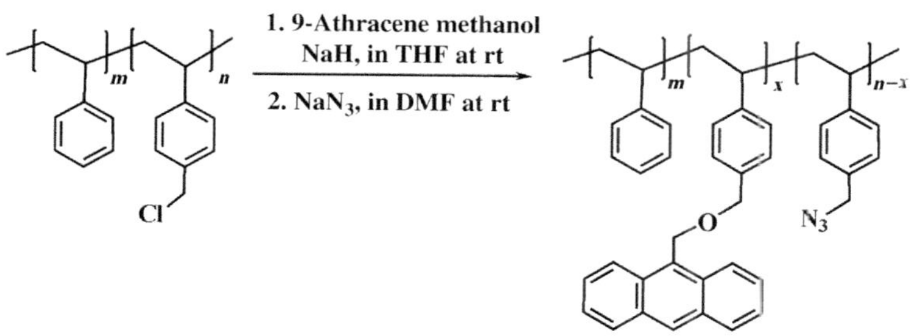

SCHEME 5.4 Synthesis of polystyrene with anthryl and azide pendant groups.[26] *Reprinted with permission from Dag, A.; Durmaz, H.; Demir, E.; Hizal, G.; Tunca, U.* J. Polym. Sci., Part A: Polym. Chem. ***2008****, 46, 6969–6977. Copyright 2008 John Wiley and Sons.*

Riva et al.,[27] combining ring-opening polymerization and click chemistry, reacted R-chloro-ε-caprolactone with sodium azide either prior to polymerization or afterward in order to prepare clickable materials (Scheme 5.6). The introduction of the azide group to the polymer backbone was complete

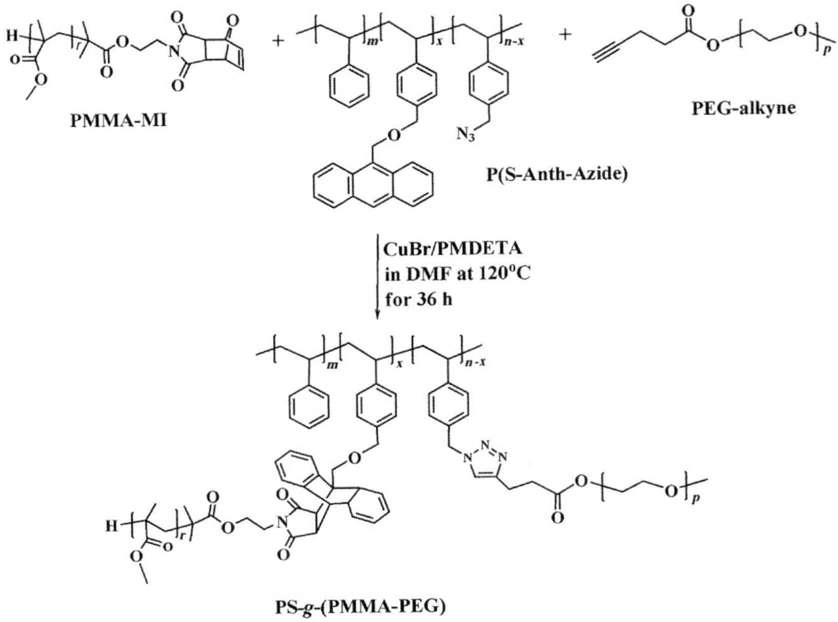

SCHEME 5.5 One-pot preparation of PS-*g*-(PMMA–PEG) heterograft copolymer by using double-click reactions of P(S-Anth-Azide), PMMA–MI, and PEG-alkyne in presence of CuBr/PMDETA in DMF at 120°C.[26] *DMF*, Dimethylformamide; *PEG*, poly(ethylene-glycol); *PMMA*, poly(methyl methacrylate); *PS*, polystyrene. *Reprinted with permission from Dag, A.; Durmaz, H.; Demir, E.; Hizal, G.; Tunca, U.* J. Polym. Sci., Part A: Polym. Chem. ***2008***, 46, *6969–6977. Copyright 2008 John Wiley and Sons.*

within 24 h, while the preparation of the azido monomer took 3 days at elevated temperatures. Subsequently, the authors functionalized the obtained materials with propargyl benzoate using water-free conditions, in order to prevent transesterification and the associated increased molecular weight distributions. Catalysis by CuI/1,8-diazabicyclo[5.4.0]undec-7-ene (DBU) in tetrahydrofuran (THF) was highly efficient, leading to complete functionalization in only 2 h at 35°C.

Narumi et al.[28] demonstrated the potential power of attaching small molecules to the chain end of a well-defined polymer: an azide-functionalized ATRP initiator was used to grow poly(*N*-isopropylacrylamide) [P(NIPAM)], having a lower critical solution temperature (LCST; the critical temperature below which the components of a mixture are miscible for all compositions) of approximately 32°C. Small acetylene derivatives with different polarities, such as phenylacetylene, (4-phenoxy)phenylacetylene, hex-1-yne, dec-1-yne, ethyl propiolate, and propargyl alcohol, were then attached to the P(NIPAM) ends via CuAAC, modifying the LCST in a temperature range of 10°C,

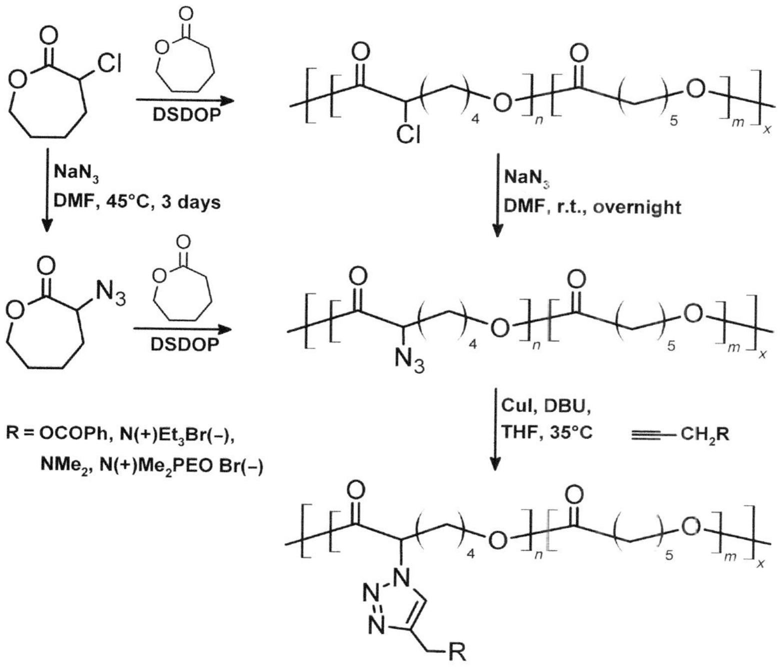

SCHEME 5.6 Derivatization of poly(ε-caprolactone) by combining ROP and click chemistry.[27] *ROP*, Ring-opening polymerization. *Reproduced by permission of The Royal Society of Chemistry from Riva, R.; Schmeits, S.; Stoffelbach, F.; Jérôme, C.; Jérôme, R.; Lecomte P.* Chem. Commun. ***2005***, 42, *5334–5336.*

which is noticeable considering the high molecular weight of the prepared polymer chain.

A common method for adding chain-end functionality to linear polymers involves modification of the ω-chain end of the polymer, through termination of the living polymer chain with a functional terminating agent or by chemically altering the radical functionality present at the chain. By far, as already exemplified, the most well-known and commonly used approach is the displacement of the terminal halogen atoms of polymers prepared by ATRP with sodium azide.[29,30] This method also prepares the way for subsequent modification of the molecules by CuAAC. In this regard, Matyjaszewski et al.[31] performed a detailed study using small-molecule azides representative of acrylic, cyanoacrylic, styrenic, and methacrylic polymer chain ends as probes of the reactivity of azide terminal polymers made by ATRP and halogen displacement. The obtained results underlined the fact that the reactivity and the rate of the click reaction depended on the nature of the monomer unit from which the terminal halogen was displaced. In particular, the authors demonstrated that the rate of triazole formation is strongly linked to both the electronic properties of the azide and the steric environment. For example, the acrylate-type model compound underwent reaction with a rate constant that is about 10 times smaller than that of the methacrylate-type model compound, which could suggest that replacement of a poly(methyl acrylate) chain end would reach 90% conversion ten times faster than would a PMMA chain end (being the constant rates in the order of 2×10^{-2} and $2 \times 10^{-3}\ s^{-1}$, respectively).

Previously, the same authors had also studied the evolution of azide–alkyne cycloadditions by varying different reaction parameters (among which are types of ligand, solvent, reducing agent, metal) by using the step-growth click-coupling of low-molecular-weight α,ω-diazido-terminated PS prepared by ATRP.[32] In their work, they found that aliphatic amine ligands showed significant rate enhancements compared to pyridine-based ligands and that tridentate ligands increased the rate of reaction with respect to tetradentate ligands. Finally, noncoordinating solvents were found to give greater rate enhancements compared to coordinating solvents. Despite the typical susceptibility of Cu^{I} complexes to oxidation, the addition of an excess of hydrazine as a reducing agent allowed click reactions to be conducted faster (probably due to the basicity of this substance) and under limited amounts of air with decreased catalyst concentrations.

Finally, they demonstrated that azide–alkyne cycloadditions could be successfully catalyzed also by oxidatively stable metal complexes, including those of Ni^{II}, Pd^{II}, and Pt^{II}, with the last one showing the highest catalytic activity with respect to those of the other metals.

Telechelic polymers are prepolymers capable of entering into further polymerization or other reactions through their reactive end groups. Telechelic PS was prepared by Matyjaszewski et al.[33] through the same strategy of azide displacement of the terminal ATRP halogen atom. In particular, a difunctional initiator was used to grow PS in both directions from a central core and, upon displacement of the bromine atoms with sodium azide, propargyl alcohol was placed onto both chain ends of the polymer by CuAAC to give a bis(hydroxyl-terminated)PS derivative (Scheme 5.7). Interestingly, the authors found that for a single polymer chain, the rate of reaction of the first chain end is three orders of magnitude higher than the reaction rate for the second chain end. To give a time scale, an alcohol group was inserted at 97% of both chain ends after a reaction time of 8 h.

The use of functional initiators in combination with replacement or substitution of dormant chain ends provides an excellent two-step route to either homobifunctional or heterobifunctional materials. The combination of CuAAC and ATRP has made excellent use of this technique for the preparation of several different complex and functional polymeric architectures from heterobifunctional telechelic materials. The general approach employs an alkyne-functionalized initiator to synthesize a polymer bearing an alkynyl group at one end and a halogen at the opposite end. Again by subsequent

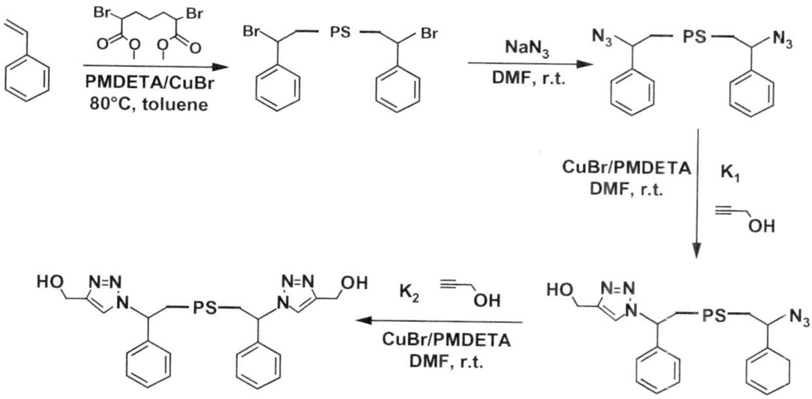

SCHEME 5.7 Synthesis of α,ω-dihydroxy-terminated PS linear polymers.[33] *PS*, Polystyrene. *Reprinted with permission from Gao, H.; Louche, G.; Sumerlin, B. S.; Jahed, N.; Golas, P.; Matyjaszewski, K.* Macromolecules ***2005***, 38, *8979–8982. Copyright 2005 American Chemical Society.*

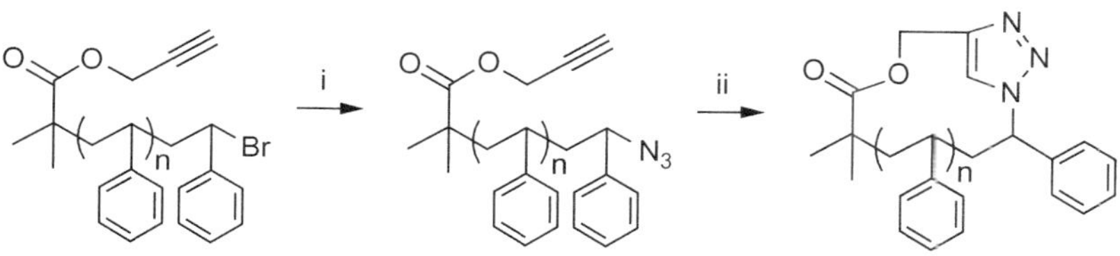

SCHEME 5.8 The terminal Azidation (i) and "click" cyclization (ii) of polystyrene prepared via ATRP.[35] (i) NaN_3 in DMF, 25°C; (ii) CuBr/Bipy, in degassed DMF, 120°C. *ATRP*, Atom transfer radical polymerization; *DMF*, dimethylformamide. *Reprinted with permission n from Laurent, B. A.; Grayson, S. M.* J. Am. Chem. Soc. ***2006***, 128, *4238–4239. Copyright 2006 American Chemical Society.*

displacement of the terminal halogen with sodium azide, a heterobifunctional telechelic polymer is obtained. Following this approach, Tsarevsky et al.[34] synthesized short PS chains with alkynyl and azido end-groups and treated them as AB monomers by subjecting them to CuAAC conditions. A significant polymeric growth was observed and at the end macromolecules with molecular weight higher than 100 kDa were obtained. About 7% of unreacted starting material was present and the authors hypothesized the formation of cyclic products. Starting from this suggestion, Grayson et al.[35] used comparable telechelic polymers for the synthesis of desired cyclic polymers by click cycloaddition (Scheme 5.8). Telechelic PS bearing an alkyne group at one chain end and an azide moiety at the other was prepared using the same synthetic strategy reported in Tsarevsky et al.[34] Subsequently CuAAC conditions were applied, but under significantly more dilute reaction concentrations than those used by Tsarevsky. In order to avoid the use of excessive amounts of solvent and to keep low the concentration of unreacted functional groups, a continuous addition method was used. In this way, cyclic PS made from polymers with molecular weights of up to 4 kDa was obtained with very high yield.

CuAAC was also used in conjunction with radical formation under atom transfer radical coupling conditions to synthesize ABC triblock copolymers in a one-pot reaction[36]: poly(*tert*-butyl acrylate)-*block*-PS-*block*-poly(ethylene oxide) (P*t*BA–PS–PEO) and P*t*BA-*block*-PS-*block*-poly(ε-caprolactone) (P*t*BA–PS–PCL) were successfully prepared by this method in DMF at 90°C. The precursors with predetermined number-average molecular weight and low polydispersity indices, such as PS with α-alkyne and ω-bromine end groups, P*t*BA with azide end group, PEO and PCL with a 2,2,6,6-tetramethylpiperidine-1-oxyl end group, were directly prepared by living polymerization technique using the compounds with corresponding functional groups as initiators, and no further modifications of the end groups were needed, except for P*t*BA–N_3. The coupling reaction between precursors was carried out in the CuBr/*N*,*N*,*N*′,*N*″,*N*″-pentamethyldiethylenetriamine system with high efficiencies. This reaction sequence was possible due to the dual role of the copper(I) catalyst, which was active for both the CuAAC reaction and the formation of radicals by bromine abstraction from ATRP (macro) initiators. Similarly, CuAAC has been combined with Diels–Alder [4 + 2] cycloaddition to obtain in one-pot ABC triblock copolymers of PEG–PS–PMMA and polycaprolactone PCL–PS–PMMA[6] by reacting furan-protected maleimide-end-functionalized PMMA, PS with α-anthracene and ω-azide end-functionality, and PEG or PCL with an alkyne end-functional group in *N*,*N*-DMF for 36 h at 120°C.

Emrick et al.[37] reported the incorporation of alkyne moieties into δ-valerolactone in water in the presence of $CuSO_4$ and sodium ascorbate at 80°C: the polymerization of this monomer proceeded with slower kinetics than for ε-caprolactone (several hours were required), but the two monomers could be copolymerized in any ratio, with the alkyne group providing a more direct route to peptide-functionalized polyesters without requiring the use of protecting group chemistry.

Malkoch et al.[38] synthesized styrenic, acrylic, and acrylamide monomers containing tetramethylsilane (TMS)-protected alkyne groups and subsequently incorporated these monomers into various copolymers with different reactive functional groups with tandem or simultaneous reactions. For example, reaction of an alkyne-functionalized PS derivative with a linker unit containing both an azide group and a nucleophilic amine in the presence of an active ester gave in one-pot the corresponding derivative in which the linker group was attached to the backbone by triazole formation while having also undergone an amidation reaction. This pioneering work demonstrated the utility of developing orthogonal methods for building functional molecules and from that report to the time of this writing there has been a rich collection of CuAAC-based functionalization reactions for backbone derivatization, thanks to the near quantitative yield, regiospecific conversion, and compatibility with a broad range of functional groups and reaction conditions demonstrated for this click reaction.[1,38]

Combining ATRP with click chemistry allowed the preparation of conjugated block copolymers of PS and poly(3-hexylthiophene) (P3HT).[39] The polymerization of styrene was started thanks to an azide-functionalized initiator, and modified Grignard reagents were utilized for the synthesis of dialkyne-P3HT. The azido PS was click-coupled to homotelechelic P3HT to yield ABA triblock copolymers by a CuI-catalyzed cycloaddition, needing an alkyl spacer between the conjugated main chain of P3HT and the alkyne end groups to proceed.

Triblock copolymers were also prepared by one-pot synthesis using two separate click reactions.[6] In particular, the functional polymers maleimide-end-functionalized PMMA, α-anthracene-ω-azido PS, and alkyne-terminated PEG or PCL were prepared and subsequently mixed in one pot, where the coupling proceeded via both a Diels–Alder reaction between PMMA and PS, and azide–alkyne cycloaddition between PS and PEG (or PCL). Size exclusion chromatography showed the quantitative yield of the reaction, demonstrating the possibility to extend this one-pot synthesis to previously unattainable block copolymers thanks to the introduction in their precursors of functionalities suitable for both azide–alkyne cycloaddition and Diels–Alder click chemistry.

With this same approach, it was possible to prepare multisegmented block copolymers, and polymers that contain multiple repeated units of block copolymers.[34] Subsequently, the concept was extended to the click coupling of diazido-terminated tri- and pentablock copolymers in order to obtain multisegmented block copolymers with alternating hard and soft segments.[40] Polymers with 5–7 block copolymer repeated units within each chain were obtained and, interestingly, each multiblock copolymer exhibited different thermal and mechanical properties than its block copolymer predecessor, outlining the importance and potentiality of this synthetic strategy in material chemistry.

Even more complex architectures can be obtained by click chemistry. Star polymers can be obtained starting from linear polymers and a multifunctional coupling agent, each with complementary clickable functionality. Gao and Matyjaszewski[41] in 2007 prepared three- and four-arm star polymers by coupling azido-terminated PS arms onto tri- and tetraalkyne cores via CuAAC, with a final yield of 90% and 83%, respectively, in only 3 h (Scheme 5.9).

The selective nature of the CuAAC was demonstrated by Zhang et al. in the synthesis of hydrophilic ABC miktoarm star polymers by combing two successive ATRPs with click chemistry (Scheme 5.10).[42] This kind of polymer has arms with different chemical composition, showing very good

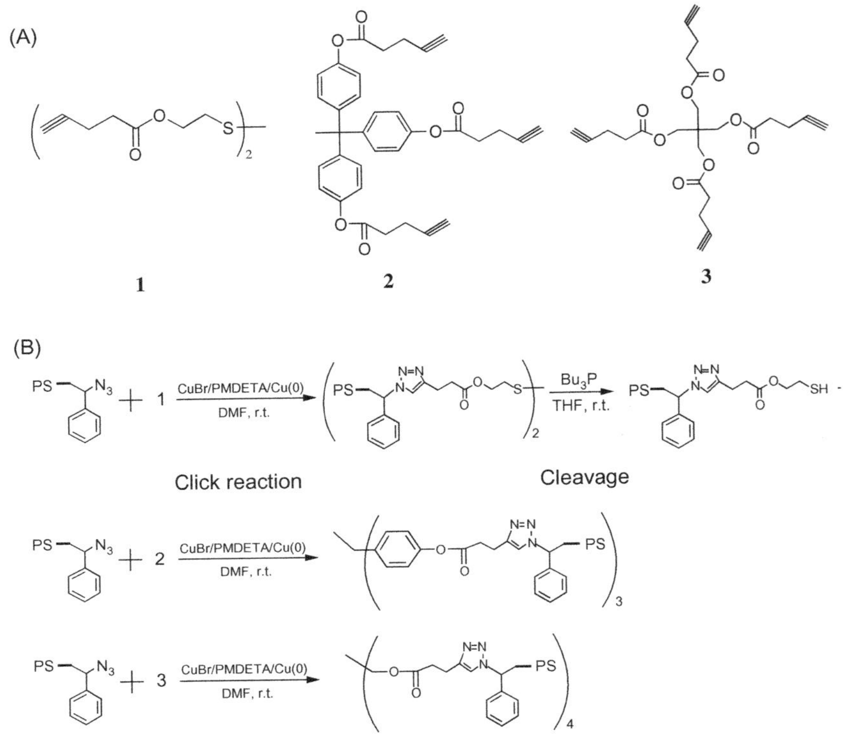

SCHEME 5.9 (A) Structures of dialkyne-containing (1), trialkyne-containing (2), and tetraalkyne-containing (3) coupling agents and (B) synthesis of linear, three-arm, and four-arm star PS polymers by click reactions.[41] *PS*, Polystyrene. *Reprinted with permission from Gao, H.; Matyjaszewski, K.* Macromolecules ***2006***, 39, *4960–4965. Copyright 2006 American Chemical Society.*

emulsifying properties and being very useful as targeted delivery agents. In this work, monopropargyl 2-bromoglutarate reacted with the terminal hydroxyl group of PEO and both an alkyne moiety and an ATRP-initiating site were introduced. Subsequently, P(NIPAM) was grown from this same site, and the alkyne moiety was used for the attachment of azide-functionalized poly(2-(diethylamino)ethyl methacrylate) (PDEA). Since P (NIPAM) is thermo-responsive and PDEA is pH-responsive, the micellization of this miktoarm star polymer could be induced by increasing either temperature or pH.

Again the combination of ATRP and CuAAC was used to obtain brush polymers.

Gao et al.[43] prepared 2-hydroxyethyl methacrylate (HEMA) by ATRP and subsequently functionalized the hydroxyl groups with alkyne moieties via carbodiimide-mediated esterification, making the HEMA backbone

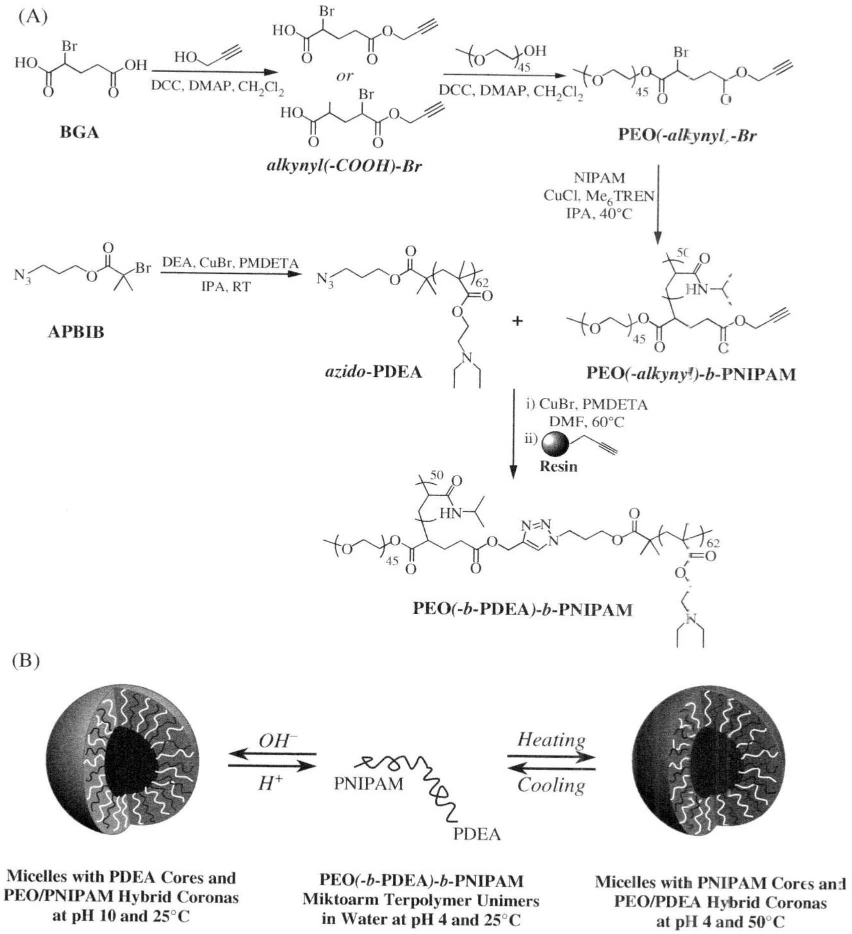

SCHEME 5.10 (A) Preparation of PEO(-*b*-PDEA)-*b*-PNIPAM double hydrophilic ABC miktoarm star terpolymer, and (B) pH- and thermo-induced supramolecular self-assembly of PEO(-*b*-PDEA)-*b*-PNIPAM ABC miktoarm star terpolymer into two types of micellar aggregates possessing hybrid coronas in aqueous solution.[42] *PDEA*, Poly(2-(diethylamino)ethyl methacrylate); *PEO*, polyethylene oxide. *Reprinted with permission from Zhang, Y.; Liu, H.; Hu, J.; Li, C.; Liu, S.* Macromol. Rapid Commun. ***2009***, 30, *941–947. Copyright 2009 John Wiley and Sons.*

"clickable". This allowed its conjugation with several azide-terminated polymers with different density and molecular weight. The grafting density was near 90% for low-molecular density polymers like PEO$-N_3$.

The CuI-catalyzed azide–alkyne cycloaddition has been extensively employed also in the synthesis of dendrimers, macromolecules that display a symmetrical, layered structure, and a large number of functional groups at the chain ends. Generally these molecules are obtained by complicated procedures, allowing their growth in a convergent (from the chains ends toward the inside of the structure, involving first the preparation of the arms and subsequently their coupling to a multifunctional core) or in a divergent (from the core toward the external part of the molecules, involving first the preparation of the multifunctional cores and then their connection in sequence) manner. Click chemistry has strongly simplified and made faster their preparation, using both convergent[44–46] and divergent[47–49] strategies or a combination of the two.[50] Large degradable dendrimers with molecular weight higher than 35,000 g/mol containing triazaadamantane units at each branch point and a number of functional groups at the periphery larger than 80 were prepared following this approach.[50] The obtained dendrimers are sensitive to pH and could be degraded in organic solution by HCl addition. The selectivity of CuI-catalyzed azide–alkyne cycloaddition was demonstrated also in the synthesis of hyperbranched polymers. The preparation of triazole-containing chains was attempted by both thermal and catalytic polymerization of AB_2 monomers.[51] While the thermal reaction produced a mixture of 1,4- and 1,5-disubstituted poly(triazole)s, the presence of the Cu catalyst obtained only the desired 1,4-disubstituted poly(triazole). The reaction was employed also to functionalize hyperbranched polymers[52]: melt-transetherification polymerization allowed to copolymerize an AB_2 monomer with a monomer displaying B-type functionality and subsequently the acetylene groups of both of them reacted with fluorescent azidomethylpyrene via CuAAC.

In 2012 Dove et al.[53] prepared a polycarbonate scaffold having norbornene side chains (Scheme 5.11) that can undergo three different reactions depending on the working conditions; that is, 1,3-cycloaddition in presence of azides, inverse-electron-demand Diels–Alder (iEDDA) reaction in presence of tetrazines, and radical addition by thiols under UV radiation (Scheme 5.12). Due to the orthogonality of these three reactions, they can be performed in a sequential one-pot manner. A large number of functionalized polymers can be prepared thanks to this approach, simply by varying the postprocessing conditions.

In 2016 Bradley et al.[54] followed this same strategy for the postpolymerization modification of PEGs.

Also macroporous foams like poly(dicyclopentadienes) can be functionalized by iEDDA. These materials are characterized by highly interconnected pores at a microcellular level and are utilized in a range of applications such as mechanical dampeners; thermal, acoustic, and electrical insulators; solid supports for catalysis and separations; and medical devices.[55] Slugovc et al.[56] added tetrazine to poly(dicyclopentadiene) and heated the mixture for 2 days. At the end, they verified by elemental analysis that the functionalization took place every fourth double bond, maintaining the open porosity of the polymer. The use of different tetrazine derivatives for the postfunctionalization of such kind of materials is under study.

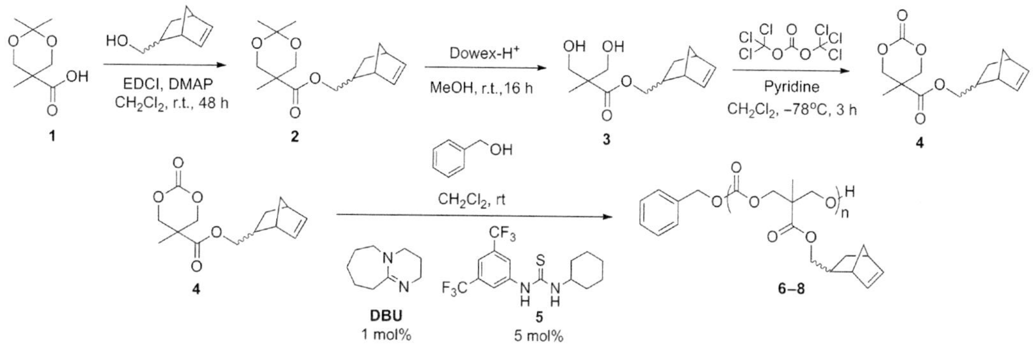

SCHEME 5.11 Synthesis and ring-opening polymerization of norbornene-functional cyclic carbonate monomer.[53] *Reprinted with permission from Williams, R. J.; Barker, I. A.; O'Reilly, R. K.; Dove, A. P.* ACS Macro Lett. ***2012***, 1, *1285–1290. Copyright 2012 American Chemical Society.*

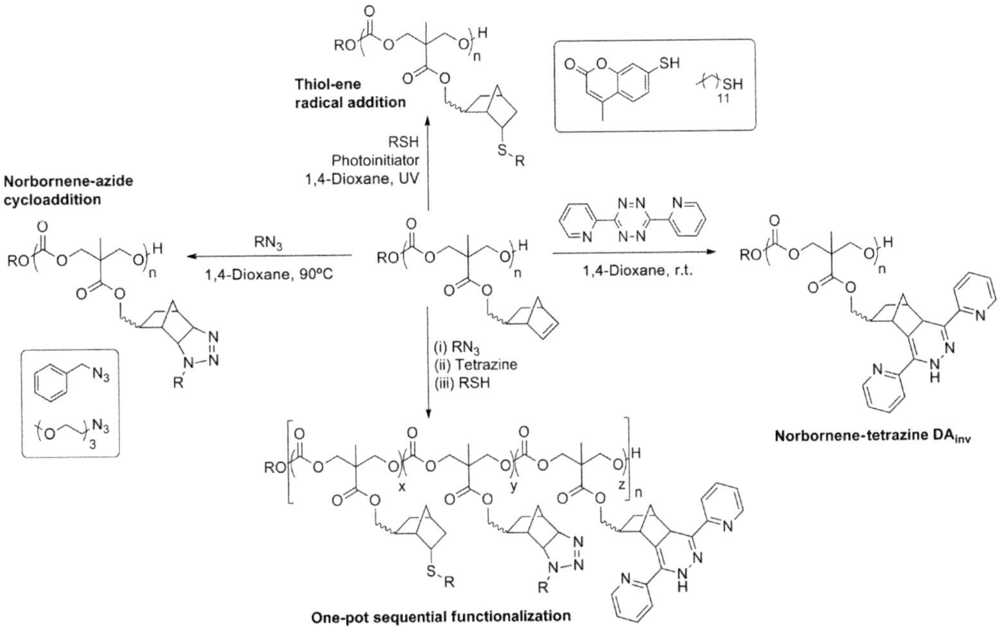

SCHEME 5.12 Individual and one-pot functionalizations of norbornene-functional poly(carbonate) scaffolds.[53] *Reprinted with permission from Williams, R. J.; Barker, I. A.; O'Reilly, R. K.; Dove, A. P.* ACS Macro Lett. ***2012***, 1, *1285–1290. Copyright 2012 American Chemical Society.*

Hydrogels and liquid crystal gels

Click chemistry has been demonstrated to be very useful also for the synthesis of hydrogels with biological applications and liquid crystal gels for optoelectronic applications, since the strategy can be applied in water, is compatible with several functional groups useful and suitable for these applications, and obtains few or no side products and a rigorous control over the structure and the mechanical properties, requirements strongly needed in these kinds of applications. Concerning the biological field, Ossipov and Hilborn[57] prepared poly(vinyl alcohol) (PVA)-based hydrogels cross-linked by triazole units by CuI-catalyzed click chemistry between PVA gels prepared by carbonyldiimidazole-mediated couplings of azide and alkyne-functionalized amines to the polymer backbone and PVA with a low weight percentage (1%–5%) of alkyne and azide groups. A hydrogel with different mechanical and elastic properties was obtained by using diazido-PEG as the cross-linker for the alkyne groups pendant PVA.

Despite hyaluranon being generally unstable in the presence of Cu^{I}, Lamanna et al. were able to prepare hyaluranon hydrogels using the CuAAC reaction as the cross-linking mechanism.[58]

First of all, water-soluble polysaccharide derivatives bearing side chains ended with either azide or alkyne functionality have been prepared: in particular, hydrosoluble hyaluronic acid (HA) sodium salt derivatives bearing azide groups (HAAA) have been obtained by reaction between HA sodium salt in MES [2-(*N*-morpholino)-ethanesulfonic acid] buffer at pH = 4 and 11-azido-3,6,9-trioxaundecan-1-amine (Scheme 5.13) while HA–propargylamide hydrosoluble derivatives have been prepared reacting HA in MES acid buffer with propargylamine (Scheme 5.14).

When the latter two types of derivatives were mixed together in aqueous solution at room temperature, they gave rise to a 1,3-dipolar cycloaddition reaction resulting in fast gelation—in the presence of catalytic amounts of Cu(I)—with formation of the HA-based click-gels (Scheme 5.15).

The structurally simplest polymer networks are termed model networks (MNs) and are typically comprised of linear telechelic polymers, or macromonomers, covalently cross-linked through their end groups with multifunctional small molecules. α,ω-Diazido-poly(*t*BA) (P*t*BA, **1** in Scheme 5.16) was prepared by ATRP of *t*BA and subsequent substitution of bromine end groups with azide. Cross-linking of the linear polymer with tri- and tetraalkyne coupling agents (**3** and **4** in Scheme 5.16) via click chemistry led to the formation of the first *tert*-butyl acrylate MNs reported in literature (**2a** and **2b**, respectively).[59]

The use of tough hydrogels as biomaterials is limited as a consequence of time-consuming fabrication techniques, toxic starting materials, and large strain hysteresis under deformation. Dove et al.[60] prepared double network hydrogels (among the strongest synthetic soft materials) by dual click

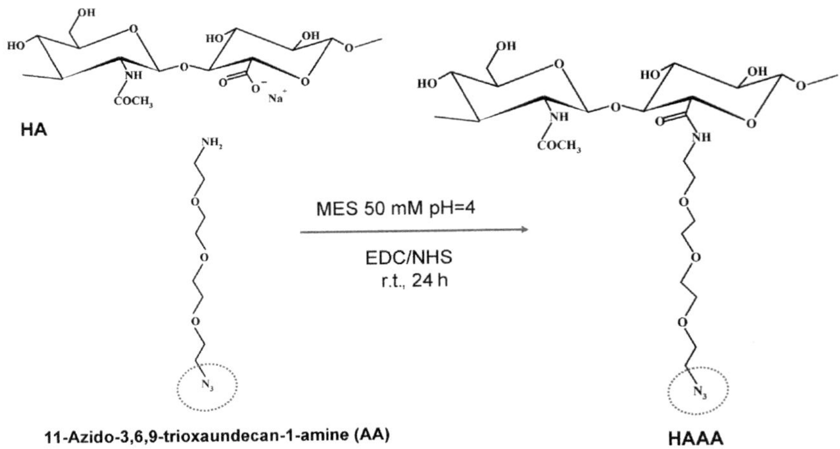

SCHEME 5.13 Reaction between HA sodium salt and AA to give hydrosoluble HA derivatives bearing azide groups (HAAA). MES = [2-(*N*-morpholino)-ethanesulfonic acid buffer]; EDC · HCl = [*N*-(3-dimethylaminopropyl)-*N'*-ethylcarbodiimide hydrochloride]. *AA*, 11-Azido-3,6,9-trioxaundecan-1-amine; *HA*, hyaluronic acid; *NHS*, *N*-hydroxysuccinimide.[58] *Reprinted with permission from Crescenzi, V.; Cornelio, L.; Di Meo, C.; Nardecchia, S.; Lamanna, R.* Biomacromolecules ***2007***, 8, *1844–1850. Copyright 2007 American Chemical Society.*

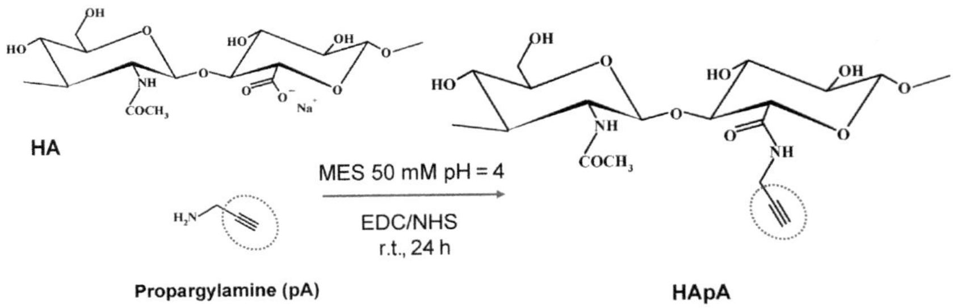

SCHEME 5.14 Reaction between HA and pA to give HApA hydrosoluble derivatives.[58] *HA*, Hyaluronic acid; *HApA*, HA–propargylamide; *pA*, propargylamine. *Reprinted with permission from Crescenzi, V.; Cornelio, L.; Di Meo, C.; Nardecchia, S.; Lamanna, R.* Biomacromolecules ***2007***, 8, *1844–1850. Copyright 2007 American Chemical Society.*

chemistry (thiol–alkyne addition and tetrazine-norbornene iEDDA reaction) in a simple one-step procedure (Scheme 5.17) under physiological conditions obtaining a material with a water content of 90% and a very high mechanical strength (compressive stresses of 14–15 MPa at 98% compression without fracture or hysteresis upon repeated load). The obtained hydrogel networks can be spatially and temporally postfunctionalized via radical thiylation and/or iEDDA addition to residual functional groups within the network. Furthermore, gelation occurs rapidly under physiological conditions, enabling encapsulation of human cells.

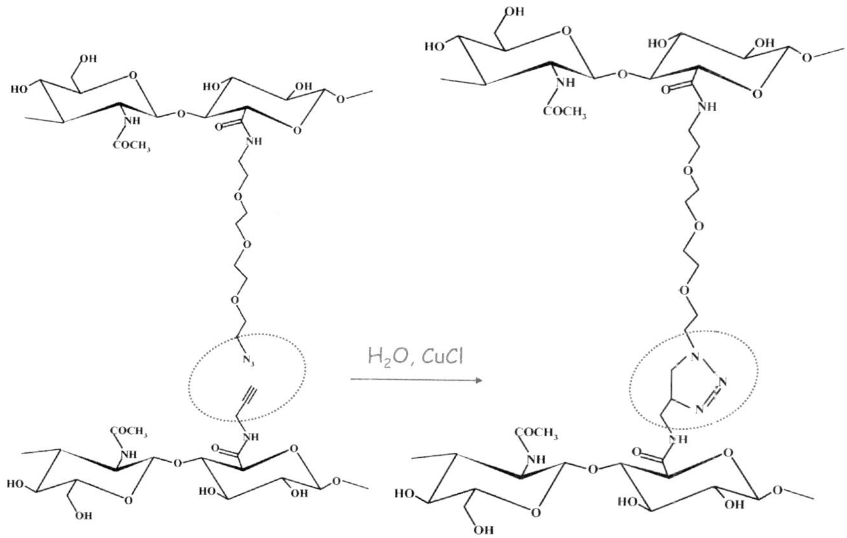

SCHEME 5.15 Formation of HA-based click-gels.[58] *HA*, Hyaluronic acid. *Reprinted with permission from Crescenzi, V.; Cornelio, L.; Di Meo, C.; Nardecchia, S.; Lamanna, R.* Biomacromolecules ***2007***, 8, *1844–1850. Copyright 2007 American Chemical Society.*

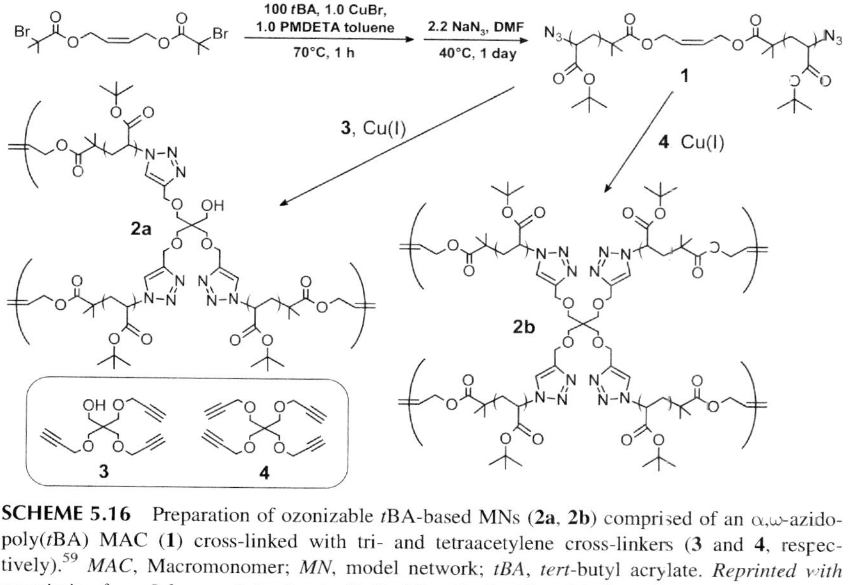

SCHEME 5.16 Preparation of ozonizable *t*BA-based MNs (**2a**, **2b**) comprised of an α,ω-azido-poly(*t*BA) MAC (**1**) cross-linked with tri- and tetraacetylene cross-linkers (**3** and **4**, respectively).[59] *MAC*, Macromonomer; *MN*, model network; *tBA*, *tert*-butyl acrylate. *Reprinted with permission from Johnson, J. A.; Lewis, D. R.; Diaz, D. D.; Finn, M. G.; Koberstein, J. T.; Turro, N. J.* J. Am. Chem. Soc. ***2006***, 128, *6564–6565. Copyright 2006 American Chemical Society.*

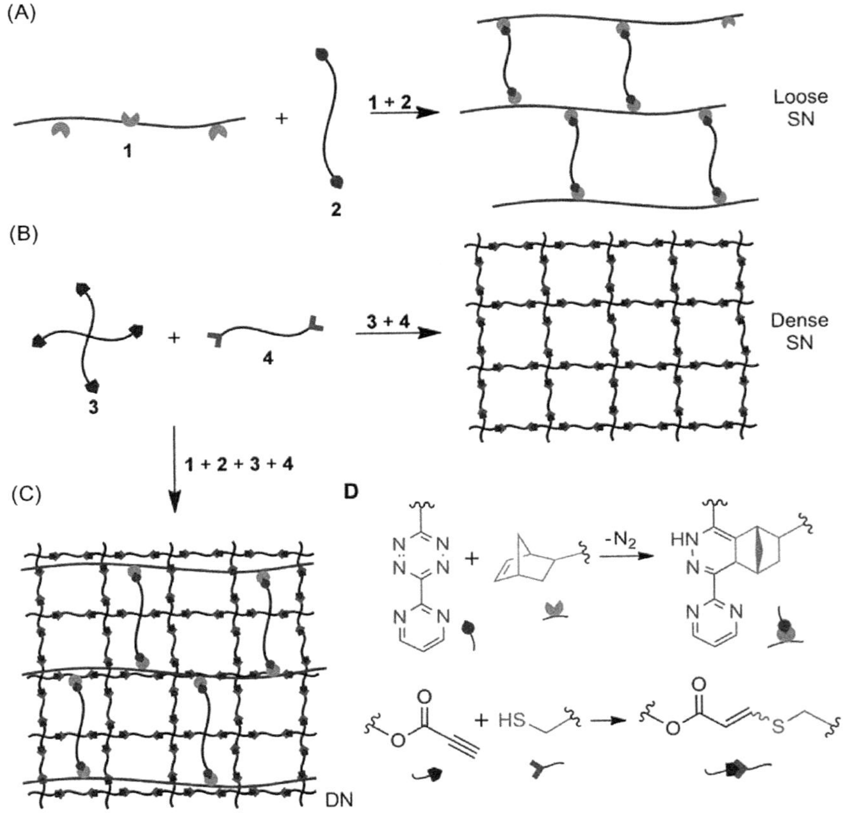

SCHEME 5.17 DN hydrogel fabrication. (A) Loose network formed by norbornene-tetrazine addition. (B) Dense network formed by nucleophilic thiol-yne addition. (C) DN gel formed by mixing all four components. (D) Tetrazine-norbornene and nucleophilic thiol-yne addition reactants used for cross-linking.[60] *DN*, Double network. *Reprinted with permission from Truong, V. X.; Ablett, M. P.; Richardson, S. M.; Hoyland, J. A.; Dove, A. P.* J. Am. Chem. Soc. ***2015***, 137, *1618–1622. Copyright 2015 American Chemical Society.*

Kornfield et al.[61] prepared gels with fast, reversible, and low-threshold optic switching under applied electric fields by click cross-linking of telechelic polymers produced by ring-opening metathesis polymerization (ROMP). Cyclooctene monomers were functionalized with either one or two mesogenic groups (functionalities that induce the liquid crystalline state) as outlined in Scheme 5.18: cyanobiphenyl mesogens were coupled via a six-carbon spacer to either dihydroxy-cyclooctene to yield monomer **5** or hydroxy-cyclooctene to give monomer **6**. This monomer pair varied the mesogen density on the final polymer backbone.

Second, the monomers were polymerized by ROMP using the highly active ruthenium catalyst ($(IMesH_2)(PCy_3)RuCl_2(CHPh)$) in dichloroethane at

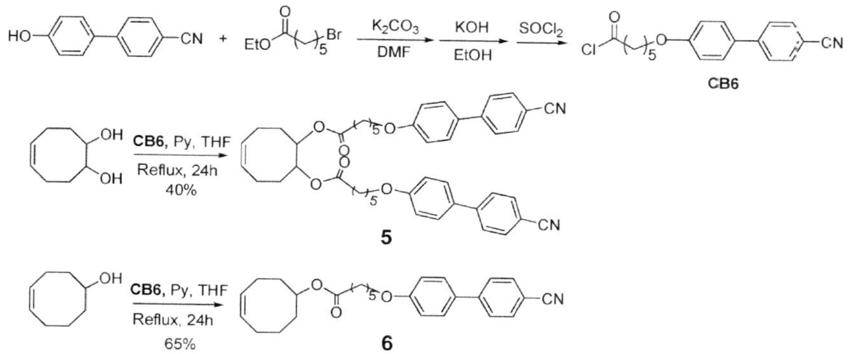

SCHEME 5.18 Synthesis of the monomers **5** and **6** by coupling of the cyanobiphenyl mesogens to either dihydroxy-cyclooctene or hydroxy-cyclooctene coupled via a six-carbon spacer.[61] *Adapted with permission from Xia, Y.; Verduzco, R.; Grubbs, R. H.; Kornfield, J. A.;* J Am. Chem. Soc. ***2008***, 130, *1735–1740. Copyright 2008 American Chemical Society.*

55°C for 24 h in presence of 1,8-dibromo-4-octene as chain transfer agent (Scheme 5.19) to obtain the telechelic polymers **7** and **8** with primary bromide end groups (the ruthenium catalyst is known to be incompatible with azide groups). The bromide end groups of **7** and **8** were transformed with quantitative yields into azide groups by nucleophilic substitution with sodium azide in DMF. Polymer 7 was nematic at room temperature, with a nematic to isotropic transition temperature (T_{NI}) between 50°C and 60°C, as measured by differential scanning calorimetry. On the other hand, polymer **8** was not liquid crystalline and did not show any phase transitions in the range 20°C–100°C. Both polymers were soluble in the small-molecule liquid crystal 5CB (*p*-*n*-pentyl-*p*′-cyanobiphenyl) up to the highest concentration tested (10 wt.%).

Controlled click cross-linking of these telechelic polymers was achieved by reacting a triacetylene species, tripropargylamine, with the polymer azide end groups in presence of CuBr as catalyst and PMDETA as a ligand in DMF at 50°C (Scheme 5.20). The solution was allowed to react for two more days after gelation had occurred to ensure complete reaction. The resulting gels were extracted in DMF and THF repeatedly to remove copper catalyst and uncross-linked fraction, and elastomers were obtained after drying under vacuum. All the elastomers readily swelled in 5CB to form LC gels with high swelling ratios, including the one derived from the non-LC polymer **8**. The swelling ratios showed a dependence on the molecular weight of the strands: larger network strands gave LC gels with higher swelling ratios. In the same way, gels prepared with several different functionalized poly(cyclooctenes) exhibited swelling ability in the molecular liquid crystal 5CB.[62,63]

Carbon-based materials

In the past two decades, carbon-based materials like graphene, fullerene, and highly porous activated carbon (biochar) are attracting great attention in

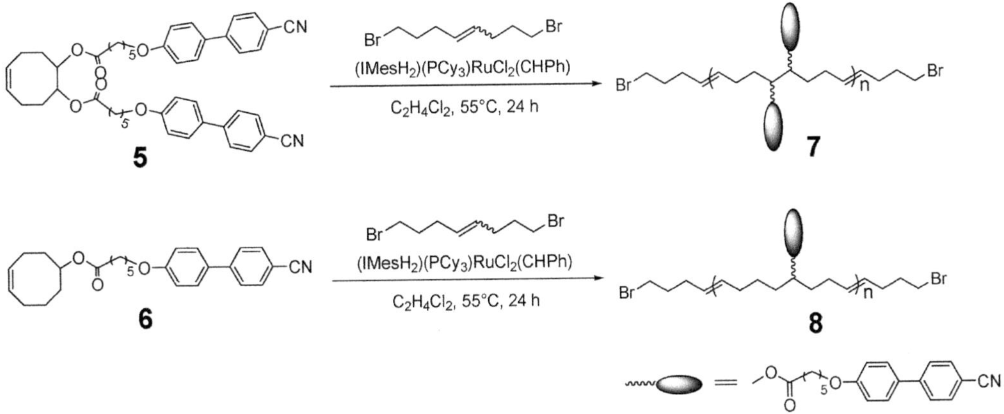

SCHEME 5.19 ROMP of functionalized cyclooctene monomers **5** and **6** to give the telechelic polymers **7** and **8**.[61] *ROMP*, Ring-opening metathesis polymerization. *Adapted with permission from Xia, Y.; Verduzco, R.; Grubbs, R. H.; Kornfield, J. A.* J. Am. Chem. Soc. ***2008***, 130, *1735–1740. Copyright 2008 American Chemical Society.*

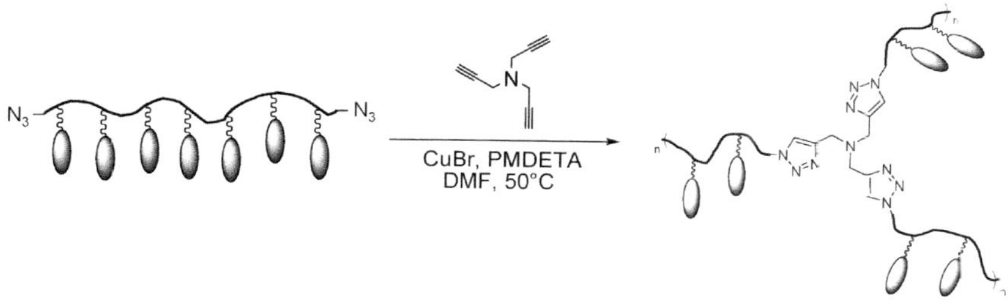

SCHEME 5.20 Cross-linking of the telechelic polymers **7** and **8** by click chemistry, leading to the formation of gels with swelling abilities in the 5CB liquid crystal.[61] *Reprinted with permission from Xia, Y.; Verduzco, R.; Grubbs, R. H.; Kornfield, J. A.* J. Am. Chem. Soc. ***2008****,* 130*, 1735–1740. Copyright 2008 American Chemical Society.*

material science and technology, thanks to their electronic and structural properties, great versatility, lightness, and eco-friendly character. In the last few years, they have found several applications as materials for energy storage.[64–68]

Cycloaddition reactions have been widely applied for the functionalization of the surface of these materials in order to modify their chemical reactivity.

Graphene is an allotrope form of carbon consisting of a single layer of carbon atoms arranged in a hexagonal lattice, and it is a very promising and fashionable material for the realization of electrodes for solid state and ions batteries, and for chemical sensors preparation. Its synthesis from graphite oxide (GO) often gives inhomogeneity among the different production lots due to reaggregation of the layers after chemical reduction. van der Wiel et al.[69] were able to obtain exfoliation of graphene oxide with formation of single layers graphene in situ by CuI-catalyzed azide–alkyne cycloaddition between azide-functionalized graphene oxide and alkyne-terminated oligomers.

Zhang et al.[70] prepared two types of graphene-based hybrid materials, namely graphene-tetraphenylporphyrin and graphene-palladium tetraphenylporphyrin, directly from pristine graphene through one-pot cycloaddition reactions. The presence of the porphyrin molecules in the hybrid materials was detected by thermogravimetric analysis (18% and 20% in mass, respectively) and UV–Vis spectroscopy, while the presence of covalent links between graphene and porphyrin was confirmed by FTIR and Raman spectroscopy.

Single-walled carbon nanotubes (SWCNTs) were reacted with *p*-aminophenyl propargyl ether in order to make their surface functionalized with alkyne groups and subsequently mixed in DMF with azide-terminated PS prepared by ATRP and nucleophilic substitution in presence of CuI-catalyst[71]: the click reaction led to PS-functionalized nanotubes highly dispersible in THF, $CHCl_2$, and $CHCl_3$ (differently from the pristine materials). The same research group was able to obtain water dispersible functionalized SWCNTs by sulfonating the PS units (up to 34 mol%) by reaction with acetyl sulfate.[72]

Wu et al.[73] prepared alkyne groups surface modified multiwalled carbon nanotubes by acylation chemistry and suspended them, in presence of CuI-catalyst, in solutions containing, respectively, poly(AzMA) and poly(propargyl methacrylate) prepared by CRP polymerization. A uniform increasing in polymer grafting density was verified by thermogravimetric analysis after each suspension step. At the end, three layers of polymers were attached to the nanotube surface, but the pendant azido groups of the azidoethyl methacrylate external layer were still available for further covalent bonds with other chains, as demonstrated by their reaction with alkyne-terminated PS and alkyne-functionalized Rhodamine B fluorescent dye.

Schirrmacher et al.[74] functionalized by iEDDA reaction (performed at room temperature and pressure) carbon nanotubes modified on the surface with diene units with gold nanoparticles conjugated to tetrazine. A similar approach was later applied to functionalize graphite surface with gold nanoparticles and fluorosilanes.[75]

Fullerenes are a large class of allotropes of carbon made of balls, "cages" or tubes of carbon atoms.[76,77] The most "famous" type, buckminster fullerene (C_{60}), is a molecule composed by 60 carbon atoms arranged in a hollow sphere. Since its discovery,[78] C_{60} has attracted intense research interest thanks to its nanostructure, leading to fascinating photoelectronic and biological properties. Similar to carbon nanotubes, the poor compatibility of pristine fullerenes with other materials leads to its strong tendency toward aggregation and low solubility in common solvents and thus limits their facile derivatization and industrial applications.[77]

Click chemistry provides a straightforward and site-specific technique for modifying fullerene surfaces with macromolecules that facilitate their handling. Cheng at al.[79] prepared organic solvents dispersible end-capped PS-C_{60} by click chemistry: the pristine fullerene molecules were functionalized with hydroxyl groups and subsequently fullerene (i.e., alkyne-functionalized fullerenes) was obtained by their esterification with propiolic acid. Azido-PS molecules prepared by ATRP and subsequent reaction with sodium azide conjugated to the alkyne groups of fulleryne giving the polymer-functionalized fullerene product, that was shown to be free from unreacted PS chains.

Several authors reported that the Cu-catalyzed Huisgen [3 + 2] cycloaddition reaction between fulleryne and azido-functionalized polymers proceeded efficiently at room temperature with no side reactions between azide and C_{60} core.[80–85] Allard et al.[81] were able to obtain water-soluble fullerene derivatives having a poly(amidoamine) (PAMAM) dendron with peripheral carboxylic groups on one side and an alkyne moiety on the other side. PAMAM

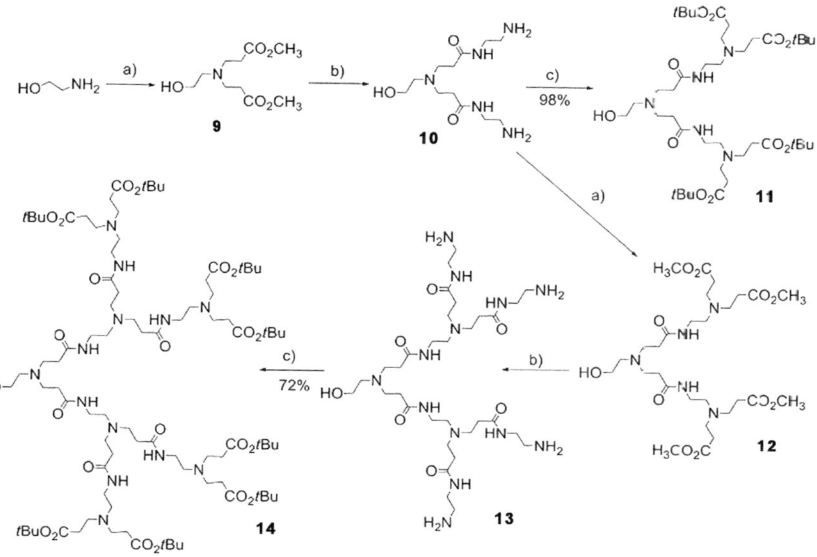

SCHEME 5.21 Synthesis of PAMAM dendrons **11** and **14**. Reagents and conditions: (a) methyl acrylate, MeOH, RT, 3 days; (b) ethylenediamine, MeOH, 0°C to RT, 3 days; (c) *tert*-butyl acrylate, MeOH, RT, 3 days.[81] *PAMAM*, Poly(amidoamine); *RT*, room temperature. *Adapted with permission from Rousseau, G.; Fensterbank, H.; Baczko, K.; Cano, M.; Stenger, I.; Larpent, C.; Allard, E.* Chempluschem *2013*, 78, *352–363. Copyright 2013 John Wiley and Sons.*

dendrons **11** and **14** having a hydroxyl moiety at their focal point and peripheral *tert*-butyl ester groups were synthesized by reacting *tert*-butyl acrylate with amine-terminated PAMAM dendrons **10** and **13** (Scheme 5.21). Dendrons **10** and **13** were prepared by a two-step process involving a Michael addition using methyl acrylate and an amidation with ethylenediamine starting from ethanol amine. Dendrons **11** and **14** were obtained with overall yields of 98% and 72%, respectively, and their structures were confirmed by NMR spectroscopy and high-resolution mass spectrometry.

Asymmetrical malonates **16** and **17** were obtained from the PAMAM dendrons **11** and **14** and the monoalkyl malonate **15** (prepared by reaction of 2,2-dimethyl-1,3-dioxane-4,6-dione with 10-undecyn-1-ol at 80°C to prevent the decarboxylation of the monoalkyl malonate) and subsequently reacted with C_{60} in DBU/I_2 solution giving the fullerodendrimers **18** and **19** with *tert*-butyl ester groups at the periphery (Scheme 5.22). Subsequently, the *tert*-butyl esters were cleaved with formic acid giving the corresponding carboxyfullerene derivatives **20** and **21**. Finally, the alkyne moiety served as chemical anchors in water to immobilize fullerodendrimers on the surface of azido-coated polymer nanoparticles by the copper(I)-catalyzed azide reaction and the alkyne cycloaddition reaction. Thanks to the high density of azide

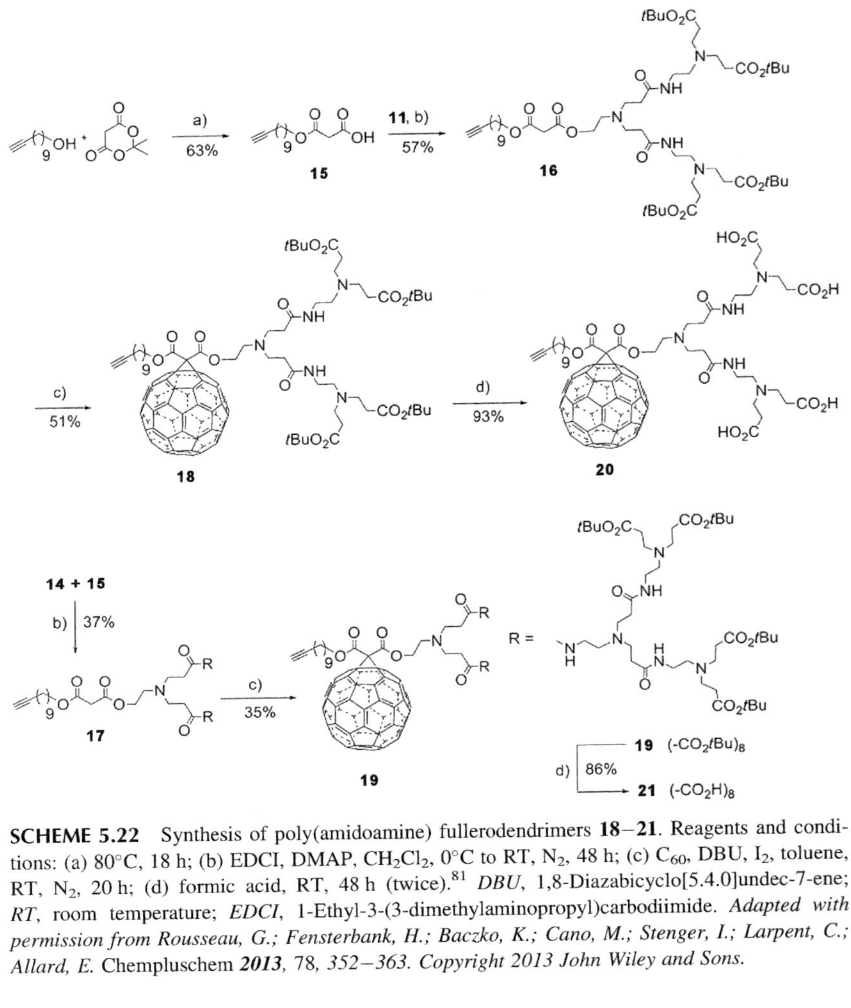

SCHEME 5.22 Synthesis of poly(amidoamine) fullerodendrimers **18–21**. Reagents and conditions: (a) 80°C, 18 h; (b) EDCI, DMAP, CH_2Cl_2, 0°C to RT, N_2, 48 h; (c) C_{60}, DBU, I_2, toluene, RT, N_2, 20 h; (d) formic acid, RT, 48 h (twice).[81] *DBU*, 1,8-Diazabicyclo[5.4.0]undec-7-ene; *RT*, room temperature; *EDCI*, 1-Ethyl-3-(3-dimethylaminopropyl)carbodiimide. *Adapted with permission from Rousseau, G.; Fensterbank, H.; Baczko, K.; Cano, M.; Stenger, I.; Larpent, C.; Allard, E.* Chempluschem ***2013****, 78, 352–363. Copyright 2013 John Wiley and Sons.*

anchoring groups on the nanoparticle surface and the size of the fullerodendrimers, unreacted azides were still available for subsequent functional arrangements.

New fullerene glycoconjugate derivatives were obtained by Pereira et al.[82] under CuAAC conditions (room temperature, $CuSO_4$ catalyst in presence of Na ascorbate, in a $H_2O-CH_2Cl_2$ solution) by reacting a bis-malonate C_{60} derivative bearing terminal alkyne groups with azido sugar molecules.

Due to the increased attention toward diketopyrrolopyrrole (DPP)-based materials, thanks to their interesting applications in organic field-effect transistors,[86] organic photovoltaics,[87] organic light-emitting diodes,[88] chemosensors,[89] and solid-state dye lasers,[90] in the past 20 years both polymers and small molecules based on DPP core have been prepared and characterized.[91,92] In particular, thanks to the appealing photovoltaic properties showed by bulk heterojunction solar cells based on fullerene and DPP derivatives, dyads and triads composed of covalently linked DPP derivatives and fullerene have received some attention.[93,94]

In this frame, Castelaín et al.[95] prepared a novel asymmetric DPP derivative poly[(9,9-dihexylfluorene)-*co*-alt-(9,9-bis-(6-azidohexyl)fluorene)] (PFA in Scheme 5.23) having azide moieties only in one of the pyrrole units of the DPP core and reacted it by the Cu-catalyzed Huisgen 1,3-dipolar cycloaddition with a fullerene derivative having an alkyne functionality[96] and alkyne-modified graphene[97] obtained by diazonium coupling in the basal planes of reduced GO (G-ALK1 in Scheme 5.23) or amidation on the carboxylic acid groups at the perimeter of GO (G-ALK2 in Scheme 5.23). The reaction with the fullerene derivative was performed with $(CH_3CN)_4PF_6$ in THF for 3 days at room temperature in presence of 1 equivalent of Cu powder with a yield of 32%. On the contrary, the synthesis of the graphene-DPP derivative, leading to the formation of the final click product PFA-G1 (Scheme 5.23), was performed in an homogenous DMF phase, thanks to the solubility of the alkine-graphene structure in this same solvent. The subsequent NMR, FTIR, and Raman analyses confirmed the evolved reactions. UV–Vis absorption analyses and cyclic voltammetry measurements showed that the DPP intramolecular contact and effective interactive cross section are stronger with flat graphene with respect to spherical fullerene due to the fact that in the DPP-clicked graphene the organic molecule is surrounded by large amounts of carbonaceous surface with which to interact whereas in the case of the clicked fullerene only a small part of the curved C_{60} surface can do it.

Inglis et al.[83] prepared for the first time star-shaped polymer-fullerene hybrids by CuAAC. In this work, PEG was functionalized with an alkyne moiety by DCC coupling with 4-pentynoic acid and subsequently six equivalents of this compound were reacted for 18 h with a sixfold azide-functionalized hexakis methanofullerene malonate crown-ether[96] at room temperature in DMF solution in presence of copper(II) sulfate and sodium ascorbate. The formation of a sixfold click product was verified by NMR spectroscopy and molecular weight distribution analysis (Scheme 5.24).

With the same approach, the authors also prepared a thermo-responsive polymer by reacting the same crown ether with the thermoresponsive P (NIPAM). Linear P(NIPAM) in water has a LCST of typically 32°C. Below this temperature, the polymer is water soluble; upon heating above this temperature the chains dehydrate and aggregate, becoming insoluble and giving rise to turbid solution and final precipitation after about 24 h. The obtained sixfold click polymer-fullerene hybrid was not at all water soluble at ambient temperature due to its complex structure consisting of a highly hydrophobic core with six hydrophilic P(NIPAM) arms radiating outward and terminating with hydrophobic dodecyl chain coming from the RAFT preparation of the

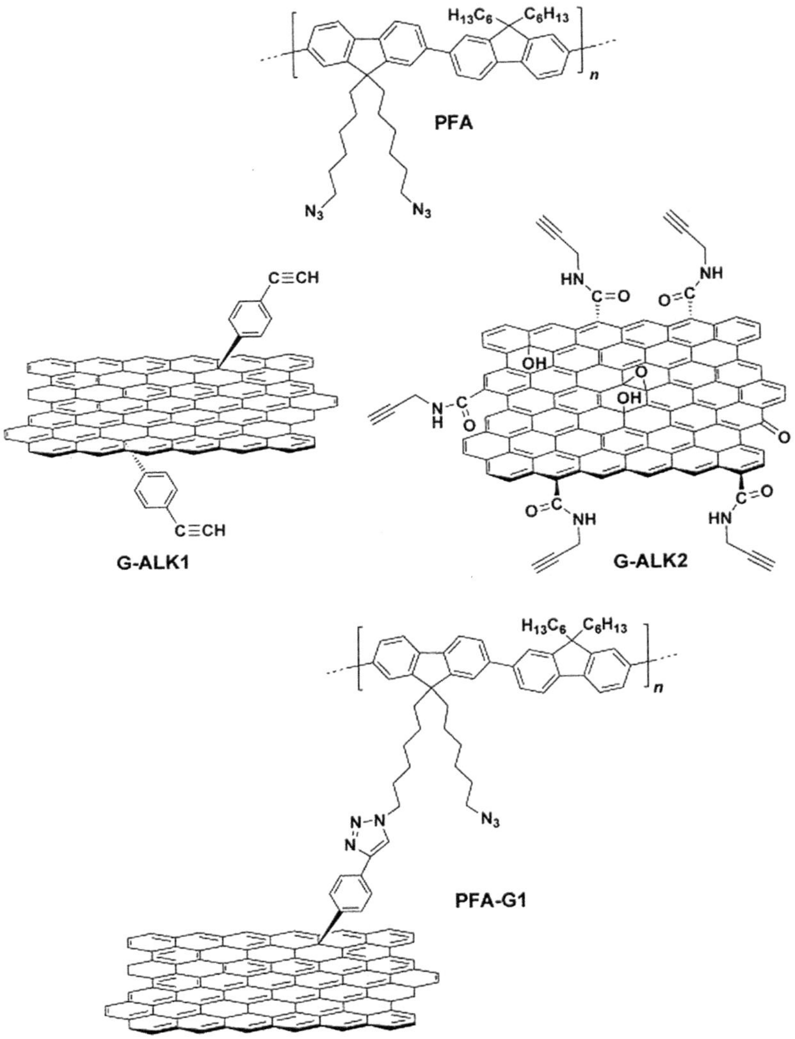

SCHEME 5.23 Chemical structure of PFA, alkyne-modified graphene obtained by diazonium coupling (G-ALK1) or amidation (G-ALK2) and the final click product PFA-G1.[95] *PFA*, Poly [(9,9-dihexylfluorene)-*co*-alt-(9,9-bis-(6-azidohexyl)fluorene)]. *Reprinted with permission from Castelaín, M.; Martínez, G.; Merino, P.; Martín-Gago, J. A.; Segura, J. L.; Ellis, G.; Salavagione, H. J.* Chem. Eur. J. ***2012****,* 18, *4965–4973. Copyright 2012 John Wiley and Sons.*

starting P(NIPAM). By heating a stirred aqueous dispersion of the click hybrid at temperature higher than 50°C, the onset of particles flocculation can be induced in few minutes. By cooling, the full reversibility of the flocculation–dispersion process was demonstrated. After removing the hydrophobic dodecyl end-groups by aminolysis, the authors obtained a

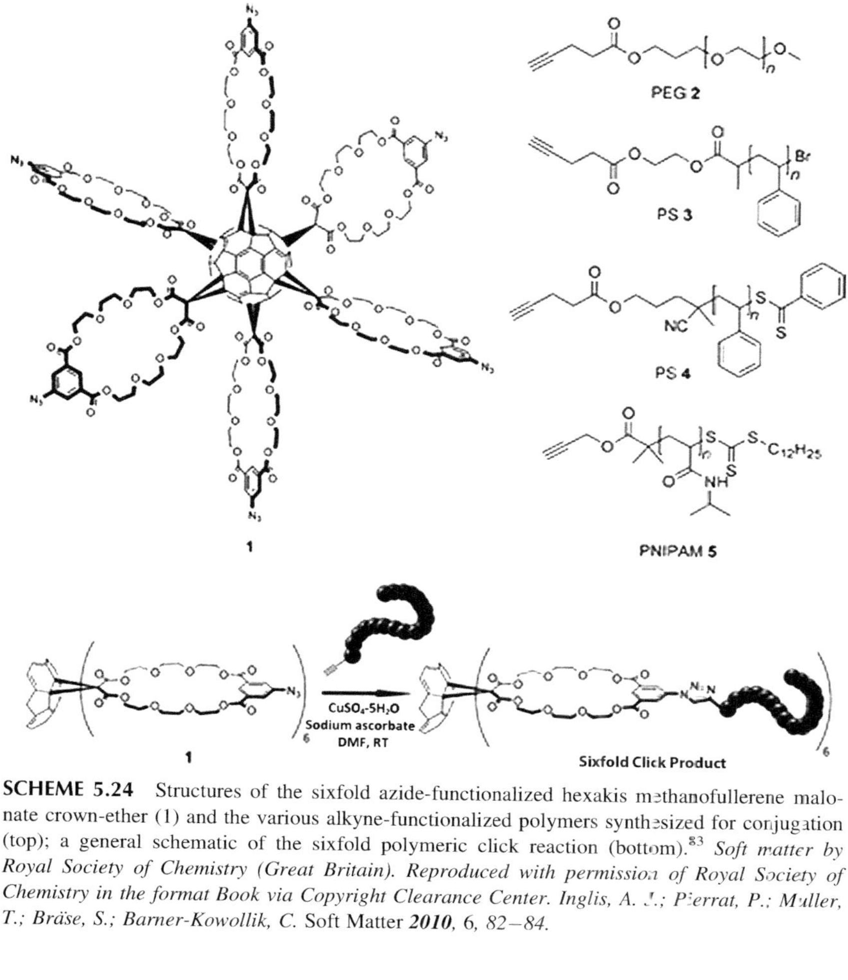

SCHEME 5.24 Structures of the sixfold azide-functionalized hexakis methanofullerene malonate crown-ether (1) and the various alkyne-functionalized polymers synthesized for conjugation (top); a general schematic of the sixfold polymeric click reaction (bottom).[83] *Soft matter by Royal Society of Chemistry (Great Britain). Reproduced with permission of Royal Society of Chemistry in the format Book via Copyright Clearance Center. Inglis, A. J.; Pierrat, P.; Muller, T.; Bräse, S.; Barner-Kowollik, C.* Soft Matter ***2010***, 6, *82–84.*

water-soluble hybrid polymer showing the same solubility characteristics of P(NIPAM).

Fullerene hexakis-adducts with a T_h-symmetrical octahedral addition pattern are a very interesting class of three-dimensional (3D) molecules: their spherical framework constitutes a very appealing platform for the preparation of innovative nanomaterials and bioactive molecules with specified functionalities.[97] The most important method for exohedral functionalization is cycloaddition to [6,6] double bonds in the fullerene core,[98] allowing for the incorporation of almost any functional groups around the fullerene core.[99,100] For example, Nierengarten et al.[100] prepared with very high yield (91%) "fullerene sugar balls" bearing 12 peripheral carbohydrate moieties by grafting unprotected sugar derivatives onto the fullerene core at room temperature in a CH_2Cl_2-water solution.

One of the proposed strategies in order to obtain fullerene hexakis-adducts bearing two or more different peripheral functional subunits is the sequential click reactions approach, consisting of the subsequent evolution of the alkyne–azide reaction with another click process.[101,102]

Iehl and Nierengarten[99] prepared a fullerene hexakisadduct bearing 12 TMS-protected alkyne groups allowing the attachment of 12 azide units under the copper-mediated Huisgen 1,3-dipolar cycloaddition conditions. In the same paper they reported the synthesis of a fullerene hexa-adduct bearing 10 azides and 2 TMS-protected alkyne groups allowing the successive attachment of 10 alkyne building blocks and 2 azide units via successive click reactions.

Again Iehl and Nierengarten coupled for the first time in literature the thiol–ene click reaction with the CuAAC reaction[103] in order to obtain a fullerene hexakis-adduct bearing three reactive subunits with different chemical nature. The thiol–ene click reaction has proven to be powerful for the functionalization of a large variety of building blocks[104] but its compatibility with fullerene derivatives is mined by the reactivity of thiyl radicals with the core of these C-based structures.[105] To solve this problem, an ad hoc functionalization of this part of the molecules was performed before the desired click reaction took place, obtaining at the end the evolution of the two desired click reactions with yields up to 91%. First, the authors prepared a structure with six identical peripheral groups (Scheme 5.25): fullerene (1 equiv.), **22** (10 equiv.), CBr_4 (100 equiv.), and DBU (20 equiv.) were reacted in *o*-dichlorobenzene (ODCB) at room temperature under stirring conditions for 72 h. The obtained solution was left to evaporate and the resulting sample was separated on a silica-gel column to obtain **23** (41% yield).

At this point, **23** (1 equiv.) was mixed with the thiol **24a** (20 equiv.) and AIBN (3 equiv.) in benzene under Ar at 80°C for 1 h giving, after purification, the compound **25a** (51% yield). The same working conditions were then applied to the thiols **24b** and **24c** leading to the formation of the thiol–ene products **25b** and **25c** (yield of 35% and 52%, respectively): the success of these reactions demonstrated that thiol–ene click reactions have success to functionalize fullerene hexa-adduct building blocks.

Later, the authors tried to graft to fullerene different groups via successive click reactions.

First, they prepared the methanofullerene **27** by reacting fullerene with **26**, iodine, and DBU under Bingel conditions, with a final yield of 50% (Scheme 5.26). The reaction of **27** with an excess of the malonate **28**, CBr_4 and DBU in ODCB led to the formation of **29** (57% yield), which finally was reacted with phenylacetylene in presence of the catalyst $CuSO_4 \cdot 5H_2O$ and sodium ascorbate giving the clicked derivative **30**. NMR measurements demonstrated that the TMS-protected alkyne unit as well as the methacrylate

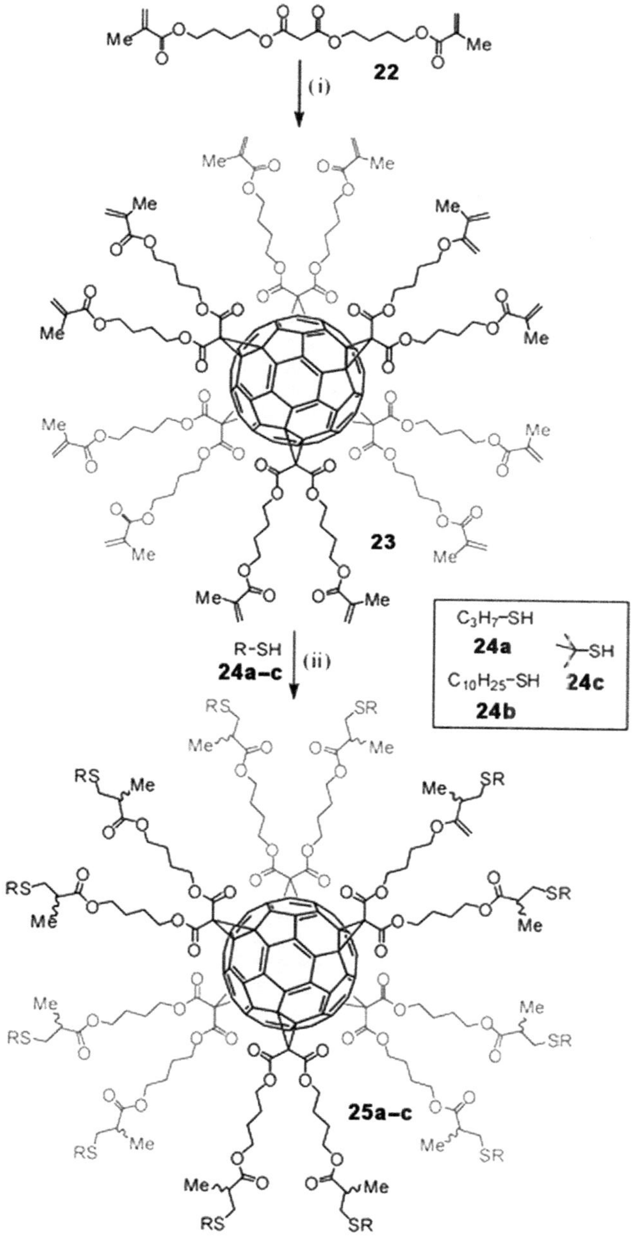

SCHEME 5.25 Functionalization of the fullerene core with reactive peripheral groups, namely methacrylate subunits. Reagents and conditions: (i) fullerene, ODCB, DBU, CBr_4, r.t. (41%); (ii) AIBN, benzene, 80°C (**25a**: 51%; **25b**: 35%; **25c**: 52%).[103] *DBU*, 1,8-Diazabicyclo[5.4.0]undec-7-ene; *ODCB*, *o*-dichlorobenzene. *Chemical communications by Royal Society of Chemistry (Great Britain). Reproduced with permission of Royal Society of Chemistry in the format Book via Copyright Clearance Center. Iehl, J.; Nierengarten, J. F.* Chem. Commun. ***2010***, 46, *4160–4162.*

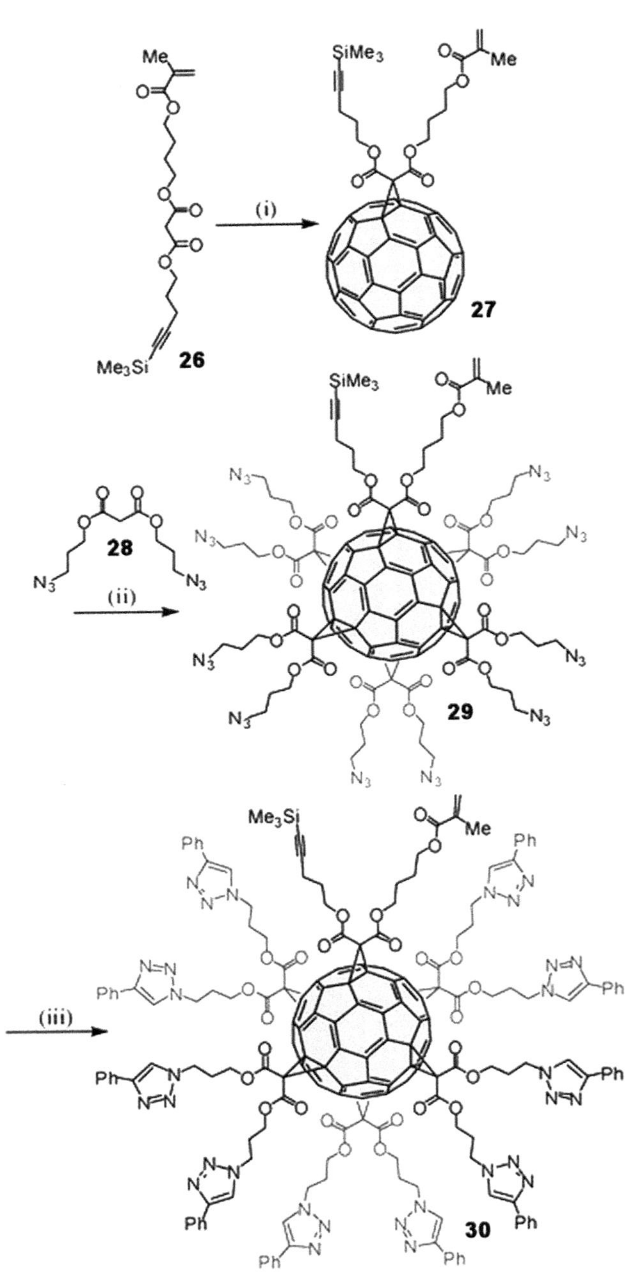

SCHEME 5.26 Preparation of a fullerene structure with a TMS-protected alkyne unit. (i) C_{60}, DBU, I_2, PhMe, r.t. (50%); (ii) DBU, CBr_4, ODCB, r.t. (57%); (iii) phenylacetylene, $CuSO_4 \cdot 5H_2O$, sodium ascorbate, $CH_2Cl_2-H_2O$, r.t. (66%).[103] *DBU*, 1,8-Diazabicyclo[5.4.0]undec-7-ene; *ODCB*, *o*-dichlorobenzene; *TMS*, tetramethylsilane. *Chemical communications by Royal Society of Chemistry (Great Britain). Reproduced with permission of Royal Society of Chemistry in the format Book via Copyright Clearance Center. Iehl, J.; Nierengarten, J. F.* Chem. Commun. ***2010***, 46, *4160–4162.*

group remained unchanged under the click reaction conditions while IR showed that no azide remains were present in **30**.

At this point (Scheme 5.27), this last structure was subjected to the thiol–ene click reaction with the thiols **24a–c** leading to the formation of the fully clicked derivatives **32a–c**, whose structures were confirmed by ^1H and ^{13}C-NMR spectra and by mass spectrometry.

Teng et al.[106] were able to obtain mono-fulleryne **35** and di-fulleryne **37** in one step (Scheme 5.28) starting from fullerene via a Bingel reaction by reacting it with terminal-alkyne-functionalized malonates [ethyl-3-butyn-1-yl malonate **34** and di(3-butyn-1-yl) malonate **36**, respectively] prepared from **33**.

Subsequently the fullerenes were reacted with azido-functionalized PS via Huisgen [3 + 2] cycloaddition click chemistry (Scheme 5.29) leading to the formation of two fullerene polymers, one with C_{60} tethered to the end of a PS chain (C_{60}-1PS, **38**) and the other with C_{60} tethered at the junction point of two PS chains of identical molecular weight (C_{60}-2PS, **39**). The fullerene polymer formation and structural differences were demonstrated by ^{13}C-NMR, FTIR, and MALDI-TOF mass spectroscopies. The thermal behavior studies confirmed that the two different fullerene polymers had different self-assembly architectures in the solid state. The polymers showed narrow polydispersity and high degree of C_{60} functionalization. UV–Vis absorption spectroscopy demonstrated that the optic property of C_{60} were not affected by the PS chains presence.

Other materials

Alkyne-terminated PS and oligoaniline were attached by click chemistry on hybrid silica nanoparticles polymerized with (6-azidohexyl methacrylate) prepared by RAFT polymerization.[107] This reaction was shown to be faster and with higher yield with respect to the corresponding reactions of free azido polymers, thanks to the highest local concentration of CuI catalyst, as a consequence of its strong interaction with the forming triazoles.[21]

Electrochromism is the phenomenon displayed by some materials of reversibly changing color stimulated by redox reactions. Electrochromic devices are based on any material that changes color depending on the applied potential: various types of inorganic and organic materials and structures can be used to realize such systems.[108,109] Some of them can be realized through the donor–acceptor strategy; that is, by alternating electron donors and acceptors, leading to the formation of low-bandgap polymers.

Xu et al. prepared pyrrolo[3,4-*d*]pyridazine-5,7-dione,[110] pyrrolophthalazinedione,[111] and pyrroloacenaphthopyridazinedione[112] starting from perquisite tetrazine by an iEDDA reaction, taking place with elimination of nitrogen. These reactions proceed without catalyst but need a temperature as high as 160°C due to the electron deficiency of both the participating

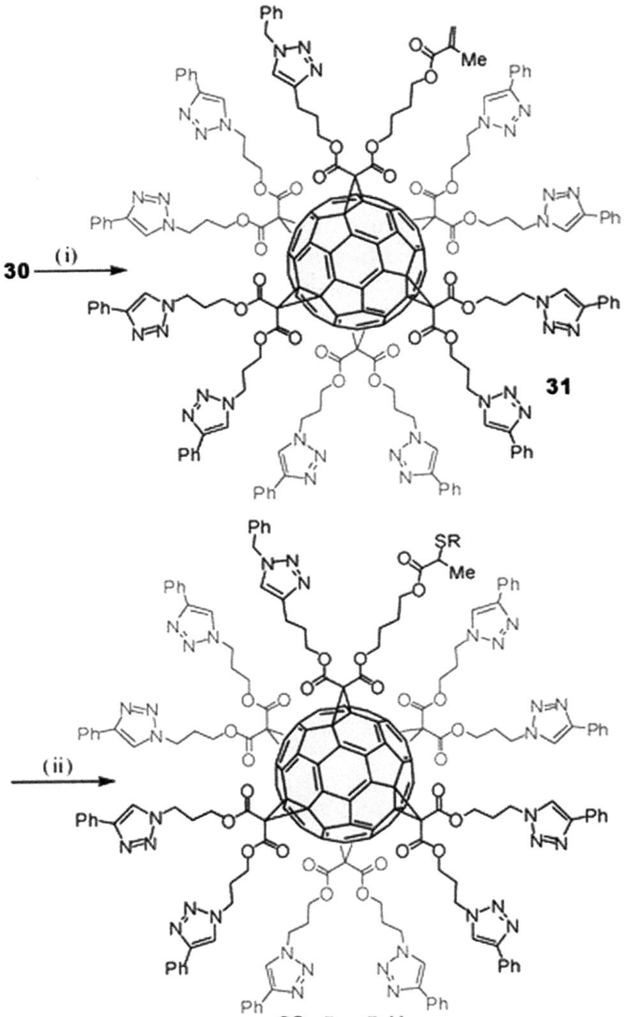

32a R = C_3H_7
32b R = $C_{10}H_{21}$
32c R = *t*Bu

SCHEME 5.27 Preparation of a fullerene structure with different functional groups by subsequent click reactions. (i) Benzyl azide, TBAF, $CuSO_4 \cdot 5H_2O$, sodium ascorbate, CH_2Cl_2–H_2O, r.t. (87%); (ii) **24a–c**, AIBN, THF, Δ, 1 h (**32a**: 70%, **32b**: 91%, **32c**: 62%).[103] *THF*, Tetrahydrofuran; *TBAF*, tetra-n-butylammonium fluoride. *Chemical communications by Royal Society of Chemistry (Great Britain). Reproduced with permission of Royal Society of Chemistry in the format Book via Copyright Clearance Center. Iehl, J.; Nierengarten, J. F.* Chem. Commun. ***2010***, 46, *4160–4162*.

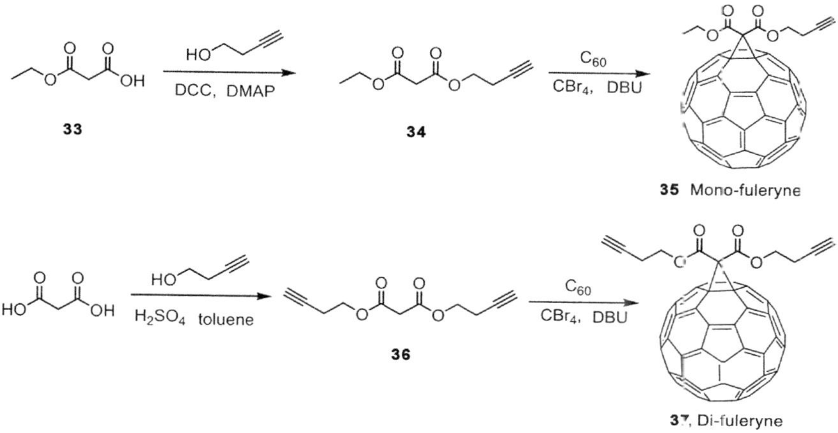

SCHEME 5.28 Preparation of mono-Fulleryne **35** and di-Fulleryne **37**.[106] *Reprinted with permission from Teng, F.-A.; Guo, Y.; He, J.; Zhang, Y.; Han, Z.; Li, H.* Des. Monomers Polym. ***2017**, 20, 283–292.*

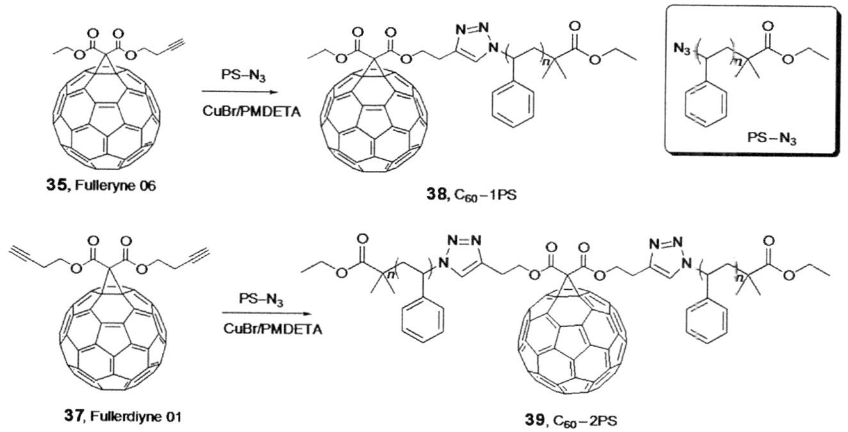

SCHEME 5.29 Synthesis of C_{60}-PS **38** and C_{60}-2PS **39**. The structure of PS–N_3 is shown in the box. *PS*, Polystyrene. *Reprinted with permission from Teng, F.-A.; Guo, Y.; He, J.; Zhang, Y.; Han, Z.; Li, H.* Des. Monomers Polym. ***2017**, 20, 283–292.*

dienophile and diene. The products show interesting electrochromic properties, good switching speeds, and high optical contrast.

Metal-organic frameworks (MOFs) are crystalline compounds consisting of metal ions or clusters coordinated to organic ligands to form one-, two-, or three-dimensional structures characterized by voids and a certain degree of porosity, which makes them particularly suitable for catalysis and gas adsorption applications.[113] Although the 3D structure and internal

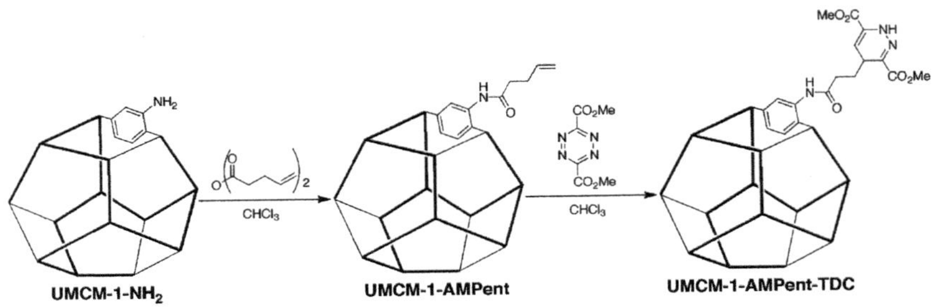

SCHEME 5.30 iEDDA reaction for the functionalization of an MOF.[115] *iEDDA*, Inverse-electron-demand Diels–Alder; *MOF*, metal-organic framework. *Reprinted with permission from Chen, C.; Allen, C. A.; Cohen, S. M.* Inorg. Chem. ***2011***, 50, *10534–10536. Copyright 2011 American Chemical Society.*

environment of the pores in theory can be controlled through proper selection of nodes and organic linking groups, the direct synthesis of such materials with the desired functionalities can be difficult due to the high sensitivity of MOF systems. Thermal and chemical sensitivity, as well as high reactivity of reaction materials, can make the obtainment of the desired products challenging to achieve. The postsynthetic exchange of organic linkers and metal ions is an expanding area of the field and opens up possibilities for more complex structures, increased functionality, and greater system control.[114] iEDDA reactions are an interesting strategy for the postfunctionalization of these structures. Cohen et al.[115] modified UMCM-1-NH_2 (University of Michigan Crystalline Material)[116] by pentenoic anhydride to introduce an alkene group and subsequently added tetrazine. The iEDDA reaction took place at room temperature giving the desired UMCM-1-AMPent-TDC structure (Scheme 5.30). Rieger et al.[117] followed the same strategy for the functionalization of a Zn-based MOF.

Surface and interface modification

The utilization of 1,3-dipolar cycloadditions, such as CuAAC, for the functionalization and modification of surfaces, interfaces, and nanoconstructs is a rapidly expanding area, as indicated by the multitude of papers reported over the last 15 years. The potential to functionalize orthogonally in a regiospecific manner, to conduct the functionalization reactions in water, and to tolerate biologically relevant groups such as peptides, proteins, and a multitude of functional handles, suggests that the 1,3-dipolar cycloadditions are an essential member of the nanoscale manipulation toolbox.[1]

O'Reilly et al.[118] prepared shell cross-linked knedel-like nanoparticles having a mixture of alkyne/COOH or azide/COOH (see Scheme 5.31) groups on the surface via transformation of the initial nanostructure utilizing carbodiimide-based amidation reactions. Subsequently, these functionalities

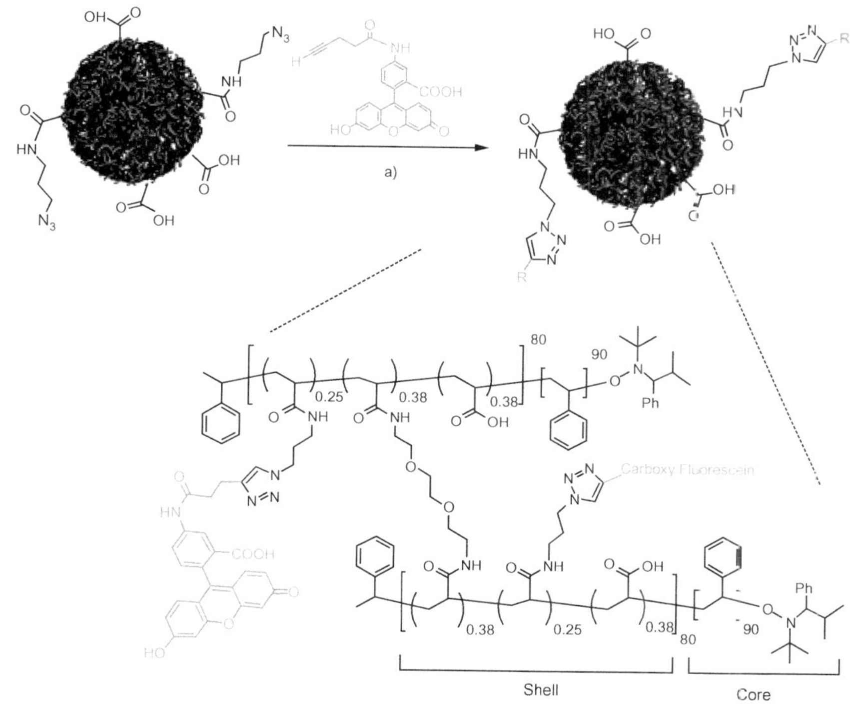

SCHEME 5.31 Shell click reaction of the nanoparticles carrying click reactive functionalities in the hydrophilic shell with an alkynyl-functionalized fluorescein dye. (a) $CuSO_4 \cdot 5H_2O$ (0.25 equiv.), sodium ascorbate (5 wt.% solution in water, 0.50 equiv.), alkynyl-functionalized fluorescent dye, RT, followed by dialysis against pH 7.3 phosphate buffered saline.[113] *RT*, room temperature. *Reprinted with permission from O'Reilly, R. K.; Joralemon, M. J.; Wooley, K. L.; Hawker, C. J.* Chem. Mater. ***2005***, 17, *5976–5988. Copyright 2005 American Chemical Society.*

were selectively modified by reaction with azido/alkynyl fluorescein dye while retaining the COOH for further amidation-based modifications (Scheme 5.31 and Scheme 5.32). Moreover, the selective introduction of azido groups in the hydrophobic core of the nanoparticles through a functional monomer strategy allowed for orthogonal 1,3-dipolar cycloadditions in the core and amidation-based strategies in the shell of these structures.

Later, the same authors used azido functional dendrimers of different generations for shell cross-linking, and residual azide functionalities were employed in a subsequent reaction with fluorescent alkynes in the shell.[119] This ability to control the reactivity of nanoparticles and the location of functional groups within a 3D structure is at the heart of the click chemistry philosophy,[1] as it has been exemplified in the earlier studies and in the selective functionalization of nanoparticle cores[120] and surfaces thanks to reactions at the chain end sites of block copolymer components.[121]

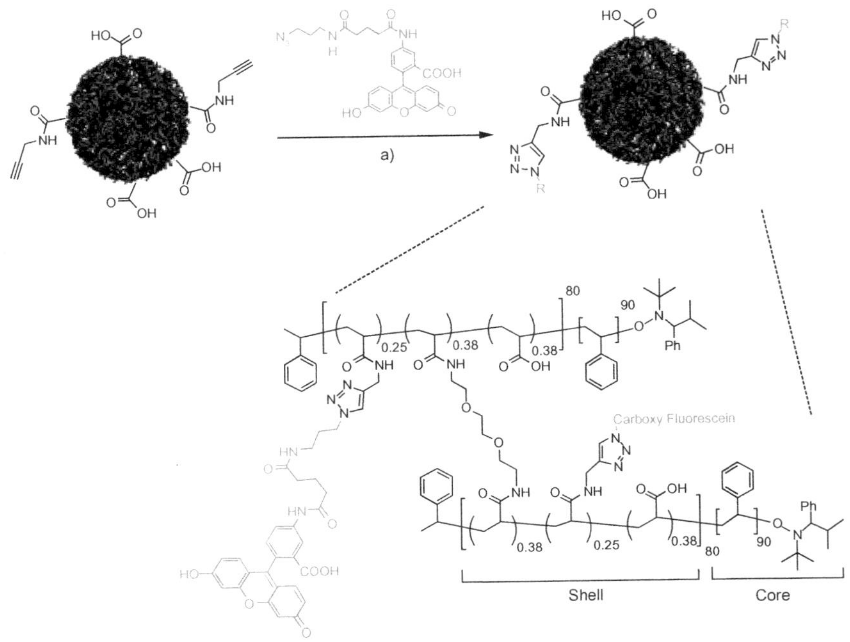

SCHEME 5.32 Shell click reaction of the nanoparticles carrying click reactive functionalities in the hydrophilic shell with an azido-functionalized fluorescein dye. (a) $CuSO_4 \cdot 5H_2O$ (0.25 equiv.), sodium ascorbate (5 wt.% solution in water, 0.50 equiv.), azido-functionalized fluorescent dye, RT, followed by dialysis against pH 7.3 phosphate buffered saline.[118] *RT*, room temperature. *Reprinted with permission from O'Reilly, R. K.; Joralemon, M. J.; Wooley, K. L.; Hawker, C. J.* Chem. Mater. ***2005***, 17, *5976–5988. Copyright 2005 American Chemical Society.*

An et al.[122] combined a Michael addition reaction (the nucleophilic addition of a carbanion or another nucleophile to an α,β-unsaturated carbonyl compound) with an 1,3-dipolar cycloaddition to selectively functionalize the interior of poly(dimethylacrylamide) nanoparticles with fluorescein units, and subsequently to couple azido groups in P(NIPAM) shell of the nanoparticles with dansyl units.

The use of the copper-catalyzed 1,3-dipolar CuAAC cycloaddition reaction has found particular utility also in the modification of self-assembled monolayers (SAMs), with many studies employing click cycloadditions to selectively and efficiently derivatize a planar substrate.

Collman et al.[123,124] reported the decoration of a SAM made of azido undecanethiol and decane thiol with redox-active ferrocene molecules containing acetylene (Scheme 5.33). The determination of the surface coverage was done by the employment of a redox-active probe, while the consumption of the azide could be followed by IR spectroscopy. This work demonstrated that surfaces presenting organic azides provide an excellent handle for surface modification and functionalization.

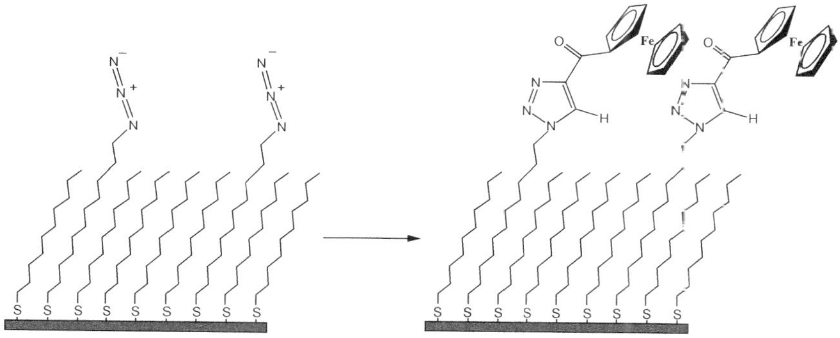

SCHEME 5.33 Decoration of a SAM made of azidoundecanethiol and decane thiol with redox-active ferrocene molecules containing acetylene.[123] *SAM*, Self-assembled monolayer. *Reprinted with permission from Collman, J. P.; Devaraj, N. K.; Chidsey, C. E. D.* Langmuir ***2004****, 20. 1051–1053. Copyright 2004 American Chemical Society.*

Also alkyne-functionalized surfaces optimally serve as platforms for click modifications: Drockenmuller et al.[125] reported the grafting of different polymers to an alkyne-functionalized SAM (Scheme 5.34). The surface of a silicon-based substrate SAM presenting alkyne functionalities was grafted with ω-azido polymers (PEG$-N_3$, PMMA$-N_3$, and PS$-N_3$) via the copper-catalyzed Huisgen 1,3-dipolar cycloaddition to yield well-defined polymer brushes with a thickness of ca. 6 nm and grafting densities of ca. 0.2 chains/nm^2. This approach, thanks to the selectivity of this click reaction, provides a general method for densely grafting onto a variety of polymer brushes bearing different functional groups onto the surface of SAMs.

Cycloadditions under nonclassical reaction conditions

Although, as underlined at the beginning of this chapter, click reactions should be very effective under simple reaction conditions (they should ideally be performed at room temperature in an open flask), there are several examples for which "the intensification of the processes" by different strategies such as applying microwave heating, using ultrasound irradiation, or employing ionic liquids (ILs) as reaction medium, is advantageous or even required.[3] Common cases are the reactions evolving in a long time and with slow kinetics under standard conditions. In this section, some examples or click reactions performed under nonclassical conditions in the material science field are briefly presented.

Microwave chemistry is based on the ability of a reaction mixture to efficiently absorb microwave energy, thanks to the so-called "microwave dielectric heating" phenomena such as dipolar polarization or ionic conduction mechanisms.[126] Controlled microwave heating in sealed vessels has been shown to strongly reduce reaction times, increase product yields, and

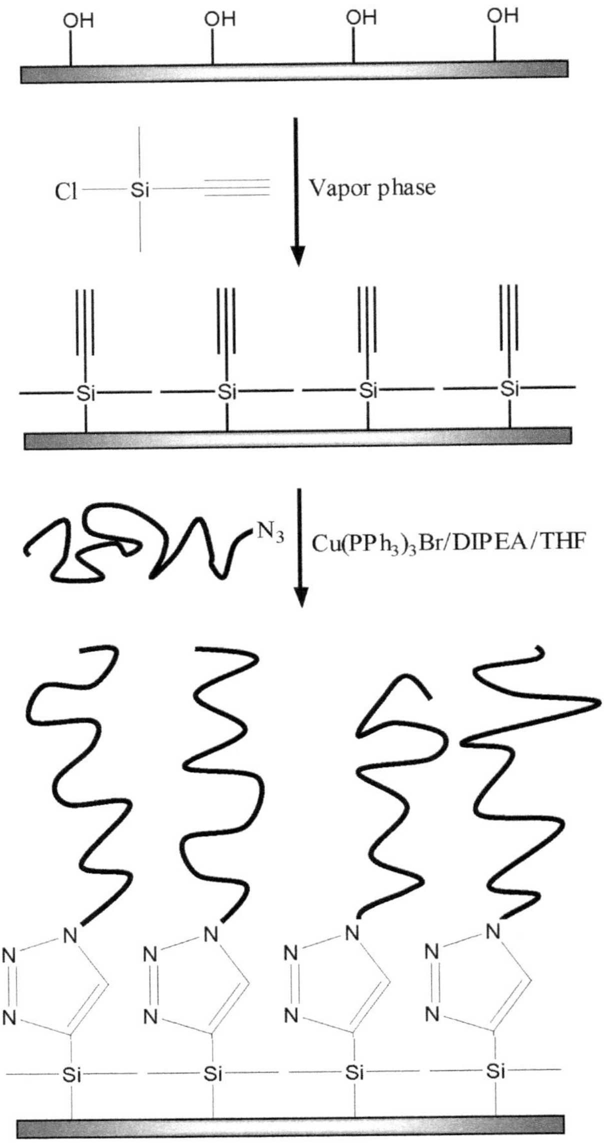

SCHEME 5.34 Click chemistry grafting of polymer brushes to "passivated" silicon substrates. *Reprinted with permission from Ostaci, R.-V.; Damiron, D.; Capponi, S.; Vignaud, G.; Leger, L.; Grohens, Y.; Drockenmuller, E.* Langmuir ***2008****, 24, 2732–2739. Copyright 2008 American Chemical Society.*

enhance the purities of the products thanks to the minimization of unwanted side reactions with respect to the conventional preparative methods.[127–129] The first applications of microwave heating to cycloaddition, namely CuAAC processes, were reported in 2004 by Van der Eycken[129] and Kappe,[130] who noticed an increase in the reaction rate and in the product yields thanks to this "intensification." From those trials forward, a huge number of 1,3-dipolar cycloaddition reactions was realized applying this alternative energy transfer technique, that in some cases was the only way to lead the reactions to their completion.[131] In some cases, microwave irradiation was able also to improve the 1,4-regioselectivity of the 1,3-dipolar cycloaddition. For example, Ritter et al.[132] performed the synthesis of the new monomer mono-(1*H*-1,2,3-triazol-4-yl)(methyl)2-methylacryl-*b*-cyclodextrin **42** via microwave-assisted CuAAC reaction at 140°C in DMF starting from propargyl methacrylate **40** and 6I-azido-6I-deoxycyclomaltoheptaose **41** with $CuSO_4$ and NaAsc as catalytic systems (Scheme 5.35). After 30 min of reaction, the complete and exclusive formation of the 1,4-disubstituted-1,2,3-triazole was observed. On the contrary, by heating under the same conditions in an oil bath, a mixture of the 1,4- (**42**) and 1,5-disubstituted (**43**) cycloadducts was obtained after a time span of 24 h. The cycloaddition proceeded also without heating: in this case, it was region-selective and a conversion degree of 74% was obtained after a duration time as long as 1 week.

Delattre et al.[133] performed the CuAAC reaction between 6I-azido-6I-deoxycyclomaltoheptaose peracetate and 1,3,5-tris(2-propynyl-oxy)-benzene with the final aim to prepare a tripod molecule containing an aromatic core bearing three peracetylated cyclodextrins, useful to study the photoluminescent properties of complexation phenomena with different pesticides. The authors showed that the reaction rate was increased by microwave irradiation with respect to the conventional click chemistry conditions.

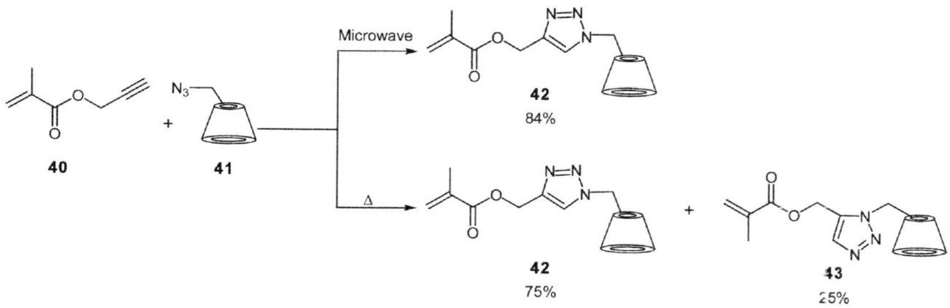

SCHEME 5.35 Influence of microwave and conventional conditions on the regioselectivity of the CuAAC reaction leading to the formation of the cyclodextrin **42**. The reactions were performed at 140°C in *N,N*-dimethylformamide, using $CuSO_4 \cdot 5H_2O$ and Na ascorbate as catalytic system. The reaction time was 30 min under microwave heating and 24 h in an oil bath.[132] *CuAAC*, Copper-catalyzed azide–alkyne cycloadditions. *Reprinted with permission from Munteanu, M.; Choi, S.; Ritter, H.* Macromolecules ***2008***, 41, *9619–9623. Copyright 2008 American Chemical Society.*

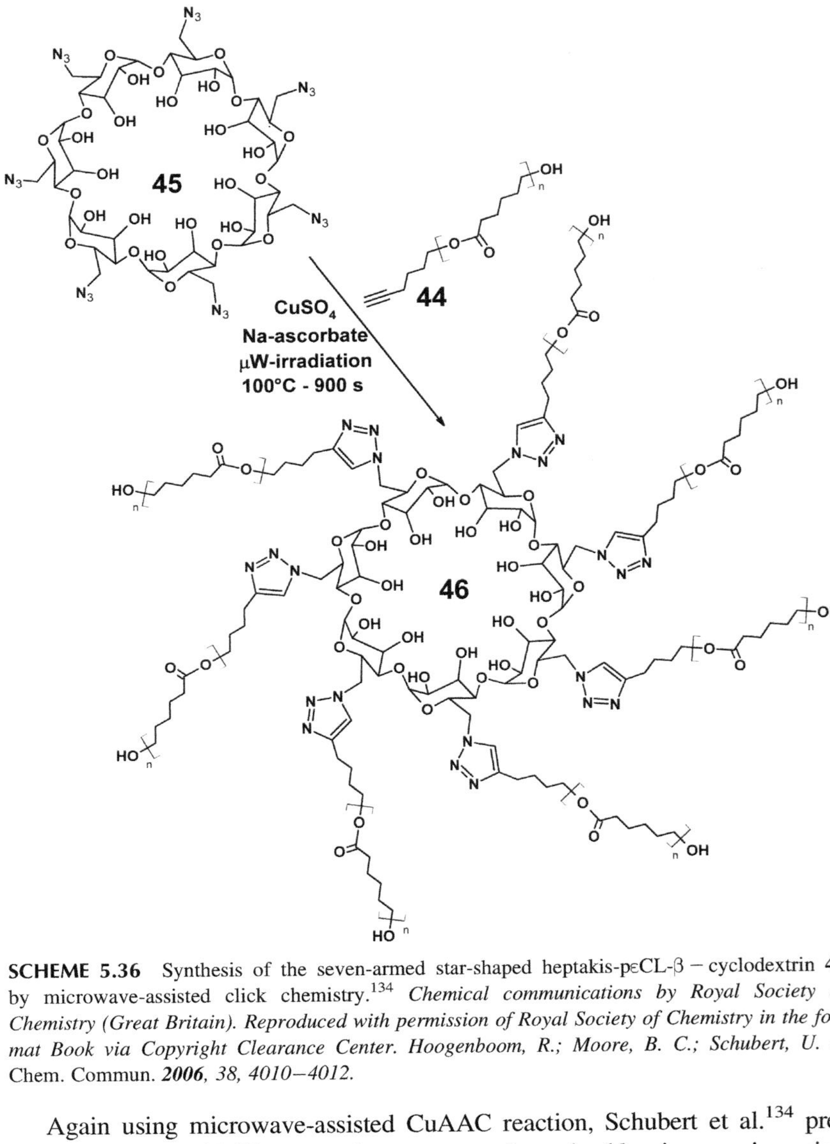

SCHEME 5.36 Synthesis of the seven-armed star-shaped heptakis-pεCL-β − cyclodextrin **46** by microwave-assisted click chemistry.[134] *Chemical communications by Royal Society of Chemistry (Great Britain). Reproduced with permission of Royal Society of Chemistry in the format Book via Copyright Clearance Center. Hoogenboom, R.; Moore, B. C.; Schubert, U. S.* Chem. Commun. ***2006****, 38, 4010–4012.*

Again using microwave-assisted CuAAC reaction, Schubert et al.[134] prepared a star-shaped PCL. ε-caprolactone was polymerized by ring opening using 5-hexyn-1-ol as initiator and tin(II) octanoate as catalyst, yielding a PCL. Subsequently, the acetylene-functionalized pεCL **44** was mixed with heptakis-azido-β-cyclodextrin **45** and heated by microwaves at 100°C, giving the seven-armed star-shaped heptakis-pεCL-β−cyclodextrin **46** (Scheme 5.36).

Concerning the applications of this assisted technique for C-based materials preparation, Schuster et al.[135] prepared 1,2,3-triazole-linked porphyrin–fullerene dyads in the attempt to study electron transfer mechanism and kinetics. Microwave-assisted CuAAC reaction was realized both between the alkynyl porphyrin and various azidobenzaldehyde components (yields up to 100%) and between azidophenylporphyrins and alkynyl benzaldehydes (yields of about 84%). Finally, the Prato 1,3-dipolar cycloaddition (frequently used for the functionalization of fullerenes and nanotubes) of azomethine ylides (derived from the porphyrinic aldehyde and sarcosine) to fullerene led to the formation of the porphyrin–fullerene dyads.

ILs are molten salts typically constituted of organic cations and inorganic anions. They present (1) low melting points; (2) no boiling point; (3) vapor pressure close to zero; (3) high chemical and thermal stability; (4) the ability to dissolve a wide range of organic, inorganic, metallic, and polymeric substances; and (5) interesting physicochemical properties such as density, viscosity, solubility, and hydrophobicity, which can be varied depending on the combination of cation and anion.[136] Their use as alternative environmentally friendly solvents for organic synthesis has found increased attention in the past two decades and also has attracted attention concerning CuAAC reactions, due to the lack of solubility of the azide and acetylene reactants in the "conventional green solvent" water.[3,137]

Liang et al.[138] performed a one-pot azidation–CuAAC reaction involving ferrocenylacetylene, benzyl halides, and NaN_3 in a 1:1 mixture of the IL 1-butyl-3-methylimidazolium tetrafluoroborate ($bmimBF_4$) and water in presence of the catalyst CuI and Na_2CO_3 at room temperature. After 8 h, the triazole product had been obtained with a yield of 99% and it was demonstrated that the IL could be recycled several times without loss of efficiency (Scheme 5.37). This one-pot strategy can be followed for the reaction of a wide range of acetylene and halide building blocks with high product yield.

Dondoni et al.[139] performed the cycloaddition of a sugar azide with a sugar acetylene (CuI, *i*-Pr_2EtN, 80°C) in 10 different ILs as well as in standard molecular solvents (toluene and DMF) to give the 1,4-disubstituted triazole-linked *C*-disaccharide. The highest yields were registered in ILs (up to 95% when $[C_8dabco][N(CN)_2]$ was used). Later that same year, they

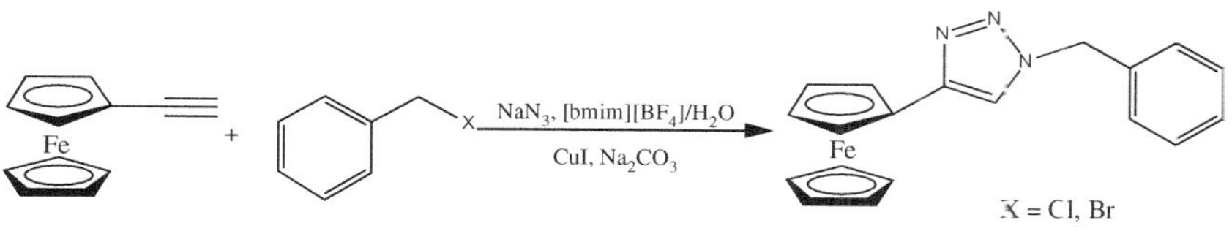

SCHEME 5.37 One-pot azidation–CuAAC reaction in a ionic liquid.[138] *CuAAC*, Copper-catalyzed azide–alkyne cycloadditions. *RSC advances by RSC Publishing. Reproduced with permission of RSC Publishing in the format Book via Copyright Clearance Center. Zhao, Y.-B., Yan, Z.-Y.; Liang, Y.-M.* Tetrahedron Lett. ***2006***, 47, *1545–1549.*

combined the use of ILs as solvents and microwave heating strategy to synthesize a triazole-linked tetra-*C*-galactosyl-calix[4]arene starting from a tetra-azido calix[4]arene derivative and ethynyl tetra-*O*-benzyl-*C*-galactoside in the presence of CuI and *i*-Pr_2EtN at 80°C.[140] Yields up to 90% were obtained and the trials made with the microwave-assisted strategy generally showed faster kinetics with respect to classic thermal treatment experiments, due to the fact that the ionic nature of the ILs allow them to be heated very efficiently by microwave dielectric heating.

Due to the already discussed low solubility of fullerene in the most commonly used organic solvents and the large amount of solvent used for the functionalization of these molecules, Mikami et al.[141] proposed an environmentally benign strategy: the Diels–Alder reaction between fullerene and anthracene was performed as a solid-state process evolving by means of photo-irradiation. The authors had previously observed that this reaction hardly took place in benzene in dark conditions at room temperature. When a 1:1 solution of the reactants was irradiated with a high-pressure mercury lamp through Pyrex vessel, the Diels–Alder products (mono- and bis-adducts) formed together with the anthracene dimer (60:10:3:26 ratio; 9 h of reaction). When functionalized anthracene molecules were used (e.g., 9-methylanthracene) as diene components, more complex mixtures were obtained. Subsequently, the same reaction was studied in solid state: a well-ground mixture of fullerene and 9-methylanthracene was irradiated with the high-pressure mercury lamp and the temperature was kept at 28°C through a cooling jacket for the 9 h reaction time. The reaction products were dissolved in C_6D_6 and CS_2 and characterized by ^{1}H-NMR using triptycene as an internal standard: only the mono- and bis-Diels–Alder adducts were identified, with a yield of 30% and 19%, respectively. Unfortunately, anthracene did not show reaction in the solid state. The authors explained this behavior with the ionization potential of the anthracene derivatives: the smaller this value is (7.33 eV for anthracene, 7.19 for methyl-anthracene),[142] the more readily the reaction with fullerene proceeds, despite the increase in the steric hindrance due to the presence of the methyl groups. So, the Diels–Alder reaction in solid state proceeds via photoinduced electron transfer from anthracene to the triplet excited state of fullerene, followed by stepwise bond formation. Even if the electron affinity of the triplet excited state of C_{60} (4.27 eV, calculated considering its electron affinity, 2.7 eV, and the excitation energy of the triplet state, 1.57 eV)[143,144] is lower than the ionization potential values of the anthracene molecules, the Coulombic interactions between the anthracene radical cation and the fullerene radical anion in the solid state (expected to be larger than the value of 2.08 eV obtained in PhCN solution by considering a charge transfer transition energy of 2.55 eV) are strong enough to compensate for the energy difference in the electron transfer.

Mikami et al.[145] in 2000 demonstrated that the photochemical Diels–Alder reaction between fullerene and Danishefsky's dienes **47a** and **47b** (Danishefsky's diene, also called Kitahara diene, is an organosilicon compound and a diene with the formal name trans-1-methoxy-3-trimethylsilyloxy-1,3-butadiene) proceeded in solution via an electron transfer with formation of radical ion pairs, followed by stepwise (open-shell) bond formation. In particular, a stereochemically defined (1*E*,3*Z*)-1-methoxy-2-methyl-3-[(trimethylsilyl)oxy]-penta-1,3-diene was used as a stereochemical probe for the reaction with fullerene, which was realized in deaerated benzene by irradiation with high-pressure mercury lamp for 9 h at −30°C (no thermal reaction occurred in the dark in the same temperature conditions). The treatment of the primarily formed intermediate silylenolether with SiO_2 and triethylamine leaves the methoxy group unaffected, and yields a mixture of the two cis/trans isomers **48/49**, with the major product being the trans-adduct **49** (Scheme 5.38).

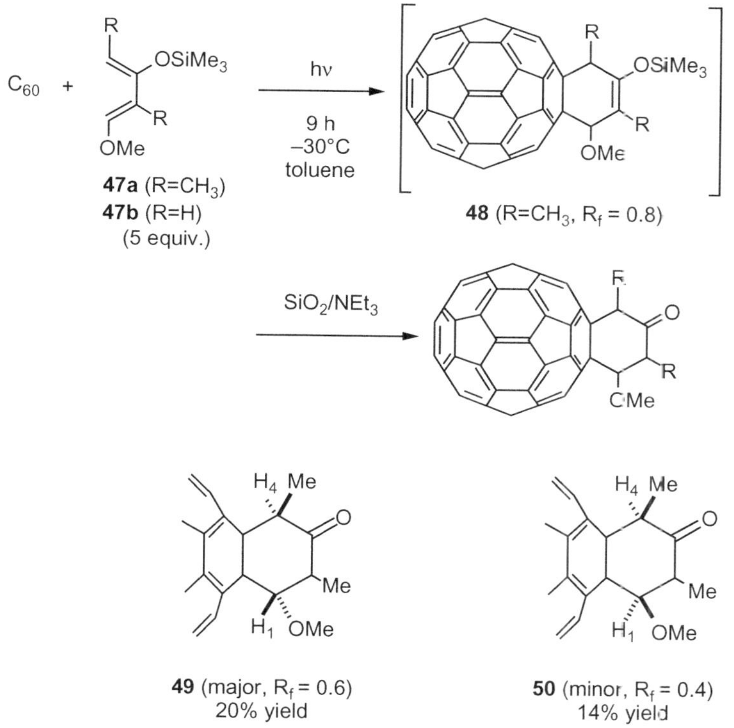

SCHEME 5.38 Photochemical Diels–Alder reaction between fullerene and Danishefsky's dienes.[145] *Adapted with permission from Mikami, K.; Matsumoto, S.; Okubo, Y.; Fujitsuka, M.; Ito, O.; Suenobu, T.; Fukuzumi, S.* J. Am. Chem. Soc. ***2000***, 122, *2236–2243. Copyright 2000 American Chemical Society.*

The transient absorption spectrum of the radical anion $C_{60}^{\bullet-}$ was detected in the visible and near-IR region using laser flash photolysis. Considering the observed stereochemistry and the fact that the observed rate constants determined from the dependence of the quantum yields on the concentrations of the diene agree well with those for the photoinduced electron transfer from the diene itself to the triplet excited state of C_{60} ($^3C_{60}^*$), the authors suggested the photoinduced electron transfer scheme reported in Scheme 5.39.

Although less common than the [4 + 2] and 1,3-dipolar cycloadditions, photochemical [2 + 2] cycloadditions of C_{60} with unsaturated compounds have also been widely studied, being the triplet excited state of C_{60}, with its high reduction potential, the principal responsible for most of these reactions. For example, preparation and isolation of [2 + 2] monoadducts of fullerene with alkynes,[146] arylalkenes,[147] dienes,[148] cycloenones,[149] acyclic enones,[150] and diones[151] have been reported in literature.

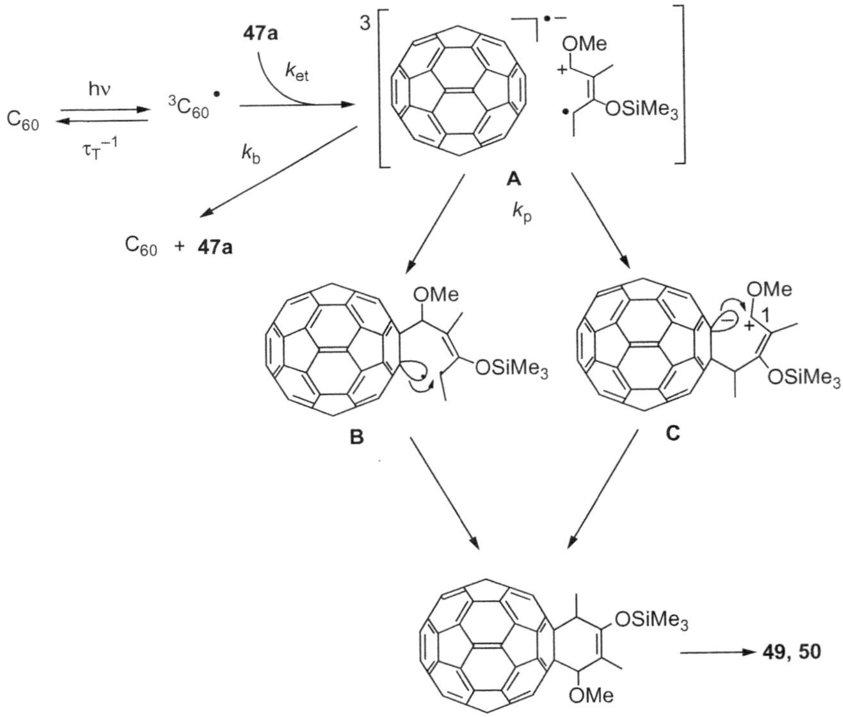

48

SCHEME 5.39 Electron transfer in the photochemical Diels–Alder reaction between fullerene and Danishefsky's dienes.[145] *Adapted with permission from Mikami, K.; Matsumoto, S.; Okubo, Y.; Fujitsuka, M.; Ito, O.; Suenobu, T.; Fukuzumi, S.* J. Am. Chem. Soc. ***2000***, 122, *2236–2243. Copyright 2000 American Chemical Society.*

As an exception, Schuster et al. reported that the photochemical [2 + 2] addition of enones to C_{60} evolves by the addition of the enone triplet excited state to the ground state of the fullerene via a triplet 1,4-biradical intermediate.[151] For all the reactions with alkenes, dienes, and enones, mechanistic studies have shown the evolution of two steps, namely the electron transfer from the double bonds to $^3C_{60}^*$ followed by the rapid collapse of the initial open intermediate to furnish the [2 + 2] adducts.

Few examples of thermal [2 + 2] cycloaddition of fullerenes have been reported in literature, including the reactions with benzyne,[152] tetraalkoxyethylenes,[146] 1,2-dicarbomethoxycyclobutadiene,[153] allenamides,[154] and morpholinocycloalkenes (enamines).[155] In particular, concerning the thermal reaction of fullerene with five- and six-membered morpholinocycloalkenes, Mikie et al.[155] reported the exclusive formation of the [2 + 2] cycloadducts in refluxing toluene, with yields of 77% and 64%, respectively. They suggested a multistep mechanism involving a secondary electron transfer with formation of the fullerene anion, radical coupling, and subsequent ion cyclization. On the contrary, the cycloheptene homologue showed negligible or null reactivity due to its increased steric hindrance.

The photocycloaddition of dienyl cyclopropanes to C_{60} allowed for stereospecific preparation of five-, seven-, and nine-membered fullerene adducts.[156] In particular, the reaction of fullerene with a 30-fold excess of 2-(trans-2′-phenylcyclopropyl)-5-methyl-2,4-hexadiene as an 80:20 *E*/*Z* mixture in deoxygenated toluene, which did not react under dark after 24 h, was irradiated by a Xenon lamp. A fast reaction was recorded, in a few minutes leading to the formation of a mixture including a major cycloproduct with a five-membered ring (60% relative yield) along with seven- and nine-membered cycloproducts. Again, formation of a biradical intermediate, which is preceded by an electron transfer step between the diene and the fullerene in its excited state, is involved in the reaction mechanism.

Tzirakis et al.[157] investigated the photochemical cycloaddition of biscyclopropyl-substituted alkenes with fullerene, finding that this reaction proceeds in a totally regioselective manner, leading to the formation of four cis-1 fullerene isomers featuring a unique 5–4–5 tricyclic fused ring system. The formation of these molecules takes place by a stepwise reaction mechanism initiated by a photoelectron transfer from the double bond of the alkene to the excited state of fullerene, leading to the formation of the geminate radical ion pair. The obtained radical cation undergoes the opening of its two cyclopropane rings, giving rise to a linear radical cation that reacts with its germinal radical anion $C_{60}^{\bullet -}$. At this point, the reaction between the radicals can follow two pathways: it can produce a 1,4-cycloadduct that, in turn, spontaneously undergoes an intramolecular [2 + 2] photocycloaddition reaction to produce the final 5–4–5 tricyclic fused ring system or it can be followed by an intramolecular nucleophilic addition to the proximal double bond, generating in situ the 1,3-dipolar intermediate that in turn can cyclize, giving the final product via an intramolecular 1,3-dipolar addition to the adjacent [6,6] double bond of the fullerene molecule.

The [4 + 2] cycloaddition reactions of *o*-quinodimethanes (*o*-xylylenes) to C_{60} have attracted much interest in fullerene chemistry, not only as efficient and versatile tools for the functionalization of fullerenes but also due to the applications of the resulting cycloadducts as C_{60}-based electroactive compounds.[158,159] In general, thermally chemo-controlled Diels–Alder reaction of C_{60} with *o*-quinodimethanes has been carried out from different precursors that, under suitable conditions, generate in situ the reactive diene intermediates, which, in turn, react with the fullerene molecules leading to the electroactive cycloadducts.[160,161] As an example, Markoulides et al.[160] investigated the optimum reaction conditions for the quantitative preparation of 3,4-di(methylene)tetrahydrothiophene-1,1-dioxide by zinc-induced 1,4-debromination of 3,4-*bis*(bromomethyl)-2,5-dihydrothiophene-1,1-dioxide in acetone by using different heating strategies; that is, the conventional heating, microwave, or ultrasonic irradiation. The obtained molecule was reacted with fullerene by [4 + 2] cycloaddition reaction leading to the efficient (the overall yield was 56%) and clean formation of the corresponding Diels–Alder monoadduct in a short reaction time.

Interestingly, the same authors developed a practical and efficient method for thermally chemo-controlled double Diels–Alder processes in one-pot, namely an innovative [4 + 2] functionalization/[4 + 2] derivatization of fullerene, starting again from 3,4-bis(bromo-methyl)-2,5-dihydrothiophene-1,1-dioxide.[162]

The [4 + 2] C_{60} adducts are generally obtained from precursors that have the dienic structure **52** (Scheme 5.40, path A) and are attached to the dienophilic C_{60} in the last step of the sequence. This strategy requires the preparation of new diene or masked-diene intermediates, which usually need multistep synthesis. As an example, attempts to prepare C_{60}-based tetrathiafulvalene adducts for photovoltaic devices by dienic chemistry were unsuccessful and different methods had to be developed.[163] Very frequently, a further derivatization of the fullerene cycloadducts is hampered by competing side reactions that lead to the formation of undesirable fullerene products, rendering the isolation of the desired C_{60} adducts quite difficult. In Markoulides et al.,[162] the authors developed a new strategy (Scheme 5.40, path B) in which the Diels–Alder reaction goes through the transient intermediate 2,2′-bisallyl diradical **53** with the diene "on" the fullerene structure.

To reach their target, the authors prepared diene-sulfone **57**[160] (Scheme 5.41) and subjected it to a [4 + 2] cycloaddition reaction with C_{60} in toluene, at 45°C–50°C. Upon completion (12 h), DMAD was added and the temperature was raised at 110°C for 30 min. TLC analysis of the reaction mixture revealed the formation of a single fullerene adduct that was purified by column chromatography followed by precipitation from $CHCl_3$–pentane. The subsequent spectroscopic analyses unambiguously supported the structure of

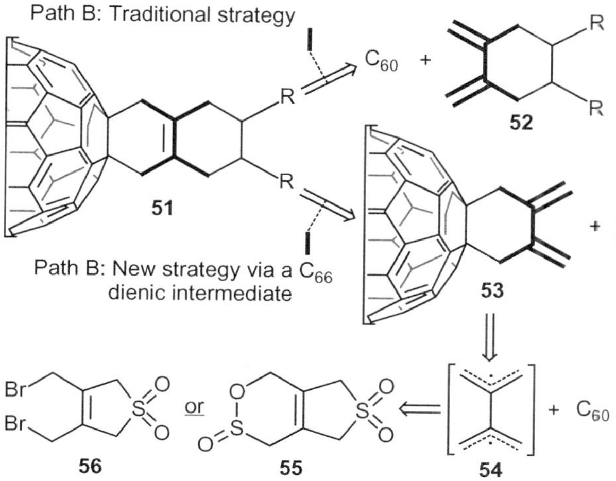

SCHEME 5.40 Strategies for C_{60} functionalization/derivatization via the Diels–Alder reaction.[162] *RSC advances by RSC Publishing. Reproduced with permission of RSC Publishing in the format Book via Copyright Clearance Center. Markoulides, M. S.; Ioannou, C. P.; Manos, M. J.; Chronakis, N.* RSC Adv. ***2013***, 3, *4750–4756.*

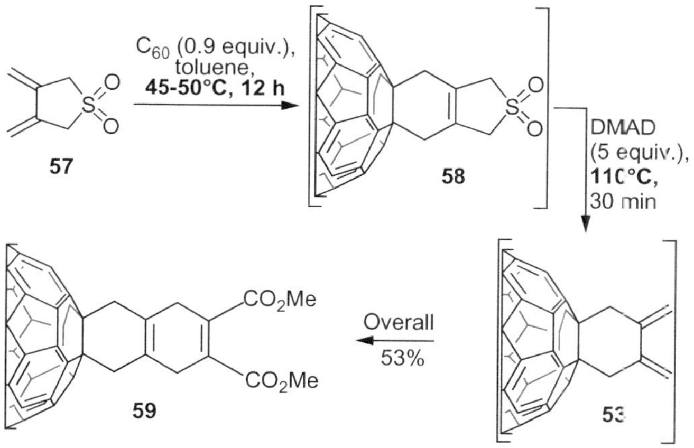

SCHEME 5.41 One-pot thermally chemo-controlled strategy for [4 + 2] functionalization/[4 + 2] derivatization of C_{60}.[162] *RSC advances by RSC Publishing. Reproduced with permission of RSC Publishing in the format Book via Copyright Clearance Center. Markoulides, M. S.; Ioannou, C. P.; Manos, M. J.; Chronakis, N.* RSC Adv. ***2013***, 3, *4750–4756.*

monoadduct **59**, which was isolated in 53% yield. This method is very promising for the functionalization/derivatization of different carbonaceous materials such as carbon nanotubes or graphene, allowing their functionalization with several different chemical groups by experimenting with different dienophiles.

References

1. Iha, R. K.; Wooley, K. L.; Nyström, A. M.; Burke, D. J.; Kade, M. J.; Hawker, C. J. *Chem. Rev.* **2009,** *109*, 5620–5686.
2. Golas, P. L.; Matyjaszewski, K. *Chem. Soc. Rev.* **2010,** *39*, 1338–1354.
3. Kappe, C. O.; Van der Eycken, E. *Chem. Soc. Rev.* **2010,** *39*, 1280–1290.
4. Kolb, H. C.; Finn, M. G.; Sharpless, K. B. *Angew. Chem. Int. Ed.* **2001,** *40*, 2004–2021.
5. Moses, J. E.; Moorhouse, A. D. *Chem. Soc. Rev.* **2007,** *36*, 1249–1262.
6. Durmaz, H.; Dag, A.; Altintas, O.; Erdogan, H.; Hizal, G.; Tunca, U. *Macromolecules* **2007,** *40*, 191–198.
7. Kim, T.-D.; Luo, J.; Tian, Y.; Ka, J.-W.; Tucker, N. M.; Haller, M.; Kang, J.-W.; Jen, A. K. Y. *Macromolecules* **2006,** *39*, 1676–1680.
8. Demko, Z. P.; Sharpless, K. B. *J. Org. Chem.* **2001,** *66*, 7945–7950.
9. Tsarevsky, N. V.; Bernaerts, K. V.; Dufour, B.; Du Prez, F. E.; Matyjaszewski, K. *Macromolecules* **2004,** *37*, 9308–9313.
10. Huisgen, R. *1,3-Dipolar Cycloaddition Chemistry;* Wiley, 1984.
11. Rostovtsev, V. V.; Green, L. G.; Fokin, V. V.; Sharpless, K. B. *Angew. Chem. Int. Ed.* **2002,** *41*, 2596–2599.
12. Tornoe, C. W.; Christensen, C.; Meldal, M. *J. Org. Chem.* **2002,** *67*, 3057–3064.
13. Braunecker, W. A.; Matyjaszewski, K. *Prog. Polym. Sci.* **2007,** *32*, 93–146.
14. Matyjaszewski, K.; Braunecker, W. A.; Matyjaszewski, K.; Gnanou, Y.; Leibler, L., Eds. *Radical Polymerization;* Wiley-VCH: Weinheim, Germany, 2007.
15. Matyjaszewski, K.; Tsarevsky, N. V. *Nat. Chem.* **2009,** *1*, 276–288.
16. Tsarevsky, N. V.; Matyjaszewski, K. *Chem. Rev.* **2007,** *107*, 2270–2299.
17. Lowe, B.; McCormick, C. L. *Prog. Polym. Sci.* **2007,** *32*, 283–351.
18. Cunningham, M. F. *Prog. Polym. Sci.* **2008,** *33*, 365.
19. Matyjaszewski, K. *J. Phys. Org. Chem.* **1995,** *8*, 197–207.
20. Lutz, J.-F.; Boerner, H. G.; Weichenhan, K. *Macromol. Rapid Commun.* **2005,** *26*, 514–518.
21. Sumerlin, B. S.; Tsarevsky, N. V.; Louche, G.; Lee, R. Y.; Matyjaszewski, K. *Macromolecules* **2005,** *38*, 7540–7545.
22. Chan, T. R.; Hilgraf, R.; Sharpless, K. B.; Fokin, V. V. *Org. Lett.* **2004,** *6*, 2853–2855.
23. Opsteen, J. A.; van Hest, J. C. M. *Chem. Commun.* **2005,** *7*, 57–59.
24. Kyeremateng, S. O.; Amado, E.; Blume, A.; Kressler, J. *Macromol. Rapid Commun.* **2008,** *29*, 1140–1146.
25. Li, Y.; Yang, J.; Benicewicz, B. C. *J. Polym. Sci., Part A: Polym. Chem.* **2007,** *45*, 4300–4308.
26. Dag, A.; Durmaz, H.; Demir, E.; Hizal, G.; Tunca, U. *J. Polym. Sci., Part A: Polym. Chem.* **2008,** *46*, 6969–6977.
27. Riva, R.; Schmeits, S.; Stoffelbach, F.; Jérôme, C.; Jérôme, R.; Lecomte, P. *Chem. Commun.* **2005,** *42*, 5334–5336.
28. Narumi, A.; Fuchise, K.; Kakuchi, R.; Toda, A.; Satoh, T.; Kawaguchi, S.; Sugiyama, K.; Hirao, A.; Kakuchi, T. *Macromol. Rapid Commun.* **2008,** *29*, 1126–1133.
29. Xiaohua, H.; Liyuan, L.; Kai, W.; Shaoliang, L.; Deyue, Y.; Yiqun, Z. *J. Appl. Polym. Sci.* **2009,** *111*, 560–565.
30. Altintas, O.; Hizal, G.; Tunca, U. *J. Polym. Sci., Part A: Polym. Chem.* **2008,** *46*, 1218–1228.

31. Golas, P. L.; Tsarevsky, N. V.; Matyjaszewski, K. *Macromol. Rapid Commun.* **2008,** *29*, 1167–1171.
32. Golas, P. L.; Tsarevsky, N. V.; Sumerlin, B. S.; Matyjaszewski, K. *Macromolecules* **2006,** *39*, 6451–6457.
33. Gao, H.; Louche, G.; Sumerlin, B. S.; Jahed, N.; Golas, P.; Matyjaszewski, K. *Macromolecules* **2005,** *38*, 8979–8982.
34. Tsarevsky, N. V.; Sumerlin, B. S.; Matyjaszewski, K. *Macromolecules* **2005,** *38*, 3558–3561.
35. Laurent, B. A.; Grayson, S. M. *J. Am. Chem. Soc.* **2006,** *128*, 4238–4239.
36. Lin, W.; Fu, Q.; Zhang, Y.; Huang, J. *Macromolecules* **2008,** *41*, 4127–4135.
37. Parrish, B.; Breitenkamp, R. B.; Emrick, T. *J. Am. Chem. Soc.* **2005,** *127*, 7404–7410.
38. Malkoch, M.; Thibault, R. J.; Drockenmuller, E.; Messerschmidt, M.; Voit, B.; Russell, T. P.; Hawker, C. J. *J. Am. Chem. Soc.* **2005,** *127*, 14942–14949.
39. Urien, M.; Erothu, H.; Cloutet, E.; Hiorns, R. C.; Vignau, L.; Cramail, H. *Macromolecules* **2008,** *41*, 7033–7040.
40. Golas, P. L.; Tsarevsky, N. V.; Sumerlin, B. S.; Walker, L. M.; Matyjaszewski, K. *Aust. J. Chem.* **2007,** *60*, 400–404.
41. Gao, H.; Matyjaszewski, K. *Macromolecules* **2006,** *39*, 4960–4965.
42. Zhang, Y.; Liu, H.; Hu, J.; Li, C.; Liu, S. *Macromol. Rapid Commun.* **2009,** *30*, 941–947.
43. Gao, H.; Matyjaszewski, K. *J. Am. Chem. Soc.* **2007,** *129*, 6633–6639.
44. Lee, J. W.; Kim, H. J.; Han, S. C.; Kim, J. H.; Jin, S.-H. *J. Nanosci. Nanotechnol.* **2008,** *8*, 4635.
45. Urbani, C. N.; Bell, C. A.; Whittaker, M. R.; Monteiro, M. J. *Macromolecules* **2008,** *41*, 1057–1060.
46. Wyszogrodzka, M.; Haag, R. *Chem. Eur. J.* **2008,** *14*, 9202–9214.
47. Kohman, R. E.; Zimmerman, S. C. *Chem. Commun.* **2009,** 794–796.
48. Elmer, S. L.; Man, S.; Zimmerman, S. C. *Eur. J. Org. Chem.* **2008,** *2008*, 3845–3851.
49. Urbani, C. N.; Lonsdale, D. E.; Bell, C. A.; Whittaker, M. R.; Monteiro, M. J. *J. Polym. Sci., Part A: Polym. Chem.* **2008,** *46*, 1533–1547.
50. Wu, P.; Malkoch, M.; Hunt, J. N.; Vestberg, R.; Kaltgrad, E.; Finn, M. G.; Fokin, V. V.; Sharpless, K. B.; Hawker, C. J. *Chem. Commun.* **2005,** 5775–5777.
51. Scheel, A. J.; Komber, H.; Voit, B. I. *Macromol. Rapid Commun.* **2004,** *25*, 1175–1180.
52. Saha, A.; Ramakrishnan, S. *Macromolecules* **2009,** *42*, 4028–4037.
53. Williams, R. J.; Barker, I. A.; O'Reilly, R. K.; Dove, A. P. *ACS Macro Lett.* **2012,** *1*, 1285–1290.
54. Jain, S.; Neumann, K.; Zhang, Y.; Geng, J.; Bradley, M. *Macromolecules* **2016,** *49*, 5438–5443.
55. Gibson, L.; Ashby, M. *Cellular Solids: Structure and Properties;* Pergamon: Oxford, 1997.
56. Knall, A.-C.; Kovacic, S.; Hollauf, M.; Reishofer, D.; Saf, R.; Slugovc, C. *Chem. Commun.* **2013,** *49*, 7325–7327.
57. Ossipov, D. A.; Hilborn, J. *Macromolecules* **2006,** *39*, 1709–1718.
58. Crescenzi, V.; Cornelio, L.; Di Meo, C.; Nardecchia, S.; Lamanna, R. *Biomacromolecules* **2007,** *8*, 1844–1850.
59. Johnson, J. A.; Lewis, D. R.; Diaz, D. D.; Finn, M. G.; Koberstein, J. T.; Turro, N. J. *J. Am. Chem. Soc.* **2006,** *128*, 6564–6565.
60. Truong, V. X.; Ablett, M. P.; Richardson, S. M.; Hoyland, J. A.; Dove, A. P. *J. Am. Chem. Soc.* **2015,** *137*, 1618–1622.

61. Xia, Y.; Verduzco, R.; Grubbs, R. H.; Kornfield, J. A. *J. Am. Chem. Soc.* **2008,** *130*, 1735–1740.
62. de Gennes, P. G.; Prost, J. *The Physics of Liquid Crystals*, 2nd ed.; Oxford University Press, 2003.
63. Hanemann, T.; Haase, W.; Svoboda, I.; Fuess, H. *Liq. Cryst.* **1995,** *19*, 699–702.
64. Frackowiak, E.; Béguin, F. *Carbon* **2001,** *39*, 937–950.
65. Zhai, Y.; Dou, Y.; Zhao, D.; Fulvio, P. F.; Mayes, R. T.; Dai, S. *Adv. Mater.* **2011,** *23*, 4828–4850.
66. Pramudita, J. C.; Pontiroli, D.; Magnani, G.; Gaboardi, M.; Riccò, M.; Milanese, C.; Brand, H. E. A.; Sharma, N. *ChemElectroChem* **2015,** *2*, 600–610.
67. Gaboardi, M.; Milanese, C.; Magnani, G.; Girella, A.; Pontiroli, D.; Cofrancesco, P.; Marini, A.; Riccò, M. *Phys. Chem. Chem. Phys.* **2017,** *19*, 21980–21986.
68. Gaboardi, M.; Sarzi Amadé, N.; Aramini, M.; Milanese, C.; Magnani, G.; Sanna, S.; Riccò, M.; Pontiroli, D. *Carbon* **2017,** *120*, 77–82.
69. Salvio, R.; Krabbenborg, S.; Naber, W. J. M.; Velders, A. H.; Reinhoudt, D. N.; van der Wiel, W. G. *Chem. Eur. J.* **2009,** *15*, 8235–8240.
70. Zhang, X.; Hou, L.; Cnossen, A.; Coleman, A. C.; Ivashenko, O.; Rudolf, P.; van Wees, B. J.; Browne, W. R.; Feringa, B. L. *Chem. Eur. J.* **2011,** *17*, 8957–8964.
71. Li, H.; Cheng, F.; Duft, A. M.; Adronov, A. *J. Am. Chem. Soc.* **2005,** *127*, 14518–14524.
72. Li, H.; Adronov, A. *Carbon* **2007,** *45*, 984–990.
73. Zhang, Y.; He, H.; Gao, C.; Wu, J. *Langmuir* **2009,** *25*, 5814–5824.
74. Zhu, J.; Hiltz, J.; Lennox, R. B.; Schirrmacher, R. *Chem. Commun.* **2013,** *49*, 10275–10277.
75. Zhu, J.; Hiltz, J.; Mezour, M. A.; Bernard-Gauthier, V.; Lennox, R. B.; Schirrmacher, R. *Chem. Mater.* **2014,** *26*, 5058–5062.
76. Hirsch, A.; Brettreich, M. *Fullerenes: Chemistry and Reactions;* WILEY-VCH Verlag Gmbh & Co. KGaA: Weinheim, 2005.
77. Langa, F.; Nierengarten, J.-F. *Fullerenes: Principles and Applications;* The Royal Society of Chemistry: Cambridge, 2007.
78. Krätschmer, W.; Lamb, L. D.; Fostiropoulos, K.; Huffmann, D. R. *Nature* **1990,** *347*, 354–358.
79. Zhang, W.-B.; Tu, Y.; Ranjan, R.; Van Horn, R. M.; Leng, S.; Wang, J.; Polce, M. J.; Wesdemiotis, C.; Quirk, R. P.; Newkome, G. R.; Cheng, S. Z. D. *Macromolecules* **2008,** *41*, 515–517.
80. Segura, J. L.; Castelaín, M.; Salavagione, H. *Curr. Org. Synth.* **2013,** *10*, 724–736.
81. Rousseau, G.; Fensterbank, H.; Baczko, K.; Cano, M.; Stenger, I.; Larpent, C.; Allard, E. *Chempluschem* **2013,** *78*, 352–363.
82. Pereira, G. R.; Santos, L. J.; Luduvico, I.; Brondi Alves, R.; Pereira de Freitas, R. *Tetrahedron Lett.* **2010,** *51*, 1022–1025.
83. Inglis, A. J.; Pierrat, P.; Muller, T.; Bräse, S.; Barner-Kowollik, C. *Soft Matter* **2010,** *6*, 82–84.
84. Castelaín, M.; Salavagione, H.; Segura, J. L. *Org. Lett.* **2012,** *14*, 2798–2801.
85. Qiao, Y.; Guo, Y.; Yu, C.; Zhang, F.; Xu, W.; Liu, Y.; Zhu, D. *J. Am. Chem. Soc.* **2012,** *134*, 4084–4087.
86. Qu, S.; Tian, H. *Chem. Commun.* **2012,** *48*, 3039–3051.
87. Cao, D. R.; Liu, Q. L.; Zeng, W. J.; Han, S. H.; Peng, J. B.; Liu, S. P. *Macromolecules* **2006,** *39*, 8347–8355.
88. Qu, Y.; Hua, J.; Tian, H. *Org. Lett.* **2010,** *12*, 3320–3323.

89. Fukuda, M.; Kodama, K.; Yamamoto, H.; Mito, K. *Dyes Pigm.* **2004,** *63*, 115–125.
90. Falzon, M.-F.; Zoombelt, A. P.; Wienk, M. M.; Janssen, R. A. J. *Phys. Chem. Chem. Phys.* **2011,** *13*, 8931–8939.
91. Zhang, Y.; Kim, C.; Lin, J.; Nguyen, T.-Q. *Adv. Funct. Mater.* **2012,** *22*, 97–105.
92. Izawa, S.; Hashimoto, K.; Tajima, K. *Chem. Commun.* **2011,** *47*, 6365–6367.
93. Karsten, B. P.; Smith, P. P.; Tamayo, A. B.; Janssen, R. A. J. *J. Phys. Chem. A* **2012,** *116*, 1146–1150.
94. Bu, H.-B.; Götz, G.; Reinold, E.; Vogt, A.; Schmid, S.; Segura, J. L.; Blanco, R.; Gómez, R.; Báuerle, P. *Tetrahedron* **2011,** *67*, 1114–1125.
95. Castelaín, M.; Martínez, G.; Merino, P.; Martín-Gago, J. A.; Segura, J. L.; Ellis, G.; Salavagione, H. J. *Chem. Eur. J.* **2012,** *18*, 4965–4973.
96. Pierrat, P.; Vanderheiden, S.; Muller, T.; Bräse, S. *Chem. Commun.* **2009,** 1748–1750.
97. Hirsch, A.; Vostrowski, O. *Eur. J. Org. Chem.* **2001,** 829–848.
98. Gottis, S.; Kopp, C.; Allardand, E.; Deschenaux, R. *Helv. Chim. Acta* **2007,** *90*, 957–962.
99. Iehl, J.; Nierengarten, J.-F. *Chem. Eur. J.* **2009,** *15*, 7306–7309.
100. Nierengarten, J.-F.; Iehl, J. V.; Oerthel, V.; Holler, M.; Illescas, B. M.; Muñoz, A.; Martín, N.; Rojo, J.; Sánchez-Navarro, M.; Cecioni, S.; Vidal, S.; Buffet, K.; Durka, M.; Vincent, S. P. *Chem. Commun.* **2010,** 3860–3862.
101. Goldmann, A. S.; Walther, A.; Nebhani, L.; Joso, R.; Ernst, D.; Loos, K.; Barner-Kowollik, C.; Berner, L.; Muller, A. H. E. *Macromolecules* **2009,** *42*, 3707–3714.
102. Nurmi, L.; Lindqvist, J.; Randev, R.; Syrett, J.; Haddleton, D. M. *Chem. Commun.* **2009,** *19*, 2727–2729.
103. Iehl, J.; Nierengarten, J. F. *Chem. Commun.* **2010,** *46*, 4160–4162.
104. Hoyle, C. E.; Bowman, C. N. *Angew. Chem. Int. Ed.* **2010,** *49*, 1540–1573.
105. Borghi, R.; Lunazzi, L.; Placucci, G.; Cerioni, G.; Plumitallo, A. *J. Org. Chem.* **1996,** *61*, 3327–3331.
106. Teng, F.-A.; Guo, Y.; He, J.; Zhang, Y.; Han, Z.; Li, H. *Des. Monomers Polym.* **2017,** *20*, 283–292.
107. Li, Y.; Benicewicz, B. C. *Macromolecules* **2008,** *41*, 7986–7992.
108. Granqvist, C. G. *Handbook of Inorganic Electrochromic Materials;* Elsevier, 2002ISBN 978-0-08-053290-5.
109. Mortimer, R. J. *Annu. Rev. Mater. Res.* **2011,** *41*, 241–268.
110. Ye, Q.; Neo, W. T.; Cho, C. M.; Yang, S. W.; Lin, T.; Zhou, H.; Yan, H.; Lu, X.; Chi, C.; Xu, J. *Org. Lett.* **2014,** *16*, 6386–6389.
111. Ye, Q.; Neo, W. T.; Lin, T.; Song, J.; Yan, H.; Zhou, H.; Shah, K. W.; Chua, S. J.; Xu, J. *Polym. Chem.* **2015,** *6*, 1487–1494.
112. Cho, C. M.; Ye, Q.; Neo, W. T.; Lin, T.; Lu, X.; Xu, J. *Polym. Chem.* **2015,** *6*, 7570–7579.
113. Cejka, J., Ed. *Metal-Organic Frameworks Applications From Catalysis to Gas Storage;* Wiley-VCH, 2011ISBN 978-3-527-32870-3.
114. Burrows, A. D.; Frost, C. G.; Mahon, M. F.; Richardson, C. *Angew. Chem. Int. Ed.* **2008,** *47*, 8482–8486.
115. Chen, C.; Allen, C. A.; Cohen, S. M. *Inorg. Chem.* **2011,** *50*, 10534–10536.
116. The Chemistry of Metal-Organic Frameworks. In *Synthesis, Characterization and Applications;* Kaskel, S., Ed.; Vol. 1; Wiley-VCH, 2016ISBN 978-3-527-33874-0.
117. Hindelang, K.; Kronast, A.; Vagin, S. I.; Rieger, B. *Chem. Eur. J.* **2013,** *19*, 8244–8246.

118. O'Reilly, R. K.; Joralemon, M. J.; Wooley, K. L.; Hawker, C. J. *Chem. Mater.* **2005,** *17,* 5976–5988.
119. Joralemon, M. J.; O'Reilly, R. K.; Hawker, C. J.; Wooley, K. L. *J. Am. Chem. Soc.* **2005,** *127,* 16892–16899.
120. O'Reilly, R. K.; Joralemon, M. J.; Hawker, C. J.; Wooley, K. L. *Chem. Eur. J.* **2006,** *12,* 6776–6786.
121. O'Reilly, R. K.; Joralemon, M. J.; Hawker, C. J.; Wooley, K. L. J. *Polym. Sci., Part A: Polym. Chem.* **2006,** *44,* 5203–5217.
122. An, Z.; Tang, W.; Wu, M.; Jiao, Z.; Stucky, G. D. *Chem. Commun.* **2008,** *48,* 6501–6503.
123. Collman, J. P.; Devaraj, N. K.; Chidsey, C. E. D. *Langmuir* **2004,** *20,* 1051–1053.
124. Collman, J. P.; Devaraj, N. K.; Eberspacher, T. P. A.; Chidsey, C. E. D. *Langmuir* **2006,** *22,* 2457–2464.
125. Ostaci, R.-V.; Damiron, D.; Capponi, S.; Vignaud, G.; Leger, L.; Grohens, Y.; Drockenmuller, E. *Langmuir* **2008,** *24,* 2732–2739.
126. Kappe, C. O. *Chem. Soc. Rev.* **2008,** *37,* 1127–1139.
127. Kappe, C. O.; Dallinger, D. *Mol. Diversity* **2009,** *13,* 71–193.
128. Caddick, S.; Fitzmaurice, R. *Tetrahedron* **2009,** *65,* 3325–3355.
129. Appukkuttan, P.; Dehaen, W.; Fokin, V. V.; Van der Eycken, E. *Org. Lett.* **2004,** *6,* 4223–4225.
130. Khanetskyy, B.; Dallinger, D.; Kappe, C. O. *J. Comb. Chem.* **2004,** *6,* 884–892.
131. Lietard, J.; Meyer, A.; Vasseur, J.-J.; Morvan, F. *J. Org. Chem.* **2008,** *73,* 191–200.
132. Munteanu, M.; Choi, S.; Ritter, H. *Macromolecules* **2008,** *41,* 9619–9623.
133. Mallard-Favier, I.; Blach, P.; Cazier, F.; Delattre, F. *Carbohydr. Res.* **2009,** *344,* 161–166.
134. Hoogenboom, R.; Moore, B. C.; Schubert, U. S. *Chem. Commun.* **2006,** 4010–4012.
135. Fazio, M. A.; Lee, O. P.; Schuster, D. I. *Org. Lett.* **2008,** *10,* 4979–4982.
136. Welton, T. *Chem. Rev.* **1999,** *99,* 2071–2084.
137. Plechkova, N. V.; Seddon, K. R. *Chem. Soc. Rev.* **2008,** *37,* 123–150.
138. Zhao, Y.-B.; Yan, Z.-Y.; Liang, Y.-M. *Tetrahedron Lett.* **2006,** *47,* 1545–1549.
139. Marra, A.; Vecchi, A.; Chiappe, C.; Melai, B.; Dondoni, A. *J. Org. Chem.* **2008,** *73,* 2458–2461.
140. Vecchi, A.; Melai, B.; Marra, A.; Chiappe, C.; Dondoni, A. *J. Org. Chem.* **2008,** *73,* 6437–6440.
141. Mikami, M.; Matsumoto, S.; Tonoi, T.; Okubo, Y.; Suenobu, T.; Fukuzumi, S. *Tetrahedron Lett.* **1998,** *39,* 3733–3736.
142. Fukuzumi, S.; Kochi, J. K. *Tetrahedron* **1982,** *38,* 1035–1049.
143. Yang, S. H.; Pettiette Conceicao, J.; Cheshnovsky, O.; Smalley, R. E. *Chem. Phys. Lett.* **1987,** *139,* 233–238.
144. Zeng, Y.; Biczok, L.; Linschitz, H. *J. Phys. Chem.* **1992,** *96,* 5237–5239.
145. Mikami, K.; Matsumoto, S.; Okubo, Y.; Fujitsuka, M.; Ito, O.; Suenobu, T.; Fukuzumi, S. *J. Am. Chem. Soc.* **2000,** *122,* 2236–2243.
146. Zhang, X.; Fan, A.; Foote, C. S. *J. Org. Chem.* **1996,** *61,* 5456–5461.
147. Vassilikogiannakis, G.; Hatzimarinaki, M.; Orfanopoulos, M. J. *Org. Chem.* **2000,** *65,* 8180–8187.
148. Vassilikogiannakis, G.; Chronakis, N.; Orfanopoulos, M. *J. Am. Chem. Soc.* **1998,** *120,* 9911–9920.

149. Schuster, D. I.; Cao, J.; Kaprinidis, N.; Wu, Y.; Jensen, A. W.; Lu, Q.; Wang, H.; Wilson, S. R. *J. Am. Chem. Soc.* **1996,** *118*, 5639–5647.
150. Vassilikogiannakis, G.; Orfanopoulos, M. *J. Org. Chem.* **1999,** *64*, 3392–3393.
151. Jensen, A. W.; Khong, A.; Saunders, M.; Wilson, S. R.; Schuster, D. I. *J. Am. Chem. Soc.* **1997,** *119*, 7303–7307.
152. Hoke, S. H., II; Molstad, J.; Dilettato, D.; Jay, M. J.; Carlson, D.; Kahr, B.; Cooks, R. G. *J. Org. Chem.* **1992,** *57*, 5069–5071.
153. Mehta, G.; Viswanath, M. B. *Tetrahedron Lett.* **1995,** *36*, 5631–5632.
154. Nair, V.; Sethumadhavan, D.; Nair, S. M.; Shanmugam, P.; Treesa, P. M.; Eigendorf, G. K. *Synthesis* **2002,** *12*, 1655–1657.
155. Mikie, T.; Asahara, H.; Nagao, K.; Ikuma, N.; Kokubo, K.; Oshima, T. *Org. Lett.* **2011,** *13*, 4244–4247.
156. Hatzimarinaki, M.; Orfanopoulos, M. *Org. Lett.* **2006,** *8*, 1775–1778.
157. Tzirakis, M. D.; Alberti, M. N.; Orfanopoulos, M. *Org. Lett.* **2011,** *13*, 3364–3367.
158. Martín, N.; Sánchez, L.; Illescas, B.; Pérez, I. *Chem. Rev.* **1998,** *98*, 2527–2548.
159. Martín, N.; Segura, J. L.; Wudl, F. In *Fullerenes: From Synthesis to Optoelectronic Properties;* Guldi, D. M., Martín, N., Eds.; Kluwer Academic Publishers: Dordrecht, The Netherlands, 2002; pp 95–120. Chapter 3.
160. Markoulides, M. S.; Ioannou, C. P.; Manos, M. J.; Chronakis, N. *RSC Adv.* **2012,** *2*, 12269–12277 and references therein.
161. Kotha, S.; Chavan, A. S. *J. Org. Chem.* **2010,** *75*, 4319–4322 and references therein.
162. Markoulides, M. S.; Ioannou, C. P.; Manos, M. J.; Chronakis, N. *RSC Adv.* **2013,** *3*, 4750–4756.
163. Fabre, J.-M. *Chem. Rev.* **2004,** *104*, 5133–5150.

Index

Note: Page numbers followed by "*f*" refer to figures.

B

C

E

H

I

N

O

P

Q

R

9780128152737

9780128152737